Ethics of Alaska Travel Writing since 1959

Arctic Humanities

VOLUME 1

The titles published in this series are listed at *brill.com/arch*

Ethics of Alaska Travel Writing since 1959

An Ecocritical Study

By

Benjamin Ferguson

BRILL

LEIDEN | BOSTON

Cover illustration: Yup'ik Mask, ca. 1900, on view at the Metropolitan Museum in New York. https://www.metmuseum.org/art/collection/search/751539

The Library of Congress Cataloging-in-Publication Data is available online at https://catalog.loc.gov
LC record available at https://lccn.loc.gov/2025026959

Typeface for the Latin, Greek, and Cyrillic scripts: "Brill". See and download: brill.com/brill-typeface.

ISSN 2772-9621
ISBN 978-90-04-73475-3 (hardback)
ISBN 978-90-04-73480-7 (e-book)
DOI 10.1163/9789004734807

This book is printed on acid-free paper and produced in a sustainable manner.

Contents

Acknowledgements

This book is dedicated first and foremost to my wife, Adeline, who has been unwavering in her support for my work. It is also dedicated to my two daughters, Ivis and Léonor, the most interesting people I know. I also want to thank my parents, Jim and Brenda, who didn't always understand my path but have always been there for me anyway, and my brother, Chris, who sparked my interest in Alaska and who has served as a sort of moral compass for my entire life.

Professionally, it takes a community to lift each other up, and Jan Borm has done nothing but support me since the moment I walked into his office and inquired about the Arctic graduate program he was running at UVSQ. Because of Jan, I have seen parts of the world I never imagined and have gotten the opportunity to meet some of the most interesting people along the way. I wouldn't be where I am today without his mentorship.

When times got tough, I also needed support from a broader community, and several people were always there when I needed them most. Emmanuelle Peraldo was patient with my professional growth, made sure I got integrated into SELVA, the Anglo-French research group on travel writing, and even checked in with me multiple times throughout my virtual PhD defense during a Covid lockdown. Likewise, Françoise Besson supported my research in SELVA, and through her tireless work organizing conferences and editing compilations, she guided me professionally, somehow always having time for my questions and comments. Sometimes, however, it isn't professional support that one needs but just plain camaraderie, and I am so thankful to Stéphanie Gourdon for being such a great friend who could empathize when I was struggling with my work.

Thanks so much to all of you, and to the many other friends, family, and colleagues who have been patient with me on this journey!

Illustrations

FIGURE 1 Yup'ik Mask, ca. 1900, on view at the metropolitan museum in New York https://www.metmuseum.org/art/collection/search/717572

Introduction

This work is an analysis of travel literature about Alaska in the modern era. In this case, "modern" is defined by anything published after the implementation of Alaska Statehood in 1959. Considering the breadth of Alaskan literature in comparison to its population during this period, there remains a dearth of academic literary study on the region. However, there is a particular work to which this book owes a great deal of debt: Susan Kollin's *Nature's State*.[1] Kollin's work particularly focuses on what she calls in her subtitle "Imaging Alaska as the Last Frontier." In effect, she analyzes how writers have fallen for dangerous tropes in the virtual space, or the communal imagination, that have physically affected the land itself. For Kollin, if we consider Alaska to be the "Last Frontier," a motto that sits right on the license plate of every car in the state, we are perpetuating the idea of Alaska as the last great "wild" space to be "tamed," or subjugated into an industrial world place. As a result, we will strip its resources and objectify its peoples, perpetuating the frontier ideologies of the 19th century. Concurrently, if Alaska's mountains are considered frontier lands to be preserved for aesthetic purposes, the tundra suffers, as it is often considered an ugly emptiness and is treated as such. The focus of my work builds off Kollin's ideas, analyzing some of the same works and some different ones, but focusing specifically on one question: what are the ethical responsibilities of the Outside writer who moves through Alaska, particularly when focusing on native culture and environmental themes?

This research was undertaken in the midst of the #metoo and Black Lives Matter movements, which together served as a catalyst for Postcolonialism in both academia and the world beyond. Therefore, I have dealt with many postcolonial questions "in the heat of the moment," as it were. The lines were quite literally written as revelations about misconduct across various dimensions of society were coming to light on a daily basis, and academics were rushing to understand what that meant for their fields. With hindsight in perspective, some of the important questions that potentially could have been more forcefully approached are: what especially are the ethics of the traditionally dominant demographic of the white man in travel writing?; what does the white man have left to say, and how should he proceed into the spaces of other cultures at this moment?; how should we readjust our view of his work?; and should his work be completely shaken apart, or should it be deconstructed

1 Kollin, Susan. *Nature's State: Imagining Alaska as the Last Frontier*. Chapel Hill: University of North Carolina Press, 2001.

 | DOI:10.1163/9789004734807_002

from multiple angles and viewed from present and past perspectives simultaneously, with nuance? Instead, in Part 1 I have explored multiple perspectives of different demographics in Alaska, specifically within the focus of the six canonical works of this study. Questions arise such as: how directly should the lever be pulled in favor of environmentalism and from what angle?; how should we define masculinity and femininity in Alaska?; how should Outside travel writers approach oral traditions?; how can Alaska Natives be fairly depicted?; how does geographical space affect the writer's spiritual and religious beliefs?; how can animals be fairly depicted in native-inhabited spaces?; and what does liberty mean in a space encroached upon by man?

In Part 2, these analyses are put into perspective, employing a lens of Ecocriticism and Environmental Justice, in a prescriptive manner. Essentially, I argue that Kollin's assertion of the importance of being wary of "the Last Frontier" in Alaskan literature is the most responsible and ethical way to write, but I add the caveat that Last Frontier adventure has served as a catalyst for the environmental movement. For instance, Kollin defends the writing of an important journalistic source, Tundra Times, but that source is now defunct, while works that idealize the frontier, such as Jon Krakauer's *Into the Wild*[2] and Jack London's *Call of the Wild*,[3] continue to entice people who create environmental change in Alaska. This obviously brings the destructive aspects with the conservationist and preservationist aspects, but all outside writers discussed in this work appear to be influenced to move through the region due at least to some extent to the pull of the Last Frontier.

I also discuss ways that environmental change can most be effected through writing. I suggest that, as with politics, a plethora of works targeting different demographics should be supported. Works that invigorate a convinced base of the environmental-minded should continue to be supported, but so should works that sneakily sit in the mind of what is termed "the middle third," or the status quo who is not inclined to act for change. Likewise, all sexes, sexual preferences, races, and cultural groups should be supported and united through literature. I will discuss, for instance, the unifying themes of salmon to Alaska. Furthermore, I argue that though works affecting large audiences are the most lauded, and often the most studied in academia, large quantities of local works that affect specific communities should be equally supported. The result would allow for nuance within the intricacies of specific communities that are sometimes left behind by the global zeitgeist of Environmental Studies. For instance, making ecological decisions while living on traditional

2 Krakauer, Jon. *Into the Wild*. 1996. Hampshire: Pan Macmillan, 2011.

3 London, Jack. *The Call of the Wild, White Fang, and Other Stories*. 1903, 1906, 1907. London: Penguin Books, 1981.

lands in a rainforest may look vastly different from making ecological decisions while living in an urban environment.

The point of many of these arguments is to serve as a metacritique for an environmental movement that has followed culture more broadly in leaning toward interconnected, global themes, particularly as so many people unite in fights over climate change. Peoples of the Arctic have learned from these arguments and have worked in concertation with the global movement on many issues, but they still feel uncomfortable and outcasted at times. For decades, movements such as Greenpeace and PETA have been out of step with native peoples struggling to adapt in their self-determination after having been abruptly forced into capitalist and Judeo-Christian systems. In pointing out these rifts, I am not intending to condemn environmental groups who are otherwise doing good work, but rather to spark an academic conversation with regard to how we can better serve Arctic peoples by understanding the importance of their cultures, beliefs, and lifestyles. By adapting and opening our minds to new ways of thinking we may be able to move forward in environmental discussions more effectively.

I will now introduce the six primary works chosen for analysis in this research and why they were chosen. Almost every decade since 1959 is represented, as are several different regions, with their diverse ecosystems and communities. Though I would have liked to have included more diverse voices from the travel writing itself, the representations covered are indeed quite diverse. Other considerations for inclusion were different motives for writing, different writing styles, and authors who had long careers I could investigate.

1 Peter Matthiessen's *Oomingmak*

In introducing the works here, time, geography, demographics, and basic reasons for inclusion will be given. Working chronologically, the first of the six works is Peter Matthiessen's *Oomingmak: The Expedition to the Musk Ox Island in the Bering Sea*,[4] published in 1967. Matthiessen had co-created *The Paris Review* while working undercover for the CIA and had recently moved back to the United States to travel and write.[5] He first published the eminent

4 Matthiessen, Peter. *Oomingmak: The Expedition to Musk Ox Island in the Bering Sea*. New York: Hastings House, 1967.

5 Wheelwright, Jeff. "A Writer's Controversial Past That Will Not Die." *The New York Times*. "https://www.nytimes.com/2018/02/02/books/review/peter-matthiessen-paris-review-cia.html." 2/2/2018

work, *Wildlife in America*,[6] in 1959, traveling to Alaska as part of his research. A few years later, he decided to return with a biologist friend, John Teal, and write exclusively about the musk ox of Nunivak Island, which lies just off the tip of southwestern Alaska, beyond the town of Bethel. For our environmental purposes, this is not a straightforward book. There is not much in the way of philosophizing, or navel-gazing, as we see in some other Matthiessen books, including *Wildlife in America*. *Oomingmak* was first published serially in *The New Yorker* (front matter), and it is a short work, coming in at 80 pages. In that span, it briefly describes Anchorage in the mid-1960s, just before the discovery of oil in Prudhoe Bay flipped the city upside down, and it also discusses local Alaska Native life on the island of Nunivak at the same moment. This was not long before the Alaska Native Claims Settlement Act (ANCSA) money began flowing, making Native tribes corporate entities, hence capitalists.

Peter Matthiessen had a prodigious career as an environmental writer, but his history with Zen Buddhism described throughout *The Snow Leopard*,[7] as well as his tendency toward zoological study so prominent with the musk ox in *Oomingmak*, create interesting parallels with other travel writers to be explored.

2 John McPhee's *Coming into the Country*

In 1977, John McPhee's *Coming into the Country* was first published.[8] This book turned out to be McPhee's most popular (Macfarlane, "Foreword," *Coming into the Country*, x) and would have been considered his magnum opus had he not put together a five-tome work of geological travel-writing, which he wrote from 1978 to 1998 and for which he won the 1999 Pulitzer Prize for General Nonfiction (ix). John McPhee has had a prolific career, serving as a staff writer for *The New Yorker* for the better part of half a century and publishing over 30 books. *Coming into the Country* is important to this research for many reasons, but one of the most outstanding aspects of it is the moment in which it was written. If Matthiessen offers us a fleeting glimpse of pre-oil Anchorage and Nunivak, McPhee meticulously limns Anchorage at the very moment when so much money is beginning to roll into its coffers that the government quite literally cannot find enough ways to spend it. Split into three sections, McPhee first takes the reader on a kayak trip with conservationists into the

6 Matthiessen, Peter. *Wildlife in America*. 1959. New York: Penguin Books, 1977.
7 Matthiessen, Peter. *The Snow Leopard*. 1978. New York: Random House, 2014.
8 McPhee, John. *Coming into the Country: Travels in Alaska*. 1977. London: Daunt Books, 2015.

extremely inaccessible Brooks Range. He then explores potential terrains for a new capital alongside local politicians in Book 2, often from a helicopter. In Book 3, the author offers a Robert Marshall-like[9] examination of the leftover gold miners along the Yukon River at the very moment the bureaucracy of litigation over indigenous and mining rights has come to the forefront, and low-carbon-footprint individualists building cabins in the area are getting caught up in its midst and kicked out of what has suddenly become enforced preservation. What makes *Coming into the Country* such a pillar for our purposes is that McPhee covers the environmental discussion from many angles at this important moment, and in various parts of the state. He interviews important officials from every side. Furthermore, multiple officials involved in the arguments of the book have published their own thoughts on the subject: Alaska Native politician, William Iġġiaġruk Hensley (*Fifty Miles from Tomorrow*),[10] National Park Service worker, John Kauffmann (*Alaska's Brooks Range: The Ultimate Mountains*),[11] and *Anchorage Daily Times* owner Robert Atwood. New Journalist, Joe McGinniss, also traveled through Alaska at roughly the same moment and wrote much more personalized accounts. These viewpoints offer potential for a comparative analysis that can only enrich our understanding of both McPhee and Alaska. If that were not enough to merit McPhee's inclusion in this research, the author has publicly stated that he does not consider himself an environmental writer (Weltzien, Maher, "Introduction," 11),[12] which opens up even more debate for analysis.

3 Barry Lopez' *Arctic Dreams*

After the pre-and post-oil Alaska works, Barry Lopez came along in the following decade, publishing *Arctic Dreams*[13] in 1986. This book is based on Lopez' travels through the western Arctic, primarily studying habits of animals and their environments. Even though Lopez' work is not exclusively Alaskan, he

9 Marshall, Robert. *Arctic Village: A 1930s Portrait of Wiseman, Alaska*. 1933. Alaska: University of Alaska Press, 1991.

10 Hensley, William L. Iggiagruk. *Fifty Miles from Tomorrow: A Memoir of Alaska and the Real People*. New York: Picador, 2009.

11 Kauffmann, John. *Alaska's Brooks Range: The Ultimate Mountains*. Seattle: The Mountaineers Books, 1992.

12 Weltzien, Alan O., editor. *Coming into McPhee Country: John McPhee and the Art of Literary Nonfiction*. Salt Lake City: University of Utah Press, 2003.

13 Lopez, Barry. *Arctic Dreams: Imagination and Desire in a Northern Landscape*. 1986. New York: Vintage, 2001.

indeed won the National Book Award in the US, was himself American, and resided for his entire adult life in the Northwest state of Oregon (Tredinnick, 45, 52).[14] Lopez was a direct descendant of the Transcendentalists, with both a Western education and a longstanding, close relationship to Native American culture, and he wrote extensively about Alaska in the work, having spent much time there. The 1980s timeframe fits in other ways as well. Lopez came of age in the rather socially liberal 1970s, publishing *Of Wolves and Men*[15] to decent success in 1978 before *Arctic Dreams* in the Reagan-era 1980s. Lopez' thinking in *Arctic Dreams* reflects socially liberal, US Northwest thinking during this transition. The writer is more spiritual than religious. He is scientific but spends long periods pontificating upon the various colors of ice and the meaning of man's place in the Arctic. He is a highly sensitive writer, and Jonathan Raban called him out for that fact (*Passage to Juneau*, 190–1),[16] which invites plenty of lively debate for this research. Also, Lopez has been one of the more heavily studied authors in modern-American environmental criticism, with three books of interviews employed in this research. This makes him, much like McPhee, a balance point from which to weigh other writers. And finally, the fact that Lopez' *Arctic Dreams* covers the western Arctic so broadly serves as an interesting geographic departure from the other works.

4 Tom Lowenstein's *Ancient Land, Sacred Whale*

The following decade brought us a vastly different work, about a different part of Alaska, and a different culture. None of the previous three authors discussed particular groups of Alaska Natives at length, of primary importance to their work, but Tom Lowenstein's 1993 *Ancient Land, Sacred Whale*[17] is primarily based on the Tikigagmiut of Point Hope, a village situated along the northwestern, Arctic coast of the state. The work is a piece of avant-garde, ethnographic poetry mixed with essay-esque reflections on Tikigaq culture. The British writer went to live in the region, having learned the basis of the Inuktitut language at the university in Fairbanks before immersing himself in the Arctic community. Much of the work is based around the whale hunt, which is a sacred and

14 Tredinnick, Mark. *The Land's Wild Music: Encounters with Barry Lopez, Peter Matthiessen, Terry Tempest William, and James Galvin*. San Antonio: Trinity University Press, 2005.

15 Lopez, Barry. *Of Wolves and Men*. 1978. New York: Touchstone, 1995.

16 Raban, Jonathan. *Passage to Juneau: A Sea and its Meanings*. London: Picador, 1999.

17 Lowenstein, Tom. *Ancient Land, Sacred Whale: The Inuit Hunt and its Rituals*. 1993. 2nd ed. New York: North Point Press, 1996.

spiritual part of Iñupiat society. However, the hunt is also important to subsistence living and was and still is part of a heated debate over animal rights.[18] Lowenstein fully understands the spiritual and physical importance of the whale hunt to Iñupiat communities, and so does Bill Hess, whose work on the same region, published six years later, covers many of the same ideas but from a much more modern lens (quite literally, as Hess' work is primarily photographical).[19] Lowenstein is the least known of the six authors, but his work fits right into the oeuvre. It offers something different: a different region, a different time, and a different people, being the only Alaska-Native-first viewpoint. It is also the only work of poetry.

5 Jonathan Raban's *Passage to Juneau*

Jonathan Raban's 1999 work *Passage to Juneau* is based around a journey from the writer's adopted city of Seattle up to Juneau, Alaska, aboard his personal boat. This is a straightforward travel-writing account in the chronological sense, but it contains a few, at least ostensible, drawbacks from our perspective: most of the journey is concerned with the British Columbian coast of Canada and not Alaska, and he writes extensively on the loss of his father and his own failing marriage. The personalized perspectives fit right into the work, however, equating Raban's stormy life with the physically stormy waters upon which he is sailing. And yet, though the metaphorical connection may potentially seem shallow, the work offers the only viewpoint in this research of the trip that so many migrants, fishermen, and shipping vessels have taken up to the state from Seattle, a city which has long served as a hub for Alaska and a base of its operations.[20] In fact, money and resources were long pulled straight out of the state and put into the pockets of Seattle businessmen (Haycox, 257; Coates, 32–3). During the gold rush as well, many people embarked from

18 Gupta, Rupa. "Indigenous Peoples and the International Environmental Community: Accomadating Claims Through a Cooperative Legal Process." *New York University Law Review: Vol. 74:1741*. "https://www.nyulawreview.org/wp-content/uploads/2018/08/NYU LawReview-74-6-Gupta.pdf." 12/1999.

19 Hess, Bill. *Gift of the Whale: the Iñupiat Bowhead Hunt, a Sacred Tradition*. Seattle: Sasquatch Books, 1999.

20 Seattle-based company Amazon, for instance, has become even more important as a supplier to Alaska's citizens than it has elsewhere because that company is one of the few that will ship from the Lower 48 for low cost. Likewise, the company Alaska Airlines is based in Seattle, and Alaskans generally must fly through Seattle in order to arrive at other places in the Contiguous US.

Seattle (Haycox, 203–4). This is not even to mention the fact that the author encounters Seattle fishermen experienced in Alaskan waters from the very first pages of *Passage to Juneau*, before he has even departed. He then proceeds to pass through fishing village after fishing village, comforting himself in their seedy bars, restaurants, and trading posts along the way. Having written *Soft City*[21] with the theme of being a sort of "pioneer" in seedy, pre-gentrified parts of London in his early adulthood, Raban is entirely comfortable in these sorts of environments (though even he is disgusted by certain bars in fishing villages of Alaska! – *Passage*, 364–73).

An interesting debate to be encountered along the way is Raban's spiritual side, which derives from his Anglican priest father and eventually leads to a form of humanist atheism (282) that is much opposed to the "New Age" proclivities of US Northwesterners like Barry Lopez. These thoughts are always at least on the periphery of Raban's writing, and will be discussed with regard to their contrasts with American thought, contrasts that are also a substantial theme throughout much of the British writer's œuvre. This perspective is therefore important for the diversity of the literature discussed in this research.

6 Nancy Lord's *Beluga Days*

The most recently published work chosen for this research is Nancy Lord's *Beluga Days*,[22] published in 2003. Along with Peter Matthiessen's novella-length work, *Oomingmak*, it was the toughest choice to make, as *Beluga Days* is not a straightforward piece of travel writing. Lord's works *Fishcamp*[23] and *Green Alaska*[24] could both perhaps make a case over *Beluga Days*. *Fishcamp*, though, is also not truly a piece of travel writing, but rather an introduction into the writer's life in a specific place where she has been living during the fishing season for decades. *Green Alaska* is indeed a veritable piece of travel writing, but it is much more historic, following in the footsteps of the famed Harriman Expedition. *Beluga Days*, by contrast, is very modern in its journalistic, environmentalist approach.

---

21 Raban, Jonathan. *Soft City*. London: The Harvill Press, 1974.

22 Lord, Nancy. *Beluga Days: Tracking a White Whale's Truths*. Berkeley, CA: Counterpoint Press, 2004.

23 Lord, Nancy. *Fishcamp: Life on an Alaskan Shore Island*. Press/Shearwater Books.1997 (hard cover) Berkeley, CA: Counterpoint Press, 1999 (paperback).

24 Lord, Nancy. *Green Alaska: Dreams from the Far Coast*. Berkeley, CA: Counterpoint Press, 1999.

In a sense, Alaskan non-fiction has a tendency to lend itself toward travel literature more than other places because not so many writers are actually from the state. Alaska is generally not a static place for non-Alaska Natives. In fact, it is the U.S. state with the highest rate of resident turnover.[25] The people who do tend to stay are those who choose to remain because the place has "spoken" to them, or somehow fits the aesthetic and lifestyle they are seeking.

Nancy Lord was born and raised in New Hampshire and came to the Brooks Range on a college trip and fell in love, felt that inexplicable pull that did not strike the others around her, only struck them as a memorable voyage (*Rock, Water, Wild*, "In the Giant's Hand," 29–36).[26] She moved to Alaska with her partner and took cannery jobs and a job at a sports store, which the two eventually bought together and sold to make their start as professional fishermen (*Fishcamp*, "Putting Up Boat," 41–2). In the winters, she plied her trade as a writer.

Beluga Days, as mentioned, is not straightforward, chronological travel literature, but it can nonetheless be viewed as a form of travel literature. The author indeed travels around the Cook Inlet in the work, exploring its history, introducing the reader to Kenai, with its white culture, and Tyonek, with its Dena'ina culture, and exploring Anchorage and all the political talks surrounding the waters of the Cook Inlet with regard to the disappearance of the local beluga. She also heads to Chicago and Quebec looking for answers. The book reads like a reflective investigation (perhaps an investigation of the investigation), and a journalistic memoir, with personal scenes of her local travels all along the way. One could be tempted to say this is just a book with a strong sense of place and not travel, but one must keep in mind that from the bottom of the Kenai Peninsula and the Cook Inlet at Homer where Nancy Lord lives much of the year is roughly five hours by car to Anchorage. The town of Kenai is about halfway, and Tyonek lies on the far side of the inlet, only accessible by plane or boat. The trip to her town of Homer is a common weekend getaway from Anchorage, but it is nonetheless quite the trip.

One last note should be made, and this concerns the fact that Nancy Lord is a woman, an important aspect as far as diversity of principal sources in this research is concerned. Though her addition should not be seen as merely due to a need for diversity, as her work, subjectively speaking, carries as much merit as any of the primary writers, the combined six works indeed benefited from the inclusion of a woman. It must be said as well that this research easily could

---

25 Sandberg, Eric. "Migration in Alaska." *Alaska Economic Trends*. "http://labor.alaska.gov/trends/mar18.pdf." 2/2018.

26 Lord, Nancy. *Rock, Water, Wild: An Alaskan Life*. Nebraska: University of Nebraska Press, 2009.

have included more women. Under the specific parameters I employed, only one woman fit, but women have indeed played a prominent role in the history and literature of Alaska and include some of the most important environmental writers of the 20th century, such as Margaret Murie and Lois Crisler, who will be discussed in detail. As for Nancy Lord, she served as Alaska writer laureate between 2008–10,[27] and her oeuvre is both extraordinarily diverse and creative, a strong representation of the state.

7 Anglophone Travel Writing: From Mercantilism to the Enlightenment

In order to approach the canon in question, it is important to first take in a historiographical analysis from multiple angles leading up to modernity in Alaska. Beginning with British travel writing and its influence on colonialism, we will look at how the US developed a unique political and social culture. We will then look at how the new country, with unprecedented space, created exceptional rhetoric around the idea of the "frontier" in order to avoid the stigma of colonialism. With that established, we will analyze how American environmentalism developed as a reaction to the exploitation of the frontier, with Alaska playing an important role in the development of preservationism in the 20th century before evolving toward Environmental Justice at the turn of the 21st century. With the entirety of this historiography in perspective, I am setting up a framework intended to demonstrate how Alaska has lain, and continues to lie, at the heart of several ideological battles while concurrently finding itself on the front line of a global climate crisis.

Anglophone travel writing is typically dated to Richard Hakluyt, an academic booster for Empire who understood that travel writing could inspire Britain out of its 16th century insularity and form its own empire to rival Spain. Hugh Brogan[28] begins his history of the United States roughly with Hakluyt (13–6), and *The Cambridge Companion to Travel Writing*[29] does as well (3–4). Working around the turn of the 17th century, Hakluyt would have been a contemporary

27 "Nancy Lord." *NancyLord.com.* "http://www.writernancylord.com/bio.html." Viewed on 10/12/2019.

28 Brogan, Hugh. *The Penguin History of the United States of America.* London: Penguin UK, 2001. p. 234, 244–5.

29 Hulme, Peter and Tim Youngs. Editors. *The Cambridge Companion to Travel Writing.* 2002. Cambridge, UK: Cambridge University Press, Fourth printing, 2008.

of Shakespeare, both men dying in 1616 (Kennedy, "Shakespeare: A Timeline," XVI).[30] Notably, Shakespeare and other playwrights had been taking inspiration from Venetian theatre, as that would have been a primary travel destination for the cultivated British of the age (Kennedy, "The Theatre in Shakespeare's Day," x; "Introduction," 1–2) and an integral part of the Grand Tour, during which artists often participated as guides to the wealthy (Buzard, "The Grand Tour and after [1660–1840]," *Cambridge Companion to Travel Writing*, 41–2). We could surmise, therefore, that Hakluyt succeeded in a propagandistic maneuver that combined both politics and literature to push Britain's focus away from Europe and toward the oceans, toward islands and other continents.

Yet, Hakluyt's propagandistic maneuvering, which eventually made travel writing texts canon with his collections *Divers Voyages Concerning the Discovery of America* (1582),[31] and *Principal Navigations, Voyages, Traffiques, and Discoveries of the English Nation* (1589), and which would eventually "challenge European perceptions of English inaction and ... promote new initiatives by showing that the English had been 'men full of activity, stirrers abroad, and searchers of the remote parts of the world,'" according to William S. Sherman ("Stirrings and searchings [1500–1720]," *Cambridge Companion to Travel Writing*, 19), did not have an immediate, overwhelming effect. It would take time for the literary movements of Shakespeare, Thomas Coryates, and John Taylor to shift focus away from the Eurocentrism that had so long been maintained. And even once Hakluyt's collections, spanning 1,500 years and 93 works (Sherman, 19), did indeed muster great cultural and political influence amongst the literate classes, a full-fledged, popular form of travel writing would not come into being for another century.

During that span, England was shaken by a civil war, as well as religious extremism, which perhaps curbed opportunities for secular literature. Daniel Defoe's *Robinson Crusoe*[32] (1719), and Jonathan Swift's *Gulliver's Travels*[33] (1726) would not be published until the dust settled. Jean Viviès briefly generalizes

30 Shakespeare, William. *Much Ado About Nothing*. 1623. Edited by R. B. Kennedy. London: Collins Classics, 2011.

31 Hakluyt, Richard. editor. *Divers Voyages Touching the Discovery of America, and the Islands Adjacent*. 1582. London: The Hakluyt Society, 1850.

32 Defoe, Daniel. *Robinson Crusoe*. 1719. Edited by Thomas Keymer. 2007. Oxford: Oxford University Press, reissued, 2008.

33 Swift, Jonathan. *Gulliver's Travels*. 1726. London: Jones & Company, 1826.

this idea in his "Introduction" to *English Travel Narratives in the Eighteenth Century*,[34] stating:

> If we ground our study in historical reality, the era can simply be labelled Georgian or Hanoverian. Like 1660, when Charles II was restored to the throne, 1714 is of course a key date when evoking political and historical chronology, marking the coronation of George I. Yet if one turns to literary chronology, this second date is far less significant, for Defoe and Swift had already written some of their works by this time. On the other hand, in the second half of the century, a turning point can indeed be identified. It is sometimes referred to (perhaps in a slightly hackneyed manner) as the moment when Reason gave way to the Age of Sensibility. Pope's death in 1744 is often cited as the beginning of this new era. The publication of Wordsworth and Coleridge's Lyrical Ballads in 1798 marks a decisive literary landmark at the century's end. (23)

This concise framework backs up the notion that travel writing was somewhat stifled after the death of Elizabeth I, until roughly the end of the English Civil War, rebounding a few decades later and starting to come into blossom in the early decades of the 18th century. Such a timeline is important for our purposes because we can then put Alaskan and Arctic exploration into context. The basis for wider acknowledgment of British exploration is laid in the age of Hakluyt, and with his contemporaries, essentially pirates, Francis Drake and Walter Raleigh (Brogan, 12–3). Wider literary interest, however, would not grasp hold of the country until Defoe and Swift. Georg Wilhelm Steller's record of "Second Kamchatka Expedition"[35] with Vitus Bering under the Russian flag, attempting to explore Alaska, would occur between 1733–1743, making landfall in 1741 (Haycox, 49–51). Likewise, Captain James Cook's first voyage into the Pacific would take place between 1768–1771. His third and final commanding voyage, which would lead him to Alaska before being killed shortly thereafter in Hawaii, would take place between 1776–1779 (Haycox, 62–9). Barry Lopez credits Cook with rendering the Pacific Ocean from a great open space of unknown and fantasy to a logical, mappable, and understandable space through which European man was capable of navigating ("After Cook, the old cartographer's

---

34 Viviès, Jean. *English Travel Narratives in the Eighteenth Century: Exploring Genres*. 1999. Translated by Claire Davison. Aldershot, England and Burlington, Vermont: Ashgate Publishing, 2002.

35 Steller, Georg Wilhelm. *Journal of a Voyage with Bering, 1741–1742*. Translated by Margritt Engel and O.W. Frost. 1743. California: Stanford University Press, 1993.

admission of ignorance, Here Be Dragons, disappeared from the perimeter of world maps" – Horizon, 59).[36]

The Enlightenment that Cook embodied implied the "scientific method," and an obsession with precision. Cook was highly focused on latitude and longitude, correcting the cartographical inaccuracies of his predecessors (Beaglehole, 114–27). This not only had the effect of making navigation clearer, but it no doubt saved the lives of countless explorers and their crews (Lopez, *Arctic Dreams,* 362). Likewise, Cook very often employed proto-anthropological description to serve as a guide for future navigators to know where, or where not, to stop and trade. For similar reasons, climate and the location of natural resources were prominent in his writings. Cook was also influenced, whether directly or indirectly, by Steller and other scientists. Steller's journals on Alaska were first penned thirty-five years before Cook's first voyage began,[37] and Steller was at the "cutting edge" so-to-speak of understanding the importance of animal and plant life. For instance, Steller saved multiple lives by understanding the connection between Vitamin C and scurvy, concocting drinks with comestible grasses he came across in his travels (Ford, *Where the Sea Breaks its Back,* 95–7).[38] He also sketched animals and plants at every opportunity, particularly noting the Steller's sea cow in detail before its extinction (Ford, 159–65).

The most common thread throughout Enlightenment-era voyages, however, is the great economic potential of exploration, a mindset which Hakluyt had already understood almost two centuries before Cook ever led an expedition.[39] Defoe's fiction can be viewed through this lens as well. For example, Robinson Crusoe essentially avoids an upper middle class existence, which would consist of living safely and conservatively off his family inheritance, deciding rather to make his own fortune at sea (Defoe, 5–10). He always has the itch to keep moving, throughout the work, and he keeps making money in different places along the way, simultaneously having adventures and avoiding the

36 Lopez, Barry. *Horizon.* London: The Bodley Head, 2019.

37 A comparative ethnohistory has since been published which weighs Tlingit oral histories against accounts from Bering's voyage, as well as succeeding encounters between the Tlingit and the Russians under Baranov. Black, Lydia, Nora Marks Dauenhauer, and Richard Dauenhauer, editors. *Anóoshi Lingít Aaní Ká / Russians in Tlingit America: The Battles of Sitka, 1802 and 1804.* Seattle and London: University of Washington Press/ Juneau: Sealaska Heritage Institute, 2008.

38 Ford, Corey. *Where the Sea Breaks Its Back: The Epic Story of Early Naturalist Georg Steller and the Russian Exploration of Alaska.* 1966. Portland, Oregon: Graphic Arts Books, 2003.

39 Beaglehole suggests that Cook read Hakluyt during his voyages (712).

stagnation against which Hakluyt warns (Sherman, "Stirrings and Searchings [1500–1720]," *The Cambridge Companion to Travel Writing*, 19–23).[40]

Investors with a similar mindset would have poured through Cook's Journals and other empiricist works in search of ways to make money (Haycox, 62–9). In fact, the explicit purpose of Cook's third voyage was to search for the Northwest Passage, which was expected to exist by sailing up the Bering Strait and across the Canadian archipelago, before ending up in Baffin Bay alongside Greenland. Cook was unsuccessful, as there existed no navigable transportation route unblocked of ice. Instead, he found what he named the Cook Inlet (Haycox, 62–4), sailing up to what would become Anchorage a little over a century later before turning back at what he called, and is still called today, Turnagain Arm (Beaglehole, 602).

8 Anglo-colonization, the Protestant Work Ethic, and the Last Frontier

In America, the repercussions of Hakluyt and his descendants in English travel writing were direct. Hugh Brogan has noted that Spain and Portugal began early domination of international waters beyond Europe at the end of the 15th century and divided up the western and eastern hemispheres, which led to Spain systematically pillaging Mexico and parts of the modern United States for gold, silver, and other quick riches (4–5). This made Spain the powerhouse of the Atlantic and Europe during the 16th century, but it would also lead to a somewhat precipitous decline. If we are to believe Brogan, the biggest reason for Britain's eventual dominance in America was its early establishment of permanent colonies. Whereas the French and Spanish more or less set up military forts and trade posts, they did not set up much in the way of actual towns, unlike the British (10–11).[41] For this reason, the British were able

40 Though Emmanuelle Peraldo has suggested that Crusoe's adventures, explorations, and accumulations are not selfish, but rather they are done for the "well-being of the nation, and what he called the public good." Peraldo, Emmanuelle. "From Mercantilist to Utilitarian Crusoe: the Transformative Impact of the Desert Island." *Études Anglaises. 2019/2 (Vol. 72)*. «https://www.cairn.info/revue-etudes-anglaises-2019-2-page-151.html?contenu =article.» 151–166.

41 Brogan notes here that he is supporting historian Charles M. Andrews' idea that "'gold, trade, tillage represent the three stages in the history of colonization, and the greatest of these, because fundamentally essential to performance, is tillage,'" though Brogan adds that not even the British understood the implications of what they were doing at first (10–1).

to systematically push the French and Spanish to the fringes through sheer numbers (and, of course, a stout navy with no ties to the "lure of continental hegemony" in Europe) (11–12); the Spanish were forced south into the Floridas and what is now Mexico and west into what are now Texas and California, and the French were eventually pushed into Louisiana and north into Quebec (8, 52). This allowed for the setting up of a contiguous conglomeration of colonies that would come to dominate North America and would be largely based on an anglophone culture and thought much different from that of the Spanish, French, or Portuguese (10–18). And it often came down to religion and interpretations of protestantism (30–50).

This colonial maneuvering is important to this research because it meant that the primary European population of North America became majority-British by the dawn of the American Revolution, and these Subjects of the Crown were no doubt unique in ideology. Even more importantly, many communities were founded by immigrants in search of religious freedom, which often meant a culture of highly conservative protestants with a heavy work ethic as a matter of religious principle. "Hellfire and brimstone" was the way of the day in the early 18th century (de Tocqueville, 188).[42,43] This is something we see even well into the 19th century, as John Muir's father, Daniel, is described in *The Story of My Boyhood and Youth*[44] as having brought his family to Wisconsin and having quite literally beat his children into religious submission, preaching hard, physical labor on the farm and rote memorization of *The Holy Bible*.

Importantly, anglophone humility and "protestant work ethic" set the stage for American thought. Benjamin Franklin, for instance, projected hard work as a key to his version of the "American Dream," and he simultaneously maintained a constant façade of modesty, even in his Socratic method of faux-naïve diplomacy (Isaacson, 27). This would resound in the manner of normative behavior American leaders would be expected to carry, unlike the trends of

---

42 de Tocqueville, Alexis. Revised by Francis Bowen. *Democracy in America*. 1835. Translated by Henry Reeve. Hertfordshire: Wordsworth Classics of World Literature, 1998.

43 A common complementary saying is "Idle hands make the Devil's work." The two both signify the fact that an early American engrained trait through the Massachusetts Puritans was strong work ethic as a means of being a good Christian, because keeping oneself busy with building a family and community was a way to "glorify God," or, as Benjamin Franklin would later put it in his famous Poor Richard's Almanac, God "helped those who helped themselves." This was a foreign idea to the Catholics and Anglicans of Europe who had long believed that "holiness required withdrawal from worldly economic concerns" (Isaacson, Benjamin Franklin, 9–10), and this idea should be considered one of the single most important aspects of understanding the formation of the American frontier mindset.

44 Muir, John. *The Story of My Boyhood and Youth*. Boston: Houghton Mifflin, 1913.

their counterparts in Western Europe whom de Tocqueville frequently calls "aristocratic" throughout *De la démocratie en Amerique* (*Democracy in America*). Furthermore, Franklin could also be considered a direct precursor to Emerson, as both men lauded "self-reliance" as a key American trait (Isaacson, 477–83). We would eventually see a similar dispensation in Theodore Roosevelt's writings on "the Strenuous Life" (Brinkley, *The Wilderness Warrior*, 68, 368–95). Both sides of the American political spectrum continue to maintain versions of this distinctly American idea, with its heavy roots in anglo-Protestantism and its access to centuries of homesteading, with Alaska becoming a key representation in the 20th century.

9 Jefferson versus Hamilton

How the US physically organized itself was directly related to this "American Dream" approach to utilitarianism, and it can be summed up in a dispute over America's soul between Thomas Jefferson and Alexander Hamilton: or, the rural / urban divide. Franklin, for instance, was thoroughly urban, and his middle class ideal leaves not much room for farming or wilderness. He wanted to build up society through hard work and modesty, not unlike his puritan forefathers. Yet, simultaneously, in western Pennsylvania, as well as Jefferson's Virginia and throughout most of the established southern United States at that point, farming was a way of life (Brogan, 104–7).

Meanwhile, Hamilton, from the Caribbean island of Nevis (Chernow, 7),[45] had worked his way into New York shipping and eventually into becoming President Washington's financial wizard and general right-hand man (338–62). In Washington's administration, Hamilton pulled his newly founded nation out of a severe, post-war recession that almost doomed the new country. In doing so, he established a national bank and a financial system so intricate that almost no one else fully understood it (Chernow, 137–8, 155–7, 170–1; Brogan, 258, 263).

With no major cities in Virginia, Jefferson believed in the utilitarian ideal of the small farmer being left alone by government and not having resources drained by large cities, unlike in Europe. This resonated well with the populations of newly arrived immigrants and the less established all over the new country who were constantly pushing westward in search of small farm propriety, and who wanted to be left alone to make their own way (Chernow, 311–2,

45 Chernow, Ron. *Alexander Hamilton*. New York: Penguin, 2004.

517; Ellis, 30–3; Brogan, 106, 258, 269). A journalistic war ensued, with Hamilton and Jefferson both writing fiercely in promotion of their respective ideologies (Chernow, 442–3).

The resulting push and pull repeated itself again and again on the American Frontier, and it became a primary theme of Jonathan Raban's work in the Pacific Northwest and Montana, notably in *Driving Home* and *Bad Land*. The Pacific Northwest being heavily linked to Alaska, those same fights migrated northward and became "The Last Frontier." Though The Last Frontier implies the dominance of the rural, throughout *Coming into the Country* John McPhee discusses how the urban pushes up against the rural. Likewise, Raban links the urban to the rural in his journey sailing up from Seattle in *Passage to Juneau*.

The ramifications of the rivalry between Hamilton and Jefferson are important for myriad reasons. To begin with, Hugh Brogan has this to say about the two ideologies:

> "For the weakness of the one was the strength of the other. Jeffersonism was too static, too backward-looking; Hamiltonism was too narrowly cynical about human nature. And the two creeds have this in common, that they both encouraged the economic individualism and the national ambitions of the new republic." (258)

Much of American history has been a back and forth between small-government utopianism and large-government federalism, expanding into a unique form of capitalist empire, which has euphemistically employed terms like "frontier," as Susan Kollin notes (20). In the battle of ideas, however, Jefferson and his acolytes, the Jeffersonians, dominated the first decades of the early 19th century (Brogan, 269, 377), until another populist, expansionist demagogue, Andrew Jackson and his acolytes, the Jacksonians, dominated from the mid-1820s almost up until the American Civil War in 1860 (266–7, 272–9); Jackson himself initiating "The Trail of Tears" in 1837 (a systematic mass movement of the Cherokees, and other Native Americans, westward across the Mississippi River, with "at least 4,000" dying along the way – 68).

It would be easy to say that Jefferson won out: the expansion westward was essentially an effecting of his vision of the yeoman farmer spread out across the burgeoning nation, with few major cities in geographical comparison to Europe (219–48). Hamilton's large, East Coast, urban banking structure, however, financed much of the expansion, bringing large amounts of wealth back to the East and the federal government (380). According to Brogan, Hamilton believed "'Ambition and avarice' were the most reliable pillars of the state," with Hamilton's financial structure set up around this ideology (258). The point I'm making is that wherever an ostensible, rural utopia has existed, capitalism (railroads, resource extraction, city building) has always followed just behind (380–93). In this way, Susan Kollin is absolutely right in being wary of Last Frontier ideology.

The skeptical perspective would point out that this is a generalized history of European culture's encounters with the rest of the world in recent centuries, and is nothing unique to America. However, in exploring Alaska's gold rush at the turn of the 20th century we see that there is actually a substantial difference between Jefferson's small government ideology, which becomes The American Frontier (and subsequently The Last Frontier), and a larger federal government asserting control. The Klondike gold rush, on the border of Canada's Yukon Territory and the Alaskan Interior, was highly controlled by Canadian forces and not extraordinarily dangerous from a criminal standpoint.[46] Order was kept (Haycox, 204–6, Borneman, 186–99).[47] Meanwhile, on Alaska's west coast, the Nome gold rush was a veritable extension of the Wild West, with guns, robberies, and deaths a daily occurrence (Haycox, 207–14, 221; Borneman, 206–12). Rex Beach's celebrated work *The Spoilers* is a fictionalized testament to the experience, and famed, Wild West personage Wyatt Earp notably tried his luck by opening "a tent saloon" (Haycox, 207).

Referencing back to the Brogan quote that opened this section: Jeffersonism and Hamiltonism both have strengths and weaknesses, but their overall enmeshed, back-and-forth impact on the American mindset ultimately led to a new type of Empire. If the British conquered much of the world through government-backed mercantilism (Brogan, 80–6), the individualism of the American mindset has evolved into a different, capitalist Empire (381–92). It has been no less ambitious or harmful, but that harm has been couched in euphemistic terms such as "frontier," as Americans have long maintained a policy of moral exceptionalism (Kollin, 20). Interestingly, Joseph Ellis declares the end of the Hamilton/Jefferson debate's relevancy to be the 1930s "… when Franklin Roosevelt invoked Hamiltonian methods (i.e., government intervention) to achieve Jeffersonian goals (i.e., economic equality)" (Ellis, 8–9). This may or may not be true, but it is interesting to note that a trend of Native acceptance and reconciliation (that has never come to full fruition) began within a

46 Although Soapy Smith created a quasi-mafia en route to Dawson, in Alaska, which systematically bamboozled people out of their money along the way. That is, before Soapy himself was murdered (Haycox, 211).

47 Borneman, Walter R. *Alaska: Saga of a Bold Land.* New York: Harper Collins, 2003.

few decades of the 1930s, leading to such decisions as Alaska's ANCSA (Alaska Native Claims Settlement Act) (Haycox, XII–XIII, 283–4).

10 The Rise of Environmental Awareness

Environmental Historian Roderick Nash notes that environmentalism tends to arise from urban areas first, as urbanites tend to lament what they have already lost, whereas rural areas are often still in the midst of the conquering phase (345–6). Nash goes so far as to say, "Initially, the marginal valuation of civilization is much higher than that of wilderness. Wilderness at this stage is so abundant as to constitute a threat to society" (346). He includes a diagram in which he depicts marginal valuation for nature and civilization, where appreciation for nature rises as its availability drops, in a negative correlation. Concurrently, on the previous page of *Wilderness and the American Mind* (345) Nash suggests that environmentalism in the U.S. has typically come from Harvard and San Francisco elites, with Alaska's preservation movement originally led by urbanites like Robert Marshall (son of a New York millionaire – Brinkley, *The Quiet World*, 231–2).

The 19th century is presumably when that negative correlation between valuation of nature and appreciation of nature meets in the center, as the Industrial Revolution was in full swing. In literature, the Transcendentalists were working out an ideology for this moment (Emerson's essays, "Nature" and "The American Scholar" being published in the 1830s; Thoreau's *Walden*[48] being published in 1854), while men like Frederick Law Olmsted were creating the physical representation of Transcendentalist ideals to be exported across the country. This is important because succeeding generations would continue to put pressure on open lands, and the environmentally conscious among them would eventually create a model of preservation on an enormous scale. These preserved lands tended to be cherished and viewed as America's cultural advantage over Europe (Brinkley, *The Wilderness Warrior*, 15, 35–9), but other lands were simultaneously built up into cities and surrounding natural resources were extracted to serve those cities.

Alaska would naturally become an attempted extension of this economic booster model. Furthermore, Alaska would begin to be viewed as an aesthetic extension of Olmsted's parks created for cities. Barry Lopez worried about

48 Thoreau, Henry David. *Walden and Civil Disobedience*. 1854 and 1849. New York: New American Library, 1st Signet Classics Printing, 2012.

this phenomenon in the 20th century. He warned that photographers for The Sierra Club were beginning to create what he described as an equivalent to pornography, as it creates a view of places as objects to be idealized and conquered as forms of artistic perfection: "They were consciously trying to create gorgeous, overwhelming images. They created them in response to an ideal about the beauty of landscape that was, qualitatively I think, no different from men choosing certain types of bodies and arranging them in certain kinds of poses to mirror a set of expectations that American men associate with erotic feminine beauty" (Tydeman, *Conversations with Barry Lopez*, 108–9). In doing so, Lopez is suggesting that we risk not appreciating the terrain that does not correspond to that particular notion of beauty. John McPhee suggests that Sierra Club leader David Brower falls into a similar trap in *Encounters with the Archrduid*. Brower is swayed by Charles Fraser's aesthetic view of Carolina and Georgia Islands and lets pass a proposal for environmental destruction he would under normal circumstances denigrate (143–50).

The negative effects of creating aesthetic expectations for land, against which Barry Lopez warned, can be seen in the fact that many boosters view the "empty" tundra of northern Alaska as useless for civilization. With no apparent aesthetic value, exploitation is justified. In the mid-20th century, there were even plans for the use of a Hydrogen bomb to carve out a port in Arctic Alaska and another for an oversized dam essentially serving no city (Coates, *The Trans-Alaska Pipeline Controversy*, 111–161).

What is evident in tundra exploitation is that aesthetics are not innate. When Barry Lopez is discussing his mistrust of certain types of photography, he is not railing against aesthetics in general, but rather carefully manipulated aesthetics like magazines that depict unattainable, conformist, beauty standards. As McPhee notes in *Coming into the Country* (310, 380), and Nancy Lord in *Rock, Water, Wild* as well (3–4), fifty-five gallon drums and broken down automobiles notoriously line properties throughout Alaska, for pragmatic reasons, and both writers come to view these eyesores as a welcoming aesthetic. McPhee goes so far as to explicitly state that he once considered them ugly, but that now they almost bloom in his eyes like a flower (*Country*, 380). A second wave ecocritical analysis, to be discussed shortly, would be that Alaskan land and peoples, maintaining cultural tenets divergent from those of the Lower 48, are exploited at least partially due to their unique aesthetics and leanings toward utilitarianism. A particularly strong example would be Susan Kollin's comments on the difference between Tundra Times' coverage of the Exxon Valdez oil spill (subsistence and biomes affected) and that of the Outside press (beauty affected) (4–5). The following section will move toward the turn of the 20th century with the creation of the conservation movement under Theodore

Roosevelt and its echoes, which reverberated all the way up through the modern Alaskan era, and even into the 21st century in some ways.

11 Theodore Roosevelt, Gifford Pinchot, and John Muir

In the years just after the turn of the 20th century, President Theodore Roosevelt famously agreed to the flooding of the Hetch Hetchy Valley in the Yosemite, sparking a feud with John Muir (Brinkley, *Wilderness*, 544). To Roosevelt, however, his own actions made perfect sense. The president was no doubt an economic booster, who believed in the inevitability of American westward expansion (535–6), but he also believed in preserving as much land as possible for future generations (545), notably in Alaska (806). Therefore, while the president was indeed a booster for westward expansion and urbanization, he was simultaneously a "wise use" conservationist and a preservationist, depending on the circumstance (545). This moment is particularly important to American environmental history because it is the major decision widely considered to be the original split between preservationists and conservationists (Nash, 135–40), a deeply emotional argument that eventually moves into Alaska. At the time, Roosevelt, who was personal friends with Muir (Brinkley, *Wilderness*, 544–5), was also very close to a forester named Gifford Pinchot who was in the process of developing the idea of "wise use" conservation, or exploiting forests but also maintaining them. The idea was that people need resources such as timber, but that deforestation has potentially been the downfall of multiple societies (340–2; Brinkley, *Rightful*, 65–8). Both Muir and Pinchot had the President's ear regarding Hetch Hetchy (and the need for both potable water and preservation), but eventually the reservoir won out (Brinkley, *Wilderness*, 544–5, 636). In maintaining this flexible mentality on a "case by case" basis, Roosevelt has made enemies of preservationists for many generations, even with the unprecedented quantity and quality of preservation that he accomplished for future generations (535–6). On the other hand, Roosevelt is also frowned upon by modern libertarians and individualists of the West, who are frustrated that so much land has been locked up by the federal government (631–72).[49] Pinchot, as well, was an important factor in the preservation of much land, but he was never forgiven for his "wise use" conservation (544, 803–5). Meanwhile, Muir

49 Bump, Phillip. "The Oregon occupiers' land dispute, explained in 9 maps." *The Washington Post*. "https://www.washingtonpost.com/news/the-fix/wp/2016/01/04/the-oregon-occupiers-complaint-explained-in-9-maps/?noredirect=on&utm_term=.f4dd96f56ebd." 4/1/2016.

has become the patron saint for preservationists, and creator of the Sierra Club (Colwell, 20).

This discussion is pertinent to this research because the Hetch Hetchy argument and "wise use" conservation both came along just after the US Census Bureau declared the American Frontier closed and Frederick Jackson Turner's "Frontier Research" was developed along similar lines (Nash, 145–7, 260–1). Roosevelt understood this idea of closure as much as anyone and essentially wanted to set the country geographically in place. According to Nash, Roosevelt corresponded directly with Turner (149), voicing agreement with Turner's idea that the frontier and wilderness defined the nation.[50] Roosevelt also agreed that frontier and wilderness were both diminishing in the face of encroachment by civilization. "This alarmed Roosevelt chiefly because of its anticipated effect on national virility and greatness," says Nash (149). In practice, this meant that Roosevelt believed in the development of cities across the country, while preserving as much land for later generations as possible outside those cities. Roosevelt often argued that future generations must be considered, in lieu of merely taking and conquering as much as possible in the present (Brinkley, *Wilderness*, VI). Therefore, a shift away from 19th century Jeffersonianism was inevitable. Muir, however, was at odds with the solutions of that political moment. He would have preferred to have completely preserved it all, what is now called Deep Ecology (135–40; 160–70). From a certain perspective, Muir could be considered naive, as westward immigration was never going to ebb. When we consider the constant opportunities available, and the European traditions of conquering of the North American continent, this becomes especially clear. But Muir did not take this view. Muir was not a Roosevelt politician who needed to understand the dynamics of urban and rural life, and the country as a whole. He did, however, watch firsthand as the 49ers effected one of the worst manmade environmental disasters theretofore seen on Earth in their quest for gold in California (Colwell, 144–5). He

50 Barry Lopez finds Frederick Jackson Turner's ideas important enough to highlight them in his chapter "The Country of the Mind" in *Arctic Dreams*. Lopez uses Turner's theories to highlight the former's ostensibly paradoxical idea, " … the landscapes in which history unfolds are both real, that is, profound in their physical effects on mankind, and not real, but mere projections, artifacts of human perception" (256). Lopez mentions that in the 19th century, westward expansion affected the country's perception of itself, as well as its values "derived from an understanding of its citizens' experiences on the frontier," but Americans simultaneously spent much of their time affecting the land as well, "Politicians and promoters, newspaper editors and businessmen argu[ing] hotly over the suitability of the tallgrass and shortgrass prairies for farming."

would also live to see his beloved Alaskan glaciers overrun by tourism, affecting Tlingit communities, as well as that tribe's lands and waters (Meaux, 3–6).

As the West was essentially conquered by the 20th century, Alaska, so difficult to settle on a large scale, remained a site for resource extraction (Nash, 274–5). Resource extraction would therefore become the primary issue of environmentalism throughout 20th century Alaska (273–315). Up to this point, I have worked to prove how European cultures and the physical landscapes that influenced them in the U.S. led to a unique set of circumstances that would eventually result in the modern environmental movement of the 1960s (IX), right around the time that this research marks the beginning of the modern era in Alaskan literature. The following section will delve into the roots of the environmental movement of the 20th century, particularly focusing on the rise of The Environmental Justice Movement toward the end of the 20th century, and its importance to the state of Alaska and its multitude of peoples.

12 Environmental Justice in Alaska

The entire argument heretofore discussed has been based around an American ideology, which derived from, and eventually evolved away from, a mix of its European counterparts. Even into the 20th century, and debatably the 21st, we have seen a predominantly white literature from both travelers and environmentalists to Alaska. Toward the end of the 20th century, however, the Environmental Justice movement, in connection with what Lawrence Buell calls "second wave ecocriticism" (*The Future of Environmental Criticism*, 22),[51] reared its head, and particularly gained steam with the publication of *The Environmental Justice Reader*[52] in 2002. Buell says, "for first-wave ecocriticism, 'environment' effectively meant 'natural environment.' In practice if not in principle, the realms of the 'natural' and the 'human' looked more disjunct than they have come to seem for more recent environmental critics …" (*Future*, 21). What Buell is getting at is that more and more environmental critics are exploring humanity's relationship with nature, wherever they may be. This is particularly true in cities where humanity most often finds itself, and the question that naturally arises is: who has access to what relationship with

51 Buell, Lawrence. *The Future of Environmental Criticism: Environmental Crisis and Literary Imagination*. 2005. Malden, MA, Oxford, and Victoria, Australia: Blackwell Publishing, Second Printing, 2006.

52 Adamson, Joni, Mei Mei Evans, & Rachel Stein. Editors. *The Environmental Justice Reader: Politics, Poetics & Pedagogy*. Tucson: The University of Arizona Press, 2002.

nature? Hence the Social Justice movement's merging with Ecocriticism to create Environmental Justice.

The Environmental Justice Reader itself tackles a wide variety of issues, including urban access to natural environments, African American access to natural environments, Native American water rights, and urban waste and pollution concentrated in predominantly-minority neighborhoods. The collection also includes criticism of the historically patriarchal, White paradigm and its resulting "genocide" against Native Americans (Johnson, "Saving the Salmon, Saving the People," 271). The genocide argument is tenable and based around multiple generations of physical and mental abuse, as well as pollution and quite literal kidnapping of children, along with a cultural extermination in a bid to impose gentrification on Native American communities. But, not only was there a cultural genocide: there have also been multiple instances of quite literal genocides against Native American and First Nations communities: Andrew Jackson's "Trail of Tears" (Brogan, 68), the mania surrounding the 1849 California Gold Rush (Colwell, 144–5), and Promyshlenniki enslavement (as well as indiscriminate killing) of Aleuts (Borneman, 60). There is, therefore, an alternate history that must be discussed more broadly and deeply, but, unfortunately, there is a much slimmer oeuvre of literary publication to choose from with regard to Native American degradation than there is first-wave environmentalism to critique.[53] With that said, this section will explore environmentalism in Alaska, particularly through the lens of the Environmental Justice Movement. The Environmental Justice Reader will be employed as the lodestar for the convergence of the two, but individual authors and works will also be analyzed to guide the way.[54]

One work, entitled *Kivalina: A Climate Change Story*,[55] by Environmental Justice advocate Christine Shearer, employs the tiny Alaskan island of Kivalina and its fight against the oil industry as its focus point. Shearer's multiple techniques merit discussion herein, as they are novel in their attempt at attacking such a difficult and ostensibly convoluted issue. She begins with a short, personal story, discussing the experience of physically traveling to the island of Kivalina and the negative reaction that many people, such as a local taxi driver in Anchorage, have when she tells them that she is headed to investigate the

53 Hence the importance of ethnohistories such as *Russians in Tlingit America*.

54 Though African American demographics are often invisible in Alaskan literature, *Black Lives in Alaska* was recently published after it began as an exhibit at The Anchorage Museum. This is a work of Social Justice, but shades of Environmental Justice are implied.
Hartman, Ian C. and David Reamer. *Black Lives in Alaska: A History of African Americans in the Far Northwest*. Seattle: University of Washington Press, 2022.

55 Shearer, Christine. *Kivalina: A Climate Change Story*. Chicago: Haymarket Books, 2011.

lawsuit Kivalina is filing against various oil companies (1–5). She also discusses her first impressions of the island (3–5). She then offers a brief summary of Kivalina's environmental issues and its subsequent lawsuit (7–14) before stepping completely back from the local case altogether. As Environmental Justice is about advocating for people, animals, and places that are often fallen prey to disinformation campaigns, Shearer utilizes the bulk of the book's volume not to discuss Kivalina itself, but rather those disinformation campaigns. This is a bold move, as the book is quite literally entitled Kivalina, with a small boat on ice pictured on its cover. Shearer's technique is to move through the history of Environmental Justice against offending corporations, beginning with asbestos (18–23) and lead (23–6), and even before specific issues when corporations began to take the upper hand in the judicial system in the mid-1800s (17). Tobacco is where the story truly shifts, however, as this is the moment when large corporations begin to wage organized disinformation campaigns geared toward broad populaces (26). Shearer lays this history out comprehensively in her first chapter "Blueprint for Denial." The following chapter, "Shaping Legality," explains the evolution of the legal battles of these disinformation campaigns leading toward the late 20th century, and strategies employed by Koch Industries. A history of legal influence by the fossil fuel industry, including Bush Administration connections to the industry and the industry's involvement in Iraq, are discussed in a chapter entitled "Fossil Fuels and U.S. Power." Then, a look is taken at globalization and the reactionary policies by the U.S. government with regard to climate change in the chapter, "Global What?"

Kivalina is 155 pages, and from page 15 to page 100 there is hardly a mention of either Alaska or the island of Kivalina. What Shearer does instead is insert short paragraphs preceding each chapter, with quotes by Tribal Administrator Colleen Swan describing the physical situation in Kivalina, which has the impact of maintaining the reader's focus on the primary purpose and theme of the book despite the fact that it drifts so far away for such long sections. Therefore, when learning about 19th century lawsuits and the Iraq War, the reader will be reminded that the point is always to refer back to Kivalina's lawsuit and why it is tenable. It is clear in these introductions that they not only remind the reader that the work is about the island of Kivalina, but each chapter gives a personal connection to the island in order to establish sympathy when the chapter itself steps away from emotion and toward fact-based, technocratic argument. Part 2 of this research will delve quite deeply into techniques of empathy, sympathy, and emotion, so for now I will just say that this book establishes emotional connections without harping on them in a manner grating to a political climate successfully polarized by the disinformation campaigns discussed in the work.

Shearer's *Kivalina* rounds off rather completely, as the last roughly 40 pages return to Kivalina's dedicated fight itself. The reason Kivalina is suing the fossil fuel industry to begin with (or, 24 companies), is that the island is quite literally disappearing, as sea ice is no longer there to serve as a barrier from storms (10–1). Not only that, but Kivalina was not even traditionally inhabited. It served as a hunting island, and the local Iñupiat were semi-nomadic. In the mid-20th century, the U.S. government decided that all children should attend schools, and it established a school on the island. As it became too dangerous to take boats to the island in bad weather (and school attendance was enforced by law), parents began moving to the island, which had never before been permanently inhabited. Therefore, the town of Kivalina developed against the will of its inhabitants (106–7). So now, a town of people who never wanted to live on the island in the first place is forced to move entirely due to climate change, but it cannot afford to do so. Therefore, it is suing the fossil fuel industry for the industry's sustained disinformation campaigns. This work merits such a lengthy discussion as it is so extraordinarily relevant, not only to Environmental Justice but to the plights of Alaska Natives. The American government inadvertently sucked Native peoples into inextricable problems, such as stagnant living in unsustainable habitats, in order to assimilate them to the developed, industrial world, but the government only made situations unbearable in many cases.

To be discussed in Part 1 of this research as well will be the idea that the symbolism of animals and plants, of biomes, may not turn out to be very effective for fighting against climate change, as has recently been attempted, particularly with the polar bear. Instead, the "human element" may turn out to be the most influential argument. And, as the Arctic is experiencing the worst effects of climate change in advance of most other places in the world, the increasing desolation of its human element is already visible. This is what Shearer argues in her chapter entitled "The Human Face of Global Warming" (101–24).

Shearer's work, as a whole, works within the technocratic framework that will be covered in the "McPhee Conundrum" chapter in Part 1, basing itself on provable facts through judicial cases, precedent, and historical context, while also moving in and out of a human element of Environmental Justice by employing Alaska Natives without relying too significantly on empathy, but also employing it wisely. This is a convincing method, and one which this research will broadly promote throughout.

To step away from Kivalina, by the second half of the 20th century, particularly by the 1970s after the more-or-less-global, social revolution of the 1960s, Alaska Native culture was beginning to be respected on its own merits. This included a shift not only in its dealing with the government but in literature as well. The rise of a new generation of nature writers, including several of those discussed herein, including Peter Matthiessen, John McPhee, and Barry Lopez, all approach Alaska Natives with an open mind, beyond the negative stereotypes built up over years. As we have seen recently in numerous instances, notably in the fight for the Keystone Pipeline and the treatment of Native Americans in the process,[56] there has not been a complete turnaround in treatment, but there has definitely been a broad change in acceptance in both academia (*The Environmental Justice Reader* itself being an example here) and in the legal world (the passing of ANCSA – Alaska Native Claims Settlement Act – for example), not to mention in everyday life. For instance, Norman Maclean notes in his famous work, *A River Runs Through It*,[57] that people would openly chastise the female Native American partner whom his brother, Paul, took out on dates in Montana in the 1930s (24). It is obviously socially unacceptable to openly chastise based on race in the 21st century, even if it indeed happens in private (and perhaps, increasingly, in certain nationalist circles). However, from a literary standpoint, there are essentially two options for attempting to understand the Alaska Native perspective: 1) from travel writers and their first-person accounts (as well as increasing accounts from Natives themselves!),[58] and 2) literary writers piggybacking off the work

56 Levin, Sam and Will Parrish. "'Treating protests as terrorism': US plans crackdown on Keystone XL activists." *The Guardian*. "https://www.theguardian.com/environment/2018/sep/20/keystone-pipeline-protest-activism-crackdown-standing-rock." 20/9/2018.

57 MacLean, Norman. *A River Runs Through It, and other stories*. 1976. Twenty-fifth Anniversary Edition. Chicago: The University of Chicago Press, 2001.

58 There is currently no Alaska Native literary star, though other forms of media, such as podcasting and children's television, are breaking through into the mainstream. *Molly of Denali* is a children's program on PBS in the US and CBC in Canada, and is based on a young Athabaskan girl's life running a trading post with her family. A multitude of Alaska Native and First Nations people were consulted in the process of making the cartoon, which is an admirable step forward in the inclusion of Arctic peoples in broader cultural discussions.

Likewise, Canadian First Nations writer, Tanya Tagaq has been praised for writing *Split Tooth*, and Tagaq acted in the HBO series, *True Detective: Night Country*, which took place in a fictional Alaskan town on the Arctic Ocean that was based on Utqiaġvik and starred several Native people. Other notable examples of Alaska Native literature include memoirs by Ernestine Hayes and William Iġġiaġruk Hensley.

Hayes, Ernestine. *The Tao of Raven: An Alaska Native Memoir*. Seattle: The University of Washington Press, 2019.

Hensley, William L. Iggiagruk. *Fifty Miles from Tomorrow: A Memoir of Alaska and the Real People*. New York: Picador, 2009.

Jacobs, Julia. "With 'Molly of Denali,' PBS Raises its Bar for Inclusion." *The New York Times*. "https://www.nytimes.com/2019/07/15/arts/television/molly-of-denali-alaska-native-pbs.html." 15/7/2019.

of anthropologists in order to examine oral histories, as seen in Jonathan Raban's *Passage to Juneau*. Part 1 of this research will examine both of these approaches, and examine the problems of what is lost in translation, not just in words but in the cultural context being lost by younger Native generations who no longer go out on the ice or hunt for subsistence, which for thousands of years has been a life purpose for these peoples (Lopez, *Arctic Dreams*, 277).

Of course, due to its rich cultural diversity and its vast, relatively-minimally-touched biomes, Alaska is one of the key areas that must be discussed with regard to Environmental Justice. Much of the movement, as mentioned earlier, focuses on minority communities being pushed out of their neighborhoods due to gentrification, or not having access to parks and gardens, or dealing with toxic pollution. Alaskans, however, are increasingly beginning to deal with the negative effects of the environmental movement. This is to say that in the fight for cleaner cities, which often entails avoiding carbon pollution, there is a new rush, this time not for gold but for precious metals to create less polluting transportation and energy in those ever-expanding cities. As a result, mines are being explored in Canada, Nevada, and Alaska in search of copper, lithium, and cobalt, in order to construct electric batteries and infrastructure.[59] In this research, Nancy Lord's *Beluga Days* is the primary work of her oeuvre to be discussed, but, more recently she has explored the potential opening of The Pebble Mine, a long-debated political topic in the state, which could potentially have disastrous effects for one of the largest salmon streams left in the world. This discussion hangs at the periphery of a book she edited entitled *Made of Salmon*, where the importance of subsistence salmon is outlined by a number of short stories and accounts by various people throughout the state. This issue, however, is merely a continuation of various forms of resource extraction that have negatively affected the biomes of Alaska in roughly the last three hundred years, since European entry. In the mid-to-late 20th century, the issue was the taking of oil, which affected communities in numerous ways, including struggles with sudden wealth and pollution, as well as population influx. Before that, sea otter and seal furs were taken in large numbers, which, as mentioned, included the enslavement and deaths of large percentages of the Aleut population to do the work of clubbing the mammals (Borneman, 60). This genocidal, bonanza-based history is not often viewed from the Alaska Native perspective,

Lopez, Issa and Nic Pizzolatto. *True Detective: Night Country*. HBO. Los Angeles, 2024.
Tagaq, Tanya. *Split Tooth*. New York: Viking Press, 2018.

59 Sadasivam,Naveena."Report:Going100%renewablepowermeansalotofdirtymining."*Grist*. „https://grist.org/article/report-going-100-renewable-power-means-a-lot-of-dirty-mining/." 17/4/2019.

but that perspective should be kept in mind as we move into analysis of the 20th and 21st centuries, attempting to understand the care that must be taken when writing about the state and its Native peoples.

One of the questions that may seem insensitive but should nonetheless be explicitly approached in order to have a precise answer is: why should literature be sensitive to Alaska Natives at all? In fact, Alaska Natives make up a minority of Alaskans now and are often unnoticeable, even to Anchorage Bowl locals (where the vast majority of Alaskans live). Many Alaska Natives live in separate neighborhoods and/or villages from much of the rest of the population. The subjective conclusion to which I have come to subscribe regarding the question of Native sensitivity is in the history: if cultures want to maintain a strong evolution in understandings of human rights, which Western Civilization purports to do, then they should be sensitive to the histories of peoples upon whom they have trampled, in order, particularly, that dominant demographics not make the mistake of viewing trampled communities as having put themselves in the intolerable situations into which they have been put by others. Therefore, not exploring this parallel history of subjugated peoples would be irresponsible. Alongside this reasoning, there is also the idea Barry Lopez constantly explores throughout so much of his oeuvre, which is that industrial society could learn a thing or two from Native peoples, if only they look and listen.

PART 1

Mapping the Ethical Fault Lines

FIGURE 2 Alutiiq/Sugpiaq mask, ca. 1870, on view at the metropolitan museum in New York https://www.metmuseum.org/art/collection/search/752034

CHAPTER 1

The McPhee Conundrum

Octogenarian, John McPhee, was present from the start of what has been termed 1st-wave environmental criticism (Buell, *The Future of Environmental Criticism*, 18–20) with works like 1967's *The Pine Barrens*,[1] which debatably set in motion the preservation of large swaths of diminishing New Jersey wilderness (Weltzien, "John McPhee and the Question of Environmental Advocacy," *Coming into McPhee Country*, 125–8). McPhee has never accepted the term environmentalist, suggesting, as so many writers about labels, that it has been thrust upon him due to his love of the outdoors, a penchant he picked up spending the summers of his youth at a place called Camp Keewaydin in Vermont. His outdoor books have consequently been titled his "Keewaydin Stories" by some (Weltzien, Maher, "Introduction," 11). This chapter is dedicated to analyzing the decades-long debate over McPhee's place in the environmental movement in order that his influence in both Alaska and in the context of this research may be put into perspective. John McPhee may be considered a lodestar for modern, Alaska travel writing due not only to the popularity of *Coming into the Country* and the depth of criticism dedicated to that work, but also due to a mix of chronology (writing at a pivotal moment in the state's history) and varied geography (he wrote about several places in the state during that moment). As a result, analyzing his work will inevitably lead to further questions, branching out to the works of other Alaska travel writers and framing some of the dilemmas facing the region.

To begin, despite his careful employment of journalistic objectivity, John McPhee nonetheless became a polarizing figure for environmental writing in the second half of the 20th century. This is exactly because he does not "pull the lever and vote" (interview, 78), as he likes to call overt subjectivity in writing. In other words, he does not tell the reader how to feel. In a *New York Times* review of McPhee's 20-year, 5-book project on geology, *Annals of the Former World*, which eventually won the Pulitzer Prize for General Non-Fiction in 1999, David Quammen accuses McPhee of "adamantine reserve" (Quammen, "Rocks of Age").[2] In refutation, John McPhee's 1993 interview with Jared Haynes ("The Size and Shape of the Canvas," *Coming into McPhee Country*) can be put into

1 McPhee, John. *The Pine Barrens*. New York: Farrar, Straus and Giroux, 1968.

2 Quammen, David. "Rocks of Age." *New York Times*. "https://www.nytimes.com/1998/07/05/books/rocks-of-age.html." 5/7/1998.

 | DOI:10.1163/9789004734807_003

perspective. McPhee tells Haynes, "… I'm not going to tell the reader how to think. I don't care how many people who write book reviews make the point that they wish I would. I'm not a writer of sermons. I also have extremely strong feelings that if a point is to be carried somewhere, it's going to work best in the eye of the reader, in the judgment of the reader" (78). Essentially, McPhee is unwilling to bow to pressure that he focus on narrow subjects judged pressing in a black-and-white, Manichean sort of way. He prefers to view what he calls "life in the round" (78), or what elsewhere has been called a "multiradial" viewpoint (Lounsberry, "Pieces of the Frame: McPhee and Portraiture," *Coming into McPhee Country*, 87–8). McPhee also avoids giving his overt opinion much of the time in order to let the reader form a judgment, which is "a thousand times more firm than the one the critics are talking about, which is the editorial writer telling somebody how to vote." Daniel Philippon starts off his essay "Academic Air" (*Coming into McPhee Country*) by analyzing McPhee's dismissal of Kathy Smith's interpretation that all work is inherently subjective. McPhee calls this "academic air." McPhee professes that literary journalists do the best they can at being objective, but Philippon calls this somewhat disingenuous from McPhee, suggesting that the writer's subjective ideas can be expressed "in any number of ways" (282). Philippon later goes on to compare being a teacher to being a literary journalist, proving his point that no one, "not McPhee, and certainly not me" is objective, that teachers often attempt objectivity as McPhee does, but that they also have an agenda, and that McPhee uses his subjective "fictionalizing act" through "selection of setting, plot, and character" (294–5). So, as with many writers, there is at least a slight disconnect between how McPhee portrays his writing and how he actually writes,[3] even if he has been quite open about his writing process in interviews (Howarth, "Introducing John McPhee," *Coming into McPhee Country*, 25; Haynes, "The Size and Shape of the Canvas," *Coming into McPhee Country*, 54–78) and in lectures at Princeton University where he taught a class called "The Literature of Fact" (Howarth, 22) for decades.

McPhee, of course, is attempting to maintain the facade of the objectivity of the respectable journalist. In doing so, he is also glossing over his meticulous structure that he mentions he spends months preparing before even writing a piece, a technique which generally funnels the reader's thought process without even being noticed (Tredinnick, 64; Phillippon, "Academic Air," 290, 295). In this way, the author accomplishes the trick of influencing the reader while

---

3 Barry Lopez likewise argues in his interviews with Mark Tredinnick in *This Land's Wild Music* that he goes into his work with a blank mind and has no agenda. " … I am simply present—with the thought, the place, the idea. It arrives" (64).

simultaneously maintaining what he calls "life in the round" and "multiradiality," or portraying all sides of a social and/or political issue. Barbara Lounsberry compares McPhee in this way to Guy Talese, who also focuses on "different angles" of people. But, she says of McPhee, "He does this, however, not with Talese's goal of discovering the reality behind the appearance (the truth behind the facade), but in order to achieve a complete view. McPhee is not so much in search of what is hidden (the essential Talese assumption); rather, he is always angling to get it all" ("Pieces and the Frame," *Coming into McPhee Country*, 87). Even if the writer is "get[ting] at it all," however, it is nonetheless not only possible but likely that a writer would have both conflicting views, of subjectivity and "multiradiality," in his head at once, even if at the subconscious level, or even if he is in a constant fight with himself over the two. This idea will be explored in depth in Part 2 when discussing Dan Sperber & Hugo Mercier's work *The Enigma of Reason*,[4] with the theory that rationalization is entirely a "social construct" (127).

1 The Question of Manichaeism in Environmental Writing

If John McPhee is expected to "pull the lever and vote," we must analyze what it means to do so. To begin with, literature has often been a tool for environmentalism, but it is not the only tool. Greenpeace has long been involved with fights in the proverbial trenches, along with Muir's Sierra Club and many others. Especially pertinent to this research's arguments, in the 1970s and 80s, Greenpeace led a campaign to end the sale of fur pelts (Wenzel, 36, 41).[5] From the perspective of liberal Europe and Continental America, this made perfect sense as an evolution away from animal abuse (1). The problem was that the economic model of entire Inuit communities in the Arctic had become based around the selling of these pelts (4). Ethnologist, George W. Wenzel, wrote a comprehensive work on the issue in the Canadian Arctic, entitled *Animal Rights, Human Rights* (1991). According to Wenzel, Greenpeace confronted several Inuit communities, suggesting they move forward along with the rest of humanity (5–6). Wenzel countered by suggesting, "The animals rights movement has acted to dissociate contemporary Inuit from their own culture, history and ecology" (34). The directors of Greenpeace were perhaps tone-deaf to the

4 Mercier, Hugo and Dan Sperber. *The Enigma of Reason: A new theory of human understanding*. London: Allen Lane, 2017.

5 Wenzel, George. *Animal Rights Human Rights: Ecology, Economy and Ideology in the Canadian Arctic*. 1991. Toronto: University of Toronto Press, 1997.

needs of communities outside the developed world, as this reliance on the sale of pelts was in fact caused by the developed world, who was suddenly telling the Inuit to stop (5). When, for instance, European Canadians moved north and came into contact with the Inuit, or First Nation Peoples, disease, alcoholism, and a descent into second-class citizenry resulted (30–1). In order to give First Nation peoples modern lifestyles, where they could have access to healthcare, education, the cash economy, etc., these semi-nomadic peoples were settled into communities, often far away from subsistence food sources (25). Therefore, they were almost forced to work for money to buy snow machines and gasoline to travel the long distances from their new settled homes to reach subsistence foods (17–21, 25, 34). Wenzel describes how some worked for money to buy boats and snow machines they could only use on the weekends, while others only hunted, fished and worked on broken machines but could not afford many of the machines to get to their prey (17–9, 112–23). Living a subsistence lifestyle, then, had become a maze of compromises. On top of that, mercury has increasingly been working its way up the food chain due to long-term CO2 emissions, as well as working its way into the Arctic Ocean, which serves as the Earth's sink, where concentrations are much higher than elsewhere.[6] These First Nations people naturally have a low carbon footprint and want to continue to live a subsistence lifestyle, which gives their lives meaning (Wenzel, 7, 59, 61–2, 138–9). Greenpeace, in their headstrong view of the issue, refused to listen (56–63, 35–55). This is not helpful for communities who do not deal in black-and-white issues, but in survival. Likewise, Nancy Lord suggested to me that her work *Beluga Days* (2004), dedicated entirely to saving the Cook Inlet beluga whale, faced criticism from Outside environmental activists because of her involvement in a subsistence beluga-killing festival held by Alaska Natives. She noted that these agitations always come from Outside, because Alaskans tend to understand traditional subsistence communities.[7]

Nuance in environmental writing is therefore important. Sometimes this is forgotten amidst the various existential crises that planet Earth is facing at any given moment, but demanding, for instance, that Arctic peoples adapt to Subtropic Postindustrialism is a colonial mindset. This hyper-attention to nuance, to "get[ting] it all" as a journalist, is why McPhee employs such terms as "life in the round." McPhee explores the multiple aspects of his chosen subjects, such as both the positive and negative consequences of engineering feats

6 "Mercury in Arctic Marine Ecosystems: Sources, Pathways, and Exposure." *US National Library of Medicine*. "https://www.ncbi.nlm.nih.gov/pmc/articles/PMC4142812/." Viewed on 14/4/18.

7 Unpublished interview with me, 30/3/2018.

in the face of nature (*Control of Nature*),[8] and yet he does so without overt accusation, allowing facts, ostensibly objective observation, and overarching structure to insinuate his subjectivity.

From a certain perspective, we could nonetheless fault John McPhee for not choosing direct enough subjects, as Rachel Carson chose with her celebrated work *Silent Spring*, which alerted the world to the zeitgeist-shifting revelation that there were serious health issues associated with the mosquito-deterrent chemical, DDT. Carson was also extraordinarily scrupulous with her writing of that work. Knowing that she would be counter-attacked for her own attacks on wealthy chemical companies, Carson's rhetoric, like McPhee's, was careful (writing meticulous descriptions of DDT composition and its usage – 15–37) but accessible (creating a vivid projection of the American town for the average reader – 1–3; and how the average reader's life will be affected in multiple ways – 103–27). She maintained a constant scientific approach and hid the fact that she personally had cancer so as not to be viewed as attacking an entire industry out of emotion (Tarter, "Some Live more Downstream than Others," 214–5). However, had McPhee chosen a more specific aspect of Alaska's pipeline controversy from the environmental perspective, such as the potentialities of a spill, the downsides of the pipeline's warmth in attracting caribou, or the future devastation invited by building the necessary highway right through the heart of pristine tundra, he would likely have invited the polarization he has tended to avoid throughout his career. And he still would not have stopped the pipeline, as so much money and so many jobs were involved for the state. Whereas, in the case of Carson, DDT and other chemicals had been directly affecting communities all over the nation (*Silent Spring*).

1.1 *Encounters with the Archdruid*

In 1971, McPhee had just published a book entitled *Levels of the Game*, whose structure weighed two tennis players, Arthur Ashe and Clark Graebner, against one another, in life and on the court. Finding the structure interesting, the writer wondered if he could pit multiple people against just one person and found his protagonist in David Brower, head of the Sierra Club (Lounsberry, 96). The work would become *Encounters with the Archdruid* (1971), McPhee traveling with Brower on three expeditions alongside Brower's "natural enemies": Charles Park, a mineralogist, Charles Fraser an ostensible "eco-developer" of islands in South Carolina and Georgia, and Floyd Dominy, a dam-builder in the Arid West (*Encounters*; Lounsberry, 96). Brower served as the perfect grey-area protagonist

8 McPhee, John. *The Control of Nature*. New York: Farrar, Straus and Giroux, 1989.

for McPhee, as the writer could relish in his play with "life in the round." This is because, though he was undoubtedly one of the most effective preservationists of the 20th century, and, under his leadership, membership in the Sierra Club exploded, he was eventually fired for egotism and controversial practices (*Encounters*, 209–220). McPhee shows "multiradial" views of Brower, praising him one minute and calling him "evangelical" the next, with backhanded complementary comparisons to famed televangelist Billy Graham (83) and often employing his characteristic, anonymous-quote structure similar to that of Robert Marshall in the 1933 work *Arctic Village*. Here is an excerpt from *Encounters*:

> "His concern is wilderness. He doesn't really care what happens to people."
> "There is a pre-Eden strain in Dave—no question."
> "I will say this: I prefer Dave's vices to the virtues of his enemies."
> "They are crucifying him, and they are self-congratulating bourgeoisie." (MCPHEE, 209)

This manner of presentation, which worked just as effectively for Robert Marshall (though less anonymous in Marshall's case), and almost always served as a bit of entertaining humor for both writers, allows for the journalist to step back from the scene and let others have their say (the multiradial view), so that the writer himself may proclaim objectivity while nonetheless expressing subjectivity. No person is one-dimensional, and this is especially true with mercurial characters like Brower, who often displays many dimensions of himself simultaneously (Smith, "John McPhee Balances the Act," 276).[9] This particular group of anonymous quotes is part of a longer passage that recounts Brower's loss of leadership in the Sierra Club, which he had taken from a small group originally led by John Muir to an enormously powerful force in the environmental movement. The quotations also exemplify how loose logic and equivocation, even outright lies,[10] through means-satisfy-the-end, bleeding-heart empathy over reason, among other issues, could hurt a movement by creating as many enemies as it gathers supporters (208–220). People were either drawn to or repelled by Brower, who was eventually pushed out by the club he had transformed (218). Yet, this powerful, charismatic figure's success could not be denied. Such is the importance of multiradiality.

---

9 Kathy Smith also points out that McPhee is meticulous about showing multiradial views of Brower's "natural enemies as well" (276).

10 According to McPhee, Brower put out full-page advertisements attacking corporations such as the Kennecott Copper Corporation. One claimed that the corporation had created "AN OPEN PIT, BIG ENOUGH TO BE SEEN FROM THE MOON," which was apparently not factually correct. This campaign even cost the Sierra Club its "tax-deductible status" (*Encounters*, 37–8).

1.2 *Annals of the Former World*

Annals was different altogether from *Encounters*, as that unwieldy book had no overt environmentalist subject. There was no overarching statement about preservation as with *The Pine Barrens*, and there were no environmentalists pitted against "natural enemies" as with *Encounters*. However, *Annals* gained a lot of attention for approaching geology at a transitional moment, when Plate Tectonic Theory had only recently become accepted knowledge (131–3). McPhee once again latched onto a protagonist (Kenneth Deffeyes, a geologist at Princeton University where McPhee also taught – 5) and was subsequently led to other protagonists[11] for what eventually became a 5-tome,[12] 20-year project across the American Interstate Highway, US I-80. Annals won great acclaim, as it went into depth in the field of geology, making a subject not easily accessible to the non-specialist reader come to life with personal stories of his protagonists and their careers (107–9, 19–20, 157–82, 282, 437), and drawing out the wonders of Deep Time (124, 228, 432–3) through adventurous narration mixed with technical vocabulary and explanations thereof. On the one hand, the writer keeps his readers' attention through passages detailing subjects such as Mount Everest being composed of marine limestone (124). On the other hand, he focuses on the underlying logic and techniques involved in India "[driving] the Himalaya five and a half miles into the sky" (124), including the physical sounds of technical words themselves (31–2; Van Noy, "A Plate Tectonics of Language," *Coming into McPhee Country*, 211–2). This fascination with words is perhaps not unlike the appreciation of the timbre and tone of poetry performed in a language one has not mastered. The technical aspects of subjects, the jargon, the specificities, have always been of interest to McPhee, which can serve as a barrier-to-entry to his writing (Philippon, 285–6). However, the technical can also be a draw, as the writer himself notes: "Geologists communicated in English; and they could name things in a manner that sent shivers through the bones. They had roof pendants in their discordant batholiths, mosaic conglomerates in desert pavement ... " (31).

As stated in the introduction to *Coming into McPhee Country*, though, "Annals has its critics" (3). Aside from his "adamantine reserve," McPhee has been criticized in reviews for hagiography, according to William Howarth ("Introducing

11 " ... with the sedimentologist Karen Kleinspehn, then a Princeton graduate student and now a professor at the University of Minnesota; with Anita Harris, of the United States Geological Survey; with David Love, of the Survey, and with the tectonicist Eldridge Moores, of the University of California, Davis" (*Annals*, 5).

12 Book 1: *Basin and Range* (1981); Book 2: *In Suspect Terrain* (1983); Book 3: *Rising from the Plains* (1986); Book 4: *Assembling California* (1993); Book 5: *Crossing the Craton* (1998).

John McPhee," 36). Howarth does not cite specific attacks of hagiography, but Weltzien and Maher cite Harvard geologist Stephen Jay Gould, who "has objected to what he regards as McPhee's "semihagiographic treatment of geology fieldwork in the tetralogy"[13] (Weltzien, Maher, "Introduction," 3). This assertion may or may not be the case, but the defense of field work in *Annals* indeed adds to its humanistic view from a literary standpoint. Furthermore, even if a writer could come along and dig out the interesting, perhaps "cyber-punk," aspects of technological geology, it is the field geologists as subjects who fall directly in line with McPhee's tendency toward Emerson's "representative men." Gould's criticism, then, points out how McPhee can sometimes miss the multiradial view in a subjective preference toward a particular lifestyle and/or aesthetic. His fascination with Alaskan homesteaders and miners in *Coming into the Country* is perhaps another example. However,"semihagiographic treatment of geology fieldwork" is different than hagiography of actual protagonists, which Howarth combats by pointing out the darkness that McPhee has hidden in his praise of several protagonists (36–9). Bill Bradley, for instance, is addicted to perfecting basketball shots and stays outside for hours, pushing off dinner. McPhee also presents the downsides of fatigue, injury, etc. that affected Bradley (36–7). Howarth then goes on to list tennis players Clark Graebner and Arthur Ashe, and how McPhee portrays their respective darknesses (37–8). But, none of these men approach the difficulty McPhee found in portraying Henri Vaillancourt, arrogant protagonist of *The Survival of the Bark Canoe* (Haynes, "The Size and Shape of the Canvas," 58), or the "archdruid" David Brower, who has already been discussed herein.

Whether dark or bright, this habit of leaning toward ostensible hagiography in the sense of idealizing a protagonist[14] comes from McPhee's experience with "Profile" pieces at *The New Yorker*, where personages the magazine deems interesting are chosen and analyzed (Lounsberry, 85). Obviously, the mere fact of finding a subject interesting and choosing to write about that person almost automatically invites a sort of hagiography-as-idealization to some extent, as *The New Yorker* does not typically choose to publish "Profile" pieces on characters it does not admire. Criticism is saved for other pieces, seen by the likes of Jane Mayer in her investigations of the political process, for instance.[15]

13 *Annals of the Former World* is often considered a tetralogy. Each of the first four books were published independently, and the fifth book, not a full-length work, was published in 1998 with the collection.

14 "2: idealizing or idolizing biography." *Merriam-Webster*. "https://www.merriam-webster.com/dictionary/hagiography." Viewed on 27/3/2020.

15 "Contributors: Jane Mayer." *The New Yorker*. "https://www.newyorker.com/contributors/jane-mayer." Viewed on 27/3/2020.

Lounsberry indeed notes, however, that McPhee "has taken the *New Yorker* 'Profile' and turned it frontal, indeed made it three-dimensional, multi-radial, and panoptic" (79). What she means is that, whereas the profile was already an established form of journalism, McPhee developed a mastery of showing multiple sides of an individual, both critical and praising at once, while spending months painstakingly working out a structure to fit that multiradial view (Haynes, 57–8).

1.3 *Coming into the Country*

If *Annals of the Former World* is somewhat of an opus for McPhee, *Coming into the Country* is not far off in ambition, employing as its subject the entire state of Alaska. *Coming into the Country* was published in 1978, the year after the Trans-Alaska Pipeline System first pumped oil from Prudhoe Bay on the Arctic Ocean 800 miles south to Valdez in Prince William Bay (Coates, 255–62). By the late 1970s, McPhee had already published *The Pine Barrens* on his native New Jersey, coastal plain, pine forest and the encroachment of civilization thereof, and he had also published *Encounters with the Archdruid* about his travels with David Brower, head of the Sierra Club, with "three of his [Brower's] natural enemies." The environmental movement was bourgeoning but had not developed to the extent that it since has (Buell, *The Future of Environmental Criticism*, 1), so we must understand the context in which he is working. McPhee is toying with questions not as developed and scenarios of climate change, overpopulation, and pollution not quite as imminent as they currently are (Kolbert, *The Sixth Extinction*;[16] Lord, *Early Warming*).[17]

Coming into the Country is composed of three parts: "At the Northern Tree Line: The Encircled River," "In Urban Alaska: What They Were Hunting For," and "In the Bush: Coming into the Country." The first part obviously takes place "at the Northern Tree Line" of the Arctic Circle, which is a geographic destination where the forest ends and the land becomes largely bare, with sporadic stunted, or miniature, trees underfoot mixed with equally minuscule Arctic willow and moss (6, 48–53). McPhee is on a canoe trip with a small group of men, including National Park Service Planner, John Kauffmann, who would go on to write *Alaska's Brooks Range: The Ultimate Mountains* (1992).[18] This

16 Kolbert, Elizabeth. *The Sixth Extinction: An Unnatural History*. New York: Picador, 2015.

17 Lord, Nancy. *Early Warming: Crisis and Response in the Climate-Changed North*. Berkeley, Ca: Counterpoint Press, 2011.

18 The Brooks Range is coveted by climbers and outdoorsmen for its extreme isolation. It is very expensive to reach, as it is geographically distant from Anchorage and Arctic in conditions (John Kauffmann, *The Ultimate Mountains*). For that reason, this large mountain range is still relatively little known. However, it has inspired many writers, including John

part sets up the wilderness stereotype that we have of aesthetic wild Alaska, with wild bears and salmon packing the water "like a sky full of zeppelins" (6) and no one around for miles. Except, a convoluted mess of caveats is soon revealed to the reader, toppling the construction of wilderness that McPhee has just put in place. The writer suddenly mentions a brief story of recent occurrence wherein he had caught an aluminum soda can in a stream near Anchorage while fishing (9). He then states that the cost of getting to the crystal clear water, with all the salmon and bears to be feared, is exorbitant. The average Anchorite cannot afford to "air charter" all the way to the actual wild,[19] beyond the tamed environs of their city (9–10). Going even further, McPhee quotes one of his acquaintances on the trip, Bob Fedeler, who says, " ... that country near Anchorage, where many people have summer cottages, has long since been virtually fished out and is now supplied with stocked rainbows from a state hatchery. 'That is the myth of Alaska,' he says" (8–9). This is a classic multiradial McPhee maneuver, setting up a scenario and tearing it down.

A comparison can be made to the 1993 interview with Jared Haynes, wherein McPhee describes being enraptured with his protagonist, Henri Vaillancourt (*The Survival of the Bark Canoe*), who initially comes off as the perfect Emersonian "representative man" when in his element as a builder of traditional custom birch-bark canoes (Haynes, 58–9). Vaillancourt, however, turns out to be overly-arrogant and completely out of his element when McPhee actually goes out on a canoe trip with him; a batch of spoiled, homemade jerky serving the same rhetorical purpose as the can of soda that opens the reader's eyes to the truth in *Coming into the Country*. With *Survival*, McPhee could have pulled his protagonist apart piece by piece as some critics suggest he do more of, but instead, he deliberately shows the reader the positives he sees in Vaillancourt before mentioning Vaillancourt's prevalent

McPhee, Robert Marshall (*Arctic Village*), and Nancy Lord (*Rock, Water, Wild*, 29–36). As discussed in the essay, "In the Giant's Hand," from the book just cited, Lord decided to move to Alaska permanently after a summer trip to the Brooks Range while in school in 1971. She has never left.

19 The term "wild" is used here for lack of more precise vocabulary. It could be argued that there is no "wild" left in Alaska, as people leave traces everywhere (27), and the advent of the pipeline system means that trucks are constantly passing through even the Brooks Range, and planes are flying over head. Finding somewhere truly separate from society is, then, difficult. From another angle, however, McPhee argues that one goes to wild Alaska to quite literally face one's own mortality. For him, the danger posed by bears is representative of the wild (368), but he could just as easily be referring to any mortal danger one finds outside civilization. At another point, he calls the Brooks Range wild because of its lack of trails and the facility of getting lost (47), which is another way of facing mortality.

personality flaws. Likewise, as McPhee begins *Coming into the Country* with a glimpse of the pristine, the salmon, the clear streams, the bears (bears being an obsession throughout the book), he then informs us that Alaska is in fact not as pristine as conceived in the popular imagination (10). With this rhetorical maneuver in mind, it becomes more difficult to accuse McPhee of reservation; page 9's revelations through Bob Fedeler are much too clear to be considered reserved.

Part 1 does indeed end with McPhee playing devil's advocate to Kauffmann, who proclaims the need for truly wild spaces for humanity. McPhee asks, " … why do you sneakerfaces, you ecocentrics, think you need so much of it? Why do you need eighty million acres? … Entire mountain ranges?" (78). Calling Kauffmann a "sneakerface" and an "ecocentric" is a concise and highly-weighted maneuver that enables McPhee's journalistic ethics of objectivity the space to allow Kauffmann to subsequently preach about scale in Alaska for multiple paragraphs, clearly laying out the environmentalist viewpoint, allowing it to fully breathe without the writer himself appearing subjectively for the cause (78–9). This conversation is also stated after a conversation with local booster, Robert Atwood (76), strengthening the environmentalist viewpoint through narrative structure by allowing it the last word (though McPhee admittedly does no name-calling of Atwood).

Part 2, "In Urban Alaska: What They Were Hunting For," is about Urban Alaska, mostly Anchorage. The section opens describing a helicopter trip to visit different spots for a potential move of the state capital, which has just received a boon of roughly $900 million in oil leases after the private company Atlantic Richfield (ARCO) discovered vast reserves in the Arctic (Coates, 164, 179). McPhee explores the rationality of the thought process of moving capitals, "as most of them have done at one time or another" (88) and discusses whether Juneau, the extraordinarily difficult-to-reach city (111), should keep the title, whether Anchorage, the economic center of the state (89), should take it, whether Fairbanks, the centrally-located city, should get it (91), or whether a new capital should be built altogether (90–1). The plot shifts laterally back and forth between explaining the odd phenomenon that is Anchorage and explaining the capital crisis. Anchorage itself is another perfect multiradial McPhee subject in that it is essentially akin to poorly-designed suburbs of the Lower 48, yet it is unique in certain ways, including proximity to numerous mountains, rivers, and the Cook Inlet. In order to get his point across about Anchorage, McPhee employs the anonymous quote structure discussed in the *Encounters with the Archdruid* section above. This time, however, the author randomly inserts content from Anchorage commercial signs into his descriptions of the city. This extra layer added to the technique serves as a form of humor while also clearly limning the absurdity of establishing a Lower 48 suburb in subarctic wilderness:

> Almost all Americans would recognise Anchorage, because Anchorage is that part of any city where the city has burst its seams and extruded Colonel Sanders.
> 'You can taste the greed in the air.'
> BELUGA ASPHALT (112–3)

McPhee continues on with his explanation, randomly inserting the names of restaurants and small businesses. Don Delillo would employ this technique throughout his 1985 National Book Award-winning work of fiction, *White Noise*,[20] less than a decade later and to much the same purpose, in an avant-garde vein of environmental justice literature (Reed, "Toward an Environmental Justice Ecocriticism," *The Environmental Justice Reader*, 149–51). The technique can seem a bit chaotic and unorganized. Daniel Philippon notes that this was the reaction of several of his own students when they read McPhee's work ("Academic Air," 285), but there is also a certain visceral clarity anyone who has passed through such American suburbs would understand, which would not otherwise be fully accomplished were the writer to explain the issue with straightforward description. Partly because of the humor in the technique, in the indirect nature of it, this style is more in line with New Journalism,[21] McPhee's forward-thinking side, than with the directness and sincerity of much environmental writing. McPhee's traditional, transcendentalist vein, then, tends to come more from subject matter than technique; his technique often more in line with his avant-garde contemporaries such as Don Delillo and Hunter S. Thompson. McPhee goes on to say:

> Down below, among the high buildings, are houses, huts, vegetable gardens, and bungalows with tidy front lawns. Anchorage burst out of itself and left these incongruities in the centre, and for me they are the most appealing sights in Anchorage. Up against a downtown office building I have seen cordwood stacked for winter. (114)

20 Delillo, Don. White Noise. New York: Viking Adult, 1985.

21 Hunter S. Thompson's *Fear and Loathing in Las Vegas* comes to mind, with the author testing his limits on a binge of various drugs in and around Las Vegas, with the kitsch of carpets, lights, and convertibles on full display, both enjoyed and mocked simultaneously throughout the experience.

This second passage yet again depicts multiradiality: despite the fact that Anchorage can be viewed as ugly and an ecological nightmare to the aesthetics of 1st-wave environmental criticism,[22] there is something endearing about the city as well. It is a basic American city, like El Paso, Trenton, or Daytona, as McPhee mentions just afterward, but it still maintains small houses with vegetable gardens and cordwood against a modern downtown high-rise.[23] This is something one would not find in most modern American cities, and it is the rawness itself, the lack of zoning McPhee mentions, that appeals to him in his ambivalence.

If McPhee's structure can seem a bit daunting to some (Philippon, 285), it is also effective. For example, when shifting the story laterally back to the helicopter flights, one of the delegates professes a desire for a new capital to be established at the foot of Denali, even on a pristine lake nearby where a float plane landing strip could be set up to lead right to the new mansion (128). McPhee never states outright that this is a terrible idea. In fact, he has already suggested that moving capitals itself is common, leading the reader to maintain an open mind on the subject. However, having this overwhelmingly negative passage about the downsides of fast food and "cookie cutter" consumer society is a subtle message that sticks with the reader when subsequently approaching the aesthetic desire for a palace on a pristine lake to be carved out of the wilderness.

Nowhere is safe from economic progress. Critics miss the whole point when they attack McPhee for his moderation in decisions such as giving Alaska newspaper editor and owner, and local booster, Robert Atwood, a fair interview (76, 115). Atwood explicitly blasts all conservation efforts, and McPhee has no immediate, subjective riposte. Except, as noted by multiple essays in *Coming into McPhee Country*, the author's plots and structures are extraordinarily important to his form of journalism (Weltzien, Maher, 6–7; Howarth, 22–4; Lounsberry, 91–8; Philippon, 285–297). McPhee does truly seem to want to maintain the goal of objectivity of the reporter, exploring multiple sides of issues in his writing, but, like all human beings, he has an opinion, which instead of being didactic and polemic is insinuated in the structure itself (and in the subtle irony mentioned above). Through a panoptic view, *Coming into the Country* presents pristine nature encroached upon by helicopters of greedy

22 In the essay, "Toward a Symbiosis of Ecology and Justice," in *The Environmental Justice Reader*, Tom Lynch talks about the line between real issues like exhaustive water use and environmentalist aestheticism such as a few aluminum wrappers left on the ground employed in John Nichols' *The Milagro Beanfield War*, something Lynch calls "a superficial middle class environmentalism" (253).

23 From personal experience, there are pet reindeer in yards downtown and people cross-country skiing to work in winter. Moose regularly wander into parking lots of commercial centers and eat leaves from trees.

politicians, and pipelines, and development, as McPhee prods the reader to ask to what extent these things are necessary or desirable for humanity. Like most pragmatic people, McPhee understands that some form of encroachment is necessary for human civilization, but so is "preserv[ation of] plenty of land as well" (379), as he mentions in Book 3. One cannot come out of the first two books without having thought twice about boosterism. The message is all too clear, if subtle.

Having set up the debate with analyses of pristine nature and the urban, McPhee then goes even further into the moral grey area in Book 3, the eponymous "In the Bush: Coming into the Country," by heading toward the Yukon River, not far from where Robert Marshall famously traveled in 1929 to live with the miners of Wiseman (*Arctic Village*). The original placer miners were gone by the 1970s, but some people are still digging in McPhee's work, mostly employing earthmoving Caterpillar bulldozers now[24] (200, 205). Subsistence has become the primary goal for much of the new generation living in the Interior, instead of mining (159–167, 294). McPhee's characters often start off in Anchorage or Juneau, working for a season, or even a year, to save up before heading north to towns on the Yukon River like Eagle or Circle, where they learn from more experienced trappers how to truly accomplish a subsistence lifestyle and to what extent they can or should (177–82, 212–5). They have built cabins along the river and have begun trapping, if they indeed make it that far in the extraordinarily difficult transition to subsistence (161). Amidst this process, while McPhee is visiting, the pipeline is being built, and the government is in the process of negotiating national preservations (213–15). Officials of the Bureau of Land Management fly in by helicopter to visit these cabins on the Yukon that people have made their life's work accomplishing, and the officials tell them to evacuate immediately, as they are on government property: "This is now the twentieth century. You can't just do what you want to do. You cannot play with the wilderness" (212). Once again, McPhee thrives on this type of conundrum because it blends into a moral grey area almost by definition: so many proponents of the ecocritical movement advocate for self-sufficiency, the Emersonian and Thoreauvian models so often cited as influences to this day (Buell, *Future*, 26–7; Terrell Dixon, "Environmental Justice," *The*

24 Comparatively, the two works diverge in that Marshall portrays the miners of Depression-era Wiseman as very socialist. In fact, Marshall, a socialist himself, overtly pushes Wiseman as the perfect social society in the way that its miners aid each other (*Arctic Village*, 101–16, 196–7). By contrast, McPhee's miners of the 1970s are more in line with anarcho-capitalism (which will be described in Part 2 of this research when discussing Nancy Maclean's *Democracy in Chains*), wanting the government out of their lives at all cost so that they may mine as they please.

Environmental Justice Reader, 19; Suberchicot, *Littérature et Environnement*, 10–2),[25] but these people building cabins along the Yukon, the actual models of the low carbon-footprint, self-sustaining stereotype of those Transcendentalists, are nonetheless overtrapping, cutting down too many trees, and polluting river beds on federal preserves. As stated by John Kauffmann in Book 1, there has to be some place for people to go to get away (78–9). Kauffmann advocates getting away as a temporary visit, but others like Fedeler argue that temporary visits are elitist because the average man cannot afford to get to the wilderness for a vacation (9).

From the "representative men" standpoint, these people had to work long hours just to afford to get to the Yukon to work night and day to get by without the strings of modern civilization and government oversight attached (213–5). McPhee employs the phrase Alaska critic Susan Kollin suggests so ardently that we avoid, "the last frontier" (*Nature's State*, 2–11): "The Alaskan wild was, as advertised, the last frontier – where people willing to combat its cold and run its risks could live an existence free from supererogatory rules as long as they did no harm to one another" (*Coming*, 213). McPhee devotes several pages to proving the legitimacy of the right of these people to live in the bush. In this case, he indeed subjectively and overtly "pull[s] the lever and vote[s]" by suggesting that allowing these people to remain in the bush is a better scenario than " ... some faceless corporation 'responsible' to a hundred thousand stockholders, making a crater you could see from the moon" (379). The entire argument arises, McPhee notes, because oil has been found in Prudhoe Bay, and Alaska Natives will have to be compensated with both land and money. Indeed, homesteaders and miners had been coming to settle in the Country for decades, but the Alaska Native Claims Settlement Act (ANCSA), an agreement with Alaska Natives that would allow construction of the pipeline in exchange for money and preservation of land, was suddenly kicking them out (212–5).

And yet these homesteaders and miners really were depleting resources. The poet and author John Haines talks about this phenomenon in his 1989 book of essays, *The Stars, the Snow, the Fire* (14–5).[26] Wild game is over-hunted or -trapped.[27] Too many trees are cut down for cabins, and more people keep coming, along with the pipeline, which Haines states matter-of-factly are

25 Suberchicot, Alain. *Littérature et Environnement : Pour une écocritique comparée*. Paris: Éditions Champion, 2012.

26 Haines, John. *The Stars, The Snow, The Fire: Twenty-five Years in the Alaska Wilderness*. 1977. New York: Washington Square Press, 1989.

27 In the same passage, on pages 14–15, Haines also discusses the resilience of the local game. When animals are left alone for a while, their numbers tend to rebound.

pushing out the game he has been trapping on a certain ridge (67). In recent years, around John Haines' preferred trapping territory, the environs of Fairbanks, the EPA has put restrictions on burning firewood due to sustained problems with air pollution.[28] This briefly became a micro cause célèbre on social media among politically conservative Americans because it meant that the government was even telling people on the "frontier" that they could not heat their homes with wood, another example of there being nowhere left to go to get away.[29] The two ideologies put side-by-side: should there be a place to get away to visit, and should there be a place to get away to live? The two are at conflict with one another, and nothing is clear. McPhee, however, typically ostensibly aloof, interjects himself. He openly admires what the people he meets along the Yukon are accomplishing. He also shows their flaws. For instance, McPhee spends a lot of time with a family, the Gelvins, who tear up an entire creek bed with a "Cat" (Caterpillar D–9) in order to reach leftover veins of gold. McPhee comes away in favor of "preservation, of people like the Gelvins. To be sure, a lot of land as well" (379). He wants people to see that the story in Alaska is purely grey, unlike the black and white view that people often created from Outside. There are always things for which to strive, risks to take, and concessions to be made if one is sensible enough to view the entire picture and not lie to oneself. John Muir did not see this until it was too late, and his proselytizing had led to an overabundance of vacationers to Alaska who did not treat it well and rendered the very wilderness he adored "civilized." This was a major regret (Haycox, *Alaska: An American Colony*, 197–9). Though John McPhee has had nowhere near the environmental influence that John Muir had, McPhee is more cautious and does not make this mistake. He constantly makes the reader aware that any of these repercussions are possible at any time, and, in the 21st century, the ecocritical movement is beginning to see that even the desire to take "eco-vacations" has led to degradations of cultures and increasing airline pollution, as ecocritic Allison Hawthorne Deming so poignantly notes in *The Edges of the Civilized World*[30] (50–1, 183–8). Therefore, we find ourselves much where we started but with a clearer understanding: nuance and caution slowly

28 "EPA, Alaska to Work Together on Improving Fairbanks Air Quality." *EPA*. "https://archive.epa.gov/epa/newsreleases/epa-alaska-work-together-improving-fairbanks-air-quality.html." 01/05/2017.

29 "EPA Moves To Fine Alaskans For Burning Wood To Keep Warm." *Redstate*. "https://www.redstate.com/streiff/2016/12/31/epa-moves-fine-alaskans-burning-wood-keep-warm/." December 31, 2016

30 Deming, Alison Hawthorne. *The Edges of the Civilized World: a journey in nature and culture*. New York: Picador, 1998.

chip away at an issue while straightforward proselytism creates both positive and negative consequences.

2 What Does John McPhee's Work Mean for Alaska?

Susan Kollin is careful to praise McPhee in *Nature's State*, saying that he and Barry Lopez " … dismantle notions of Alaska as a pastoral or wilderness retreat, a place somehow set off from the rest of the United States or the world" (46). She also says, "Throughout his book, McPhee takes care to distance himself from some of the more problematic aspects of the mythic Alaska that Muir and Marshall helped set in motion." She then describes a dynamic Alaska dealing with economic issues and not some "pristine" wonderland, praising McPhee's multiple views of Anchorage (46). She even goes so far as to say, "In McPhee's account, the Last Frontier desperately needs an urban ecology. Thoreau's pastoral legacy, responsible for shaping so much of Euro-American self-understanding, has become an environmentally costly one in Alaska" (47). An environmentally favorable comparison to Thoreau is a very strong remark from the cautious critic, but she goes on to criticize McPhee as well. She states that the writer attempts to " … distance himself from the stereotypical sentiments directed toward Alaska," but "at times he, too, is caught up in some of the exceptionalist rhetoric shaping the region" (49). Her most poignant point is that "McPhee contributes to the notion that the region is somehow an anomalous space unlike any other American terrain, a place fully set off from any other U.S. region" (50). Kollin is very meticulous in both her praise and criticism of McPhee. The difference in opinion between her criticism and my own is, I submit that in order to get people to read a work, a bit of aestheticism in the form of praising the stereotype of the "wild" or "The Last Frontier" is necessary. Unlike locally focused works that do not fall to this stereotype, such as Nancy Lord's *Beluga Days*, *Coming into the Country* was a national best-seller, opening the broader U.S. to Alaska's struggles and educating them. The ultimate problem in Alaskan literature, then, is to which community we are writing. Kollin notes an argument by Thomas Birkland that Alaska has suffered from its own pristine stereotype because the oil companies have been able to manipulate it. These boosters argue that the state is so massive as to be unchangeable, and that their exploration would consequently do no major harm (*Coming*, 50). This is an important point on the part of Kollin and Birkland. The job of McPhee, however, was that of an Outside travel writer. He was writing for a broad audience, piquing their interest in an informative manner. If his book was flawed, it was only slightly flawed, while still being attentive to the needs of the communities he interviewed. Writers like Nancy Lord and Tom Lowenstein, we will see, tend to write incredibly carefully, with the utmost attention to the communities they visit, but they both struggle to reach a broader audience the way that McPhee has. I submit that both styles are acceptable because there should be a multitude of voices in environmental writing, on different scales and for different audiences, who accept one another's differences, even if writers are at times critical of one another. This is a theme I will return to quite frequently from here on out in this research.

A final point to make is that McPhee has succeeded in forming a mosaic including a multitude of different communities in the Alaskan spaces he explored, and the work maintains coherence due to his meticulous structure. By setting up one part on a canoe trip with preservationists, one part with politicians in Anchorage and in a helicopter, and one part with the people on the Yukon, McPhee explores almost all possible angles. He even "profiles" the Alaska Native politician William Iġġiaġruk Hensley,[31] who was educated in the American South, as well as describing the lives of the inhabitants of a local Gwich'in community. Comparatively, the other notable pipeline-era work of Alaska travel writing is *Going to Extremes*[32] by New Journalist, Joe McGinniss. This best-selling work is much less scrupulous than McPhee's. Kollin attacks McGinniss for a lack of sensitivity to Natives, as McGinniss creates an idealistic description of an Alaska Native man carving wood for hours, very still and concentrated, before suddenly jumping up and running to watch his favorite television show (Kollin, 169; McGinniss, 111–2). From one standpoint, McGinniss makes his living by putting himself in odd situations and writing whatever happens, often in a humorous manner (this is not uncommon in New Journalism). For instance, McGinniss also rides to Alaska in a pickup truck with Duane Archer, a shady Anchorage businessman (*Going to Extremes*, 12) and spends time with a government assistant who professes spending most of her free time at bars snorting cocaine with politicians in Juneau (16–8). These are revelatory experiences and situations about which John McPhee likely would not allow himself to write, as McPhee tends to be more scrupulous, spending time with "representative men": artisans, politicians, etc. McGinniss, however, easily falls into the trap of insensitivity. The point that McGinniss appears to be making is a criticism of capitalism, insinuating that if this very traditional Alaska Native can suddenly run from his wood carving to watch television then no one is safe from low-brow capitalist culture, not even at the ends of the Earth. The juxtaposition of the staid Native person suddenly running to watch an emission,

31 Hensley would go on to write his own memoir entitled *Fifty Miles from Tomorrow* (2008).

32 McGinniss, Joe, *Going to Extremes*. New York: Alfred A. Knopf, 1980.

after hours of zen-like stillness, is also humorous. But, this passage can easily be read as "punching downward" toward an indigenous minority. Unlike McPhee, who interviews to exhaustion, and unlike Lord, who reads, interviews and studies the various angles of her communities (*Beluga Days, Early Warming, Made of Salmon*), and unlike Jonathan Raban, who is an incurable academic at heart with a passion for a blend of classic travel writing (Cook, Hakluyt, etc.) and ethnology (*Passage to Juneau*, 21–37), McGinniss merely describes what he sees. Therefore, when McGinniss sees a drunken Native wetting his pants at a bar (23–4), or a Native obsessed with television, the history of the region, the generational degradation by colonialism that has led these scenes is lost on the reader. The result is a continual buildup of stereotypical prejudices already formed against that culture. In urban Alaska, the problems of homelessness, joblessness, and alcoholism among Native communities are highly visible, and it is easy to develop judgments based on these observations, which is why ethical travel writing requires background and exhaustive interviewing to create a multiradial view.

FIGURE 3 Tolovona river valley
PHOTO BY CHRIS FERGUSON

CHAPTER 2

Masculinity v. Femininity

Aside from the geographic particularities of Alaska, among the important topics to be explored is that of sex. This is primarily because the Ecofeminist movement, dealing with a "matriarchal" perspective in contrast to that of a prevailing "patriarchal" perspective, began gaining traction toward the end of the 20th century (Gaard and Murphy, *Ecofeminist Literary Criticism*, 5)[1] and has become increasingly relevant in light of the #metoo movement.[2,3] Both Ecofeminism and the #metoo movement are no doubt equally as relevant in Alaska as they are Outside, but one of the primary questions being asked in this chapter is whether those movements have manifested themselves differently in Alaska, and if so, in what ways? Understanding the nuances of these differences could potentially shed light on another dimension of perspective with regard to environmental and ethical responsibility.[4,5]

The chapter will begin by exploring the tradition of women in Alaska through its literature. It will analyze works by Libby Beaman, Margaret Murie, and Lois

1 Gaard, Greta and Patrick D. Murphy. Editors. *Ecofeminist Literary Criticism: Theory, Interpretation, Pedagogy*. Urbana and Chicago: University of Illinois Press, 1998.

2 "History & Vision" *metoo*. "https://metoomvmt.org/about/#history." Viewed on 2/4/2020.

3 This is not a definition, but a loose description of what the movement is attempting to accomplish, according to the citation noted just above: "The 'me too' movement supports survivors of sexual violence and their allies by connecting survivors to resources, offering community organizing resources, pursuing a 'me too' policy platform, and gathering sexual violence researchers and research. 'Me Too' movement work is a blend of grassroots organizing to interrupt sexual violence and digital community building to connect survivors to resources. As the 'me too' movement affirms empowerment through empathy and community-based action, the work is survivor-led and specific to the needs of different communities."

4 However, this is not an exploration of 21st century gender studies, which has a much larger scope and is fertile ground for a future researcher. Rather, it is a look at how gender roles have been interpreted and subverted in Alaska travel writing, often deriving from frontier expectations, which can sometimes border on the stereotypical from the modern viewpoint. They are nonetheless real, and have led to horrific consequences, as Alaska's short history has a very high rate of abuse against women and even femicide. Nancy Lord approaches this topic in her short story "Marks" about a man who transports women to wild areas in a helicopter and hunts them like prey, which is based on a true story.

5 In this work, Elizabeth Beaman and Margaret Murie are known affectionately as Libby and Mardy. This is how they chose to identify themselves and were called by others, in Murie's case as a matriarchal figure guiding the country and numerous highly influential men toward preservation of the Brooks Range. While I do identify male authors by their last names, I also call Nancy Lord by her last name, as well as female researchers, such as Ann Fienup-Riordan.

 | DOI:10.1163/9789004734807_004

Crisler, while simultaneously examining how women's roles have potentially been unique in Alaska, due to the fact that many of the women it attracts have a tendency to be authoritative in a manner different from the expectations of women in the Lower 48, as different skills have traditionally been required to excel in Alaska. Ecofeminism will then be explored, in order to add another dimension of clarity when discussing "masculine" and "feminine" in modern environmental writing. Finally, this section will finish up with Nancy Lord's form of 21st century Alaskan feminism and how it pulls together multiple aspects of the traditions that preceded her. It will demonstrate how her writing, particularly her later writing, codified by her work of investigative travel journalism, *Beluga Days*, has been able to move beyond the constant theme of "breaking free" from societal norms for women. Likewise, it will discuss how modern male writing has also begun to break free from societal norms by adding dimensions of emotion rarely displayed in previous generations.

First, a brief exploration of terminology should be considered. There is a tendency to state that many white women who have come to Alaska have been "strong" due to the rigorous lifestyle required outside civilization, or at least at its logistical fringes. The term "strong" connotes both physical and mental facilities. Many women, such as Libby Beaman, displayed enormous mental fortitude in convincing others to allow them to even travel to the region, and even more strength in the face of prejudice once they were in Alaska. The problem remains, however, that the word "strong" has vague connotations. Emily Dickinson, for instance, was a strong woman, fighting off gender oppression and agoraphobia, but yet she hardly left her own property, and at times her own house. Sandra Gilbert and Susan Gubar suggest that she internalized her fears and weaknesses, as well as "her anger at female subordination," and turned them into strengths in her poetry: her "life itself, in other words, became a kind of novel or narrative poem ..." (*Madwoman in the Attic*, 84–7, 583–6).[6]

Susan Kollin notably avoids such characterization as "strong," but, at one point in her chapter on female Alaska writers, entitled "Domestic Ecologies," she employs the term "authoritative" to describe these women (94). The most accurate definition for the connotation she employs is likely the *Lexico* dictionary's second entry: "Commanding and self-confident; likely to be respected and obeyed."[7] Such a connotation allows for both mental and physical strength,

6 Gilbert, Sandra M. And Susan Gubar. Editors. *The Madwoman in the Attic: The Woman Writer and the Nineteenth-Century Literary Imagination*. New Haven: Yale University Press, 1979.

7 "Authoritative." *Lexico*. "https://www.lexico.com/en/definition/authoritative." Viewed on 1/4/2020.

but it also allows for the kind of leadership roles and mutual respect within Alaskan communities that should be projected.

One more note to mention is that in Europe, Anglophone travel writing studies have tended to maintain a particular focus on a few particular women who have been rightly recognized for such authoritative character in their travels, such as Mary Wollstonecraft in Scandinavia (*Letters Written During a Short Residence in Sweden, Norway, and Denmark* – 1796)[8] and Anna Jameson in Canada (*Winter Studies and Summer Rambles in Canada* – 1838).[9] Numerous other women have traveled widely as well, such as Robert Peary's wife, Josephine, who accompanied her husband to northern Greenland, well inside the Arctic Circle).[10] Furthermore, countless Native women have no doubt traveled through these very same spaces, but there are few-to-no examples of their travels in literature. As groundbreaking as Ernestine Hayes' works *Blond Indian*[11] and *The Tao of Raven* are, both are grounded in Juneau-based memoir. No doubt, countless other women of varied backgrounds have carved the way for, and influenced, several writers discussed herein, but this chapter will focus particularly on women who have traveled to Alaska and have written about the experience.

1 Libby Beaman

Nancy Lord has expressed in relation to this research that people tend to be judged by their capabilities in Alaska more than by their sex.[12] Examples could be given otherwise, as in McPhee's *Coming into the Country* when Dick Cook's companion, Donna Kneeland, is essentially treated as an able body to do housework while he is out trapping (159–61). It is hard to deny, however, the plethora of authoritative women who have come into the Country. This section will start out by examining the experiences of Libby Beaman, a young woman from a well-to-do background of mid-19th century Washington D.C., who made her way to Alaska and lived there over the winter of 1879–80 (John,

8 Wollstonecraft, Mary. *Letters Written During a Short Residence in Sweden, Norway, and Denmark*. London: J. Johnson, 1796.

9 Jameson, Anna. *Winter Studies and Summer Rambles in Canada*. 1838. Cambridge: Cambridge University Press, 2011.

10 Diebitsch-Peary, Josephine. *My Arctic Journal: A Year Among Ice-Fields and Eskimos (1891–1892)*. 1893. Cooper Square Press, 2002.

11 Hayes, Ernestine. *Blond Indian: An Alaska Native Memoir*. Tuscon: University of Arizona Press, 2006.

12 Interview with me, 30/3/18

ed., *Libby*). The work itself will be analyzed and compared with the dominant themes and movements of 19th and 20th century American travel writing, in order to demonstrate how Beaman, despite never being published while living, was in fact much ahead of her time; she was more in line with the New Journalists of the mid-20th century than with the Transcendentalists of the mid-19th century. Beaman's work also sets up a comparison with two 20th century, feminine, Alaskan memoirs, to be explored in the following section.

Libby, having a strong personality and being well-connected, was able to procure a personal meeting with President Rutherford B. Hayes in order to ask for a job for her husband during a post-Civil War economic depression. When the only work to be had was that of a surveyor of seal hunting on the Pribilof Islands lying between Alaska and Russia, Libby convinced the hesitant President to allow her to accompany her husband (34–38). She then had to convince both her husband (39–40) and the head surveyor who was her husband's direct supervisor (42–3, 56–62), all three of whom were against the idea of letting her go to Alaska. Libby calmly pressed on, and ended up teaching English to the local Aleutians (170–1, 174–5), before getting snowed into her cabin to the point of almost dying of hunger with her husband (182–93) while unknowingly pregnant (198). Her journals are compelling in their narrative in the sense that they carry a visible narrative arc complete with climax and denouement. Along the way, the reader also encounters a classic buildup in the plot, based around a power struggle between Libby's husband and his boss. There is constant jealousy over Libby's attention, which culminates in the near-death experience mentioned above, before a short denouement describing subsequent events, including Libby's husband's boss going on to found what is now *The Los Angeles Times* (235–6).

The suspense of men constantly worried about the state of the authoritative woman, and the dangers that face her compared with perceived dangers, is something that grows stronger throughout. Eventually, the two men can hardly stand the presence of each other without insult (221–25), though Libby does not complain, even in her private journals. Instead, she lets men argue amongst themselves while she constantly pushes forward (226–7). As such, the work presents a rare first-person point of view of an authoritative 19th century American woman and her subtle tactics in moving through a world controlled by the men around her. The journals were unfortunately not published until 1987,[13] by her granddaughter, Betty John (ix–x), but this was the start of an

13 Greta Gaard and Patrick D. Murphy credit feminism with " … its revelation of the range and quantity of women's writing that had been excluded from serious literary study

attraction to the region by authoritative, white, American women[14] with writing skills (Meaux, *In Pursuit of Alaska*, 35–41)[15] who would follow John Muir and Charles Hallocks'[16] suggestion and tour the new American region (Meaux, 3–6, 15–22).

In an era fascinated with travel accounts, particularly in the West,[17] one might expect such an interesting piece of writing to have been published earlier. Beaman, though, despite being an obviously authoritative and educated woman (John, 12–4, 19, 22–4, 28–30), was also a wife, mother and caretaker during a relatively closed era, morally. It is also difficult to decipher how much of the writing was manipulated by Betty John, who vaguely discusses "fill[ing] in some gaps by conjuring up memories of the stories she told me and by doing research into her times" (x). However, the book is put together as a casual, epistolary/diary mix with the narrative discussed above and travel observations of animals and natives (70–1, 76, 90–1, 99–101, 103, 118–9, 132, 134–6) scattered throughout.

For a period of American literature still dominated by the echoes of Transcendentalism (Buell, *Literary Transcendentalism*, 6–7),[18] which was very often opaque by nature, Beaman's writing was also potentially not "serious" enough,[19] essentially meaning not "vague" enough, not complicated enough in syntax and vocabulary, and lacking the spiritual and philosophical undertones seen in much 19th century American writing (Buell, 1–6, 95). For example, at one point Beaman describes the Aleutians this way: "Most Aleuts are stocky

---

and ignored in canon formation" (*Ecofeminist Literary Criticism*, 5). Late 20th century feminism, of which Gaard and Murphy are speaking, may have indirectly led to the surfacing of this work, due to increased interest in the subject.

14 "White women" is employed here because Alaska Native women were already in the region. Many Native women were no doubt authoritative as well, though they were not published in written form until well into the 20th century.

15 Meaux, Jean Morgan. *In Pursuit of Alaska: An Anthology of Travelers' Tales, 1879–1909*. Seattle: University of Washington Press, 2013.

16 Charles Hallock was a well-known writer and contemporary to John Muir. He founded the magazine *Forest and Stream*, which would become *Field and Stream* (Meaux, 23).

17 Libby Beaman's journals date from 1879–1880. Lawrence Buell offers a clear picture of the popularity of travel writing in mid-19th century America in Literary Transcendentalism (189–94).

18 Buell, Lawrence. *Literary Transcendentalism: Style and Vision in the American Renaissance*. 1973. Ithaca and London: Cornell University Press, Second Printing, 1974.

19 Buell notes in *Literary* Transcendentalism when discussing Ellery Channing that " ... in Concord serious literature was de rigueur" (251). And serious literature in America, with no predecessor other than religious sermons, would have meant writing like Emerson himself and/or how he suggested others write (Emerson, "The American Scholar," 43–59).

and below medium height, with coarse black hair which grows long and lank down to their shoulders. The tyone, like all the others, is copper colored and has only a hint of chin whiskers, which he never has to shave" (76). Thoreau's *Walden* tends to share a more direct form of language with Beaman's style, but he is still often much less direct than Beaman:

> "But men labor under a mistake. The better part of the man is soon plowed into the soil for compost. By a seeming fate, commonly called necessity, they are employed, as it says in an old book, laying up treasures which moth and rust will corrupt and thieves will break through and steal. It is a fool's life, as they will find when they get to the end of it, if not before. It is said that Deucalion and Pyrrha created men by throwing stones over their heads and behind them …" (5)

Here Thoreau is direct in saying "men labor under a mistake," but he goes on to give indirect, didactic suggestions in the Transcendentalist, sermonic strain by evoking "an old book" and saying "treasures which moth and rust will corrupt" instead of directly stating that material goods will not last. In fact, this is essentially a secular style of preaching, while Beaman never once offers the reader suggestions about how to live one's life, or even examine it. The subtlety, one could argue, of Beaman's personality is how she was able to impose herself on the world.

Thoreau, being slightly indirect and sermonic, nonetheless tended to be much more direct than Emerson, whose essays often purposely meandered in search of divine inspiration, or as stated by Lawrence Buell, "a staccato movement through a succession of analogies, a continuous process of statement and restatement until the topic seems finally exhausted," a case of "structural miscellaneousness" (*Literary Transcendentalism*, 157). Emerson is obviously an essayist and preacher, not unlike Thoreau, getting across his didactic suggestions in the manner of the Harvard educated, Unitarian Transcendentalist that he was (Buell, *Emerson*, 12–3). Emerson rarely directly described some concrete thing, as Beaman describes the Aleutians in *Libby*, but rather, he modeled ideas, employing poetics and "many-sidedness," through his essays (Buell, 156). Beaman's style was direct and epistolary (her writing was mostly composed of letters home), and it was more in line with the women who would come after her in Alaska.

Beaman's writing, potentially not marketable as "serious" literature at the time, reflects, if anything, the travel writing of newspapermen, journalists. Tom Wolfe has declared that before the New Journalists came along, journalists were considered the lowest form of literary writing, as they were seen as simply reporting facts (39–40).[20] For instance, whereas Melville wrote the realist, "travel and adventure" novels in *Typee*[21] and *Omoo*,[22] Beaman was simply recording an experience filled with interesting descriptions. If anything, Beaman could have attempted to publish some of her descriptions and experiences in newspapers, particularly if she wanted to avoid the sensationalism of early Melville, or Elisha Kent Kane's searches for Franklin.[23] Beaman's lack of publishing, however, should not downplay her accomplishments in managing to manipulate her way to Alaska, past pessimistic men, and experience and write with direct elegance about her experiences, which were eventually published over a century later, appearing as modern and relevant[24] as ever.

2 Margaret Murie

In the middle of the 20th century, Lois Crisler (*Arctic Wild*) and Margaret Murie (*Two in the Far North*) both wrote memorable works on their experiences in bush Alaska. The following two sections will examine these works, with Libby Beaman's writing kept in mind as well, in order to delve into the question of what three female writers can reveal about women in Alaskan culture. For Murie's part, it seemed to be her distinct motherhood and compassion that shined through in her writing and drew in so many readers.

First, both Crisler and Murie could be compared to Libby Beaman in that they each came to Alaska with their respective husbands and wrote about the experience. Margaret, or "Mardy," Murie, even brought her child into the

---

20 Wolfe, Tom. *The New Journalism*. 1975. London: Picador, 1990.

21 Melville, Herman. *Typee: A Peep at Polynesian Life*. New York: Wiley & Putnam, 1846.

22 Melville, Herman. *Omoo: A Narrative of Adventures in the South Seas*. New York: Harper & Brothers, 1847.

23 Wolfe suggests that "The sort of reporting that one now finds in the New Journalism probably begins with the travel literature of the late eighteenth and early nineteenth centuries" (66). Elisha Kent Kane, was an early example of American writing in this vein.
Kane, Elisha Kent. *The United States Grinnell Expedition in Search of Sir John Franklin*. Philadelphia: Childs & Peterson, 1856.
Kane, Elisha Kent. *Arctic Explorations: The Second Grinnell Expedition in Search of Sir John Franklin*. 1857. Cambridge: Cambridge University Press, 2012.

24 As stated earlier, the publishing of Libby may have been made possible by the feminist movement, but it also bolsters the movement in its strength. The work is simultaneously relevant in its descriptions and discomfort with seal hunting, which were important topics around the time of its publication (Wenzel, *Animal Rights, Human Rights*) as the environmental movement was also gaining momentum (Buell, *The Future of Environmental Criticism*, 3).

notoriously-difficult-to-reach Brooks Range for four months (*Arctic Wild*, 209–15), as her husband, Olaus, was sent there "to band geese" (209). The idea of having a child in the bush was not well-received by others (212–3), just as many thought Beaman was crazy for traveling to the Pribilofs, or worse when she found out she was pregnant and "half-hoped that [she] could produce America's first 'snow baby'" (*Libby*, 215). The idea of bringing a child into the Country, however, serves as a literary representation of strength on the part of the caretaking protagonist, who no doubt takes a calculated physical risk in wishing the child to become strong, independent, and ecologically-connected through an upbringing outside civilization (not to mention the risk of moral outrage at foregrounding the decision in her most renowned literary work), which may stick with a person later in life, even if only subconsciously. In fact, this particular child, Martin Murie, would grow up to become an environmental writer himself.[25] Likewise, John Muir's notorious outdoor upbringing likely contributed to his own relationship with nature (Muir, *The Story of My Boyhood and Youth*). A particularly important point to make is that caretaking in the Country is also much more of a primal experience than that of the world Outside, with death constantly around the corner: body warmth, nourishment, and even companionship take on new meanings altogether. In fact, at one point in *Two in the Far North* Olaus is gone for a few hours, and upon returning he finds his wife sobbing tears of worry, a wandering mind having set in, with anxiety about the safety of him and her both having monopolized her thoughts in isolation (181–3). This kind of emotion is likely common in most men as well, in the Country, but the phenomenon, which is limned by both Mardy and Libby (*Libby*, 186–94 – though Libby is often portrayed as calmer than her hot-blooded husband) is not approached by the likes of Rex Beach, Jack London, or any of the men in the oeuvre of this research. In fact, emotion displaying mental weakness in "frontier" travel writing is historically very rare; *Into the Wild* being a rare case of complex frontier emotion being discussed at all.[26]

25 "Obituaries: Martin Louis Murie." *Reed Magazine*. "http://www.reed.edu/reed-magazine/in-memoriam/obituaries/december2013/martin-louis-murie-1950.html." Reed College. December 2013.

26 Jon Krakauer's work in general is notable in its approach to emotion. *Eiger Dreams*, a book of essays, approaches Krakauer's long obsession with mountaineering, and the emotions that go along with it, as well as the psychological baggage that bubbles forth when one is alone facing mortality outside civilization ("Devil's Thumb," 165–88). Krakauer could be considered in the mold of Peter Matthiessen in that way, but Matthiessen's *The Snow Leopard* is not exactly mountaineering as much as it is a spiritual journey. Krakauer has in this way been influential, notably upon John Vaillant, whose works, *The Tiger* and *The*

As far as the actual plotlines of the works, all three women travel with husbands, all three of the husbands being in the Country because of animals, but the differences are clear as well. Beaman's husband wants to get the job done and get out (he is overseeing mass-slaughter of sea otters, so his sentiment is understandable) (*Libby*, 106–7). Murie begins *Two in the Far North* with the clearest part of her work, a nostalgic view of growing up in ironically-bourgeois Fairbanks in the first decade of the 20th century (16–52). The section is complete with mail arriving by sleigh (37–8) and the break-up of Chena River ice in spring becoming a city-wide celebration, meaning that contact with the outside world is once again possible (39–46) and berries will soon be available to make pies (50–2) for the houses attempting to emulate that of the bourgeois of the Lower 48. These bourgeois houses themselves inhabit a street immediately adjacent to the bars and prostitutes serving as symbolic and physical remnants of the gold rush of the previous decade (21–6). The clarity of this nostalgia to such a unique experience in the developed world is striking, an aspect that gets lost toward the end of the work when Murie leans heavily on campfire conversations with important figures such as US Chief Justice, William O. Douglas (342–52). Later sections also limn the menial aspects of camping in the bush, often with excerpts taken straight from Murie's diary (344). Though the added aspect of having her child by her side naturally maintains a level of anticipation and excitement in the reader, due to the uniqueness and potential danger of the situation (219–20), parts also become redundant and vacillate between descriptive prose highs (218) and somewhat lifeless menial moments in comparison with the earlier parts (319–21). This becomes particularly clear when we weigh these parts against Nancy Lord's essay on camping in the very same Brooks Range, which is powerful in its concision and evocation of the uniqueness of, and draw to, the region ("In the Giant's Hand," 29–36). Here is an example of the very pragmatic and uneventful descriptions in Murie's camping scenes.

> The boys took him across the lake in the boat, then carried the boat across the muskeg to the river, put him across it, and left it tied up on this side for future use. The Indians plan to go in upriver on the western side, to the head of the river, cross over to the head of the East Fork of the Chandalar, then … back to Arctic Village … (*Two in the Far North*, 314)

Golden Spruce, are both a combination of frontier writing and psychological explorations, much in the mood of Krakauer.

Daniel J. Philippon notes in his essay "Academic Air" (*Coming into McPhee Country*, 282–97) on John McPhee that Philippon's students felt overwhelmed in reading *Control of Nature* due to McPhee's propensity, most highly evident in *Annals of the Former World*, of presuming the reader understands the geography, terrain, and vocabulary of the region and field of work. Except, for McPhee, the vocabulary rolling off the tongue is almost poetic, and notably purposeful. McPhee does not expect the average reader to understand everything, but rather to get sucked into a world, partially through the literal evocation of the sounds of its vocabulary (Haynes, "The Size and Shape of the Canvas," 74–6; Howarth, "Introducing John McPhee," 47). This is where Murie falters slightly. The above selection gives much description, but it gives little evocation. In other words, the reader potentially gets lost in the geography and terrain without being sucked into the passage itself. Words like "Indians," "East Fork of the Chandalar" and "Arctic Village" may create some evocation in the reader, but it certainly is not enough to create a world the way a more professional writer would, the way Murie herself succeeds in doing with the Fairbanks portion of her work. The Fairbanks chapters involve physical sights, sounds, and odors, evocations of childhood, the anticipation of love, and what lay ahead. In general, this is what is lacking later on, which is why the latter chapters often do not stand up on their own as well as the earlier chapters. Murie's general knowledge and unique experience are thoroughly present, as she certainly has extraordinary knowledge of "the Indians," their food, the animals, and their food, how the food chain works (*Two in the Far North*, 315), etc., but Murie's style reminds us that she is not a professional writer. She is rather a competent writer who sometimes transcends her own deficiencies in her descriptive prose of rare fauna and flora,[27] but maintains much of her strength through her personality as a beloved environmentalist figure, a form of genuine compassion and positivity that emanates from every page of the work. Environmental historian Douglas Brinkley says, "Mostly it was Mardy Murie's ability to motivate people and hold them accountable by her steadfast decency of spirit that set her apart. To know Mardy was to love her; she was deeply humble" (*The Quiet World*, 263).[28] Therefore, any potential frustration regarding Murie's structure, vocabulary, etc. is likely to be disarmed by her compassion and

27 Whereas certain pages are filled with basic camping information, her descriptions can sometimes erupt with multi-dimensional life: "The tiny dwarf forget-me-not is in bloom, also a white one. Here in the couloir grow scattered sturdy willows the seedheads and leaves of which have a sweet fragrance. It is a gorgeous cumulus-cloud day with a breeze just right for carrying the mosquitoes away" (305).

28 Brinkley, Douglas. *The Quiet World: Saving Alaska's Wilderness Kingdom 1879–1960*. New York: Harper Collins, 2011.

charm, which tends not to be sermonic, unlike the Transcendentalists before her, but is always present.

3 Lois Crisler

Lois Crisler, on the other hand, was avant garde in her own way, embodying the modern feminist movement before it gained full steam in the decade after her work began to appear. This section will look at Lois Crisler's time raising wolves in bush Alaska, and the subsequent writing of the experience, *Arctic Wild*. Crisler's ability to write with the American sensibility of the respectability of the "working class" will be explored as well and compared with the other female writers of this research, particularly with Nancy Lord.

To begin, Crisler did not merely accompany her husband, Herb, but actually engaged equally in projects with him. They studied various animals together, notably filming documentaries for Walt Disney. Disney had already made the conservationist film *Bambi*[29] at this point and had also won an Oscar for production of a short film set in Alaska, entitled *Seal Island*[30] (Brinkley, *The Quiet World*, 346–51). The Crislers were eventually criticized for taking wolf cubs from a den for the making of a documentary, for staging certain scenes, and for eventually killing the wolves she raised from pups described in her work because of what captivity had done to them in preventing their ability to survive on their own, as Barry Lopez has noted (*Of Wolves and Men*, 152). But, *Arctic Wild* is also a minor staple of the 20th century environmental movement, openly appreciated by William O. Douglas, Ansel Adams, Margaret Murie, Margaret Mead, and Rachel Carson (Brinkley, 350–1), as the writer turns myths about wolves on their heads. She particularly attacks their demonization in Alaska where "To the average Alaskan, wolves were only useful for their pelts" (Brinkley, 351). Crisler's work fell chronologically between Rachel Carson's *The Sea Around Us* and *Silent Spring*, and the two writers likely inspired each other's respective works as they developed a correspondence (Brinkley, 351).

Whereas Murie's draw came from sincerity, humility, and even effusion in her authorial voice, as well as authenticity in experience, Crisler had a stronger grasp as a technical writer. Murie, in fact, was trained in business and was self-taught in writing (Murie, 75–84). Crisler, by contrast, was already an English Instructor at the University of Washington before she came to Alaska to study wolves and write *Arctic Wild* (Brinkley, 348). Her prose is much more polished

29 *Bambi*. David Hand. Burbank: Walt Disney Productions, 1942.

30 *Seal Island*. James Algar. California: Walt Disney Productions, 1948.

and her structure much more cohesive than Murie's unique descriptions. Crisler, in other words, wrote like a professional:

> In the face of the Arctic, which was unknown to Cris and me ... I kept lugging my ignorant, earnest questions to Andy, sure that if he only would he could answer. He looked at me without smiling, he was taciturn, he wasted valuable question-and-answer time, between his continual flights, sitting with a magazine, reading." (1)

This is from the first page of *Arctic Wild.* The reader is immediately introduced to the dangers of Alaska via the helicopter pilot, Andy Anderson, who warns them of the many perils they will face. The mix of anticipation and fear shine through, drawing the reader into a plot which has yet to be exposed. Crisler effects this anticipation by creating a wise, almost Buddha-like character, at once working class, whose taciturnity is unnerving but who fits with the aloof and ostensibly emotionless caricature of the frontier and its "wilderness." Wilderness, so tough to define, can in one sense be narrowed down to the following: risk to one's life outside civilization. Crisler begins her book without a specific scene of action but with the anticipation of risking her life. Likewise, she mentions how the bush pilot will risk his life for her, in what can be described as his working class nobility. Right away, through this pilot, a connection is created with readers who associate themselves with the working class. When Crisler later cuddles wolves from infancy, these readers, who have been taught to fear wolves, will be more inclined to listen.

Another point to be made is that, unlike Beaman's professed naivety bolstered by her determination, and Murie's empowerment of domestic caretakers at campsites, Crisler's strength comes from being hands-on throughout the work. This idea is noted above, but it deserves more attention. Both Murie and Crisler were writing at roughly the same moment. In fact, Murie's work was published four years after that of Crisler. Yet, both works were published around 1960, when social upheaval was beginning all over the western world, roiling and surfacing with Beat poets, (Kerouac, *On the Road*, 1957; Ginsberg, *Howl*, 1956) rock n' roll, (The Beatles, "Please Please Me," 1963) and the environmentalist likes of Rachel Carson and Lois Crisler.[31] And here, Murie chooses

31 Interestingly, in the fight for statehood, which would come to fruition in 1959, the famed author, Edna Ferber, would be recruited to write on the issue. Her book, *Ice Palace* (1958), would be based around a strong, young, female protagonist who was a mixture of education, beauty, and sourdough Alaska (she liked to wear mukluks and a fur parka) (Brinkley, *The Quiet World*, 451).

to write from the position of the empowered domestic wife. Crisler chooses to write as the equal partner in work. Murie's prose is of fauna and flora, of camp sites, of parenthood, and of a somewhat bourgeois childhood. Crisler's prose is of the type of adventure one would have previously expected to see from a man. "You and your husband will gamble your lives and I will gamble mine to help you." Whereas Murie was ready to see the new, the potentially unexpected in the wildlife, Crisler was ready to face mortality in the wild. But, though Crisler takes hold of the theretofore masculine approach, she nonetheless maintains emotion and doubt, which was for so long rare in frontier writing. There are the "tears of glory, fear and love," and "its great freedom will haunt you" (Crisler, 1), and she responds by "lugging [her] ignorant, earnest questions to Andy." In that passage, she tears down her own ego in the face of the test and the man taking her and her husband into the Country. The great male frontier writers of the early 20th century and before just did not allow this set of doubt to penetrate their metaphorically tough skins, as doing so would have shown weakness. Susan Kollin even suggests that for white men of this era, "overwhelmed by the growing numbers of new immigrants from other parts of the world, a beleaguered masculinity sought to reestablish itself through wilderness experiences" (63). Crisler, in the early 1960s, combined the sense of adventure of these men with the doubt, fear, and nervous excitement that was to come with a more emotionally open society, which the 1960s revolution brought culturally (Brogan, 656–63).

Despite any differences, however, the striking similarity between these three writers is their strength and courage to defy society and spend vast amounts of time outside civilization. Where did this precedent come from, and why? As noted above, society women were early proponents of the late 19th century cruises to the glaciers that John Muir suggested people visit in order to get closer to God. Strong women who were wealthy and educated but lacked a professional outlet in a rather oppressive society could travel to the Alaskan frontier. Once there, they could feel a sense of accomplishment at breaking free from societal constraints, while also appreciating the spiritual aspects. Such is apparently the case with Septima M. Collis, whose *A Woman's Trip to Alaska* conspicuously begins with "a note of congratulation from an old friend, General William T. Sherman," according to Jean Meaux (*In Pursuit*, 35). Collis, married to a Civil War general herself, "[set] out from her Fifth Avenue home" on a luxurious cruise, and she wrote an educated, descriptive account of her voyage, only marred from the modern perspective by her denigration of the Tlingit people, a common viewpoint of the day (Collis, "The Alaska's Barnum Dancers," *In Pursuit of Alaska*, 35–41). Aside from a few wealthy women like Collis and missionaries like Caroline Willard (Meaux 61–3), early Alaskan white women were quite often known to be vendors or prostitutes for the miners of the 1890s gold rush. Meaux indeed describes women as adventurers alongside the men, but she never describes them in the occupation of mining. The fact that women were not rare, however, is a misconception that Meaux indeed exposes (179–80, 214–6). In fact, she ends a summary of Mary E. Hitchcock's piece "Ho for the Land of Gold!" by relating that the promise of gold "… led to a friendly rivalry between the men and women on the various boats traveling the Yukon as it snaked its way across the middle of Alaska" (180). Women, then, were quite clearly seeking adventure in the same way that many of the miners were (Meaux). They would have crossed the notorious Chilkoot Pass just as the men did, and they would have faced hardships and uncertainty all along the way just as the men did.

Though it should not be surprising, Meaux's compilation is indeed a revelation for a patriarchal society which so often confined women to menial housework and childrearing. This was a mostly unflattering responsibility with rare allusions to courage or strength. All three of the women analyzed thus far in this chapter have turned this idea on its head by embracing the menial and/or childrearing (raising wolves from pups in the case of Crisler), but performing these tasks in places and circumstances where most individuals would not have the courage or strength to do so. The same can be said for many women in *Coming into the Country*. The matriarch of the Gelvin family is depicted as maintaining a household of men who are constantly out flying planes and driving bulldozers in dangerous situations, and yet is herself described as hunting and running trap lines. Furthermore, she is the better shot while her husband is better at a sewing machine (203, 258–71). In another instance, McPhee creates a multiradial profile of Donna Kneeland and her companion, Dick Cook (232–3). Kneeland is immediately described as subservient to Cook, but so is everyone else (178–9). Cook is portrayed as a strong trapper who has achieved a high level of independence outside civilization, offering advice to young people who share his dream, but he also treats everyone as though they are beneath him. This is epitomized in the subtle fact that McPhee is not allowed to sit in Cook's chair (355). Despite Cook's haughtiness, Kneeland is nonetheless portrayed by McPhee as having an important job in maintaining the household (there is also a section about how Kneeland came into Interior Alaska with just about any pick of a man she could choose, and she chose Dick Cook) (160–1). In the end, however, McPhee notes his discomfort by saying, "By and large the men seem to be waited upon to an extent that even our forefathers might not have known" (227–8). McPhee's profile of the couple is both admiring and damning at once, as is the case with so many situations.

Another nuanced example could be the role of women in Alaska Native societies. On the one hand, there is currently a crisis of missing and murdered indigenous women,[32] and yet Alaska Native women traditionally performed life-or-death roles in their communities. For instance, Ann Fienup-Riordan notes that the Yup'ik of western Alaska traditionally maintained individual, child-rearing households around a larger house where all the men slept, called a qasqiq (*Boundaries and Passages*, 34–8).[33] Fienup-Riordan notes that the women kept a constantly clean path to the house so that seals would give themselves to the family (*Boundaries*, 88–142). Native women mended clothes and shoes prepared for the Arctic cold, snow, and rains as well. The Iñupiat women also played important roles in the ceremonial cutting of the whale and its dispersing, along with careful handling of seal and other meals (Hess, *Gift of the Whale*, 42–3, 106–7; Lowenstein, *Ancient Land*, 162–72; Fienup-Riordan, *Boundaries*, 168–79). In certain cases, as Jean Malaurie has noted, female Inuit have even maintained multiple sexual partners (*Les derniers rois de Thulé*, 179–81).[34] This is a double-edged sword, as it means that infanticide due to hunger is higher among females than males because males can hunt (98, 187), but the lack of numbers gave women a certain amount of empowerment in their communities.

The menial, the household, child-rearing, all can be considered important in civilization, but in traditional Alaska Native and pre-modern homestead communities they were a matter of life and death. These duties, then, could potentially boost self-worth and were even elevated to spiritual tradition in some communities. Fienup-Riordan is careful to illuminate the importance of the roles of women in Yup'ik society throughout her work. McPhee notably expresses his discomfort with what he experiences in the bush, but he also allows Donna Kneeland and Dick Cook the last word toward the end of the work, both of them criticizing "women's lib" because survival requires each to play a role outside civilization (McPhee, 355). Yet, despite her rhetoric, Kneeland's physical accomplishments are likely more impressive than those of many masculine frontier writers.

---

32 "Missing and Murdered Indigenous People Crisis," *U.S. Department of the Interior: Indian Affairs*. https://www.bia.gov/service/mmu/missing-and-murdered-indigenous-people-crisis. Viewed on 4/1/2024.

33 Fienup-Riordan, Ann. *Boundaries and Passages: Rule and Ritual in Yup'ik Eskimo Oral Tradition*. Norman: University of Oklahoma Press, Norman, 1994.

34 Malaurie, Jean. *Les derniers rois de Thulé, avec les Esquimaux polaires, face à leur destin*. 1955. Paris: Éditions Plon, «Terre Humaine»; 5e édition définitive, 1989.

If this section has shown that women in Alaska have often performed household duties, it has also shown that women traveled independently to see the glaciers just as men did, and women came to the gold rush along with the men, experiencing the same hardships. Therefore, women in Alaska should not be pigeonholed. Yet, around the time Crisler and Murie were being published, along with Carson down in the Lower 48, there was not exactly a deluge but at least a trickling of women breaking free from their "roles" in the household. This timeframe aligns somewhat with that of the Lower 48, though one could certainly argue that Alaska gave women the opportunity to break free from the chains of the expectations of their societies back home, facing the wilderness if they chose to do so. The Iditarod, for instance, until it became a quasi-ultra-marathon in recent years, was a battle of stamina, preparation, and wits. Women could compete against men in a way they could not in more directly physical sports.[35] Alaska has also had a female governor, Sarah Palin. From a literary standpoint, Nancy Lord's experience, to be discussed later in this chapter, speaks volumes for the respectability and accomplishments of modern Alaskan women. However, this research will first attempt to dig into that notoriously difficult-to-define spectrum of masculinity and femininity, employing ideas of ecofeminism to aid in the process.

4 Ecofeminism in Alaskan Literature

Ecofeminism can be helpful in defining the terms of modern environmental criticism. Whereas, at one time the idea of masculinity in American literature may have seemed mostly clear, particularly in frontier writing (Roosevelt, London, Beach, etc.), notions blurred into more of a spectrum in the 20th century (Crisler, Murie, Raban, Lopez, etc.), and have only become more multi-dimensional in the early 21st century (Krakauer, Vaillant, Eggers, Lord, etc.). This section intends to explore the concept of ecofeminism as a methodological tool in order to discuss female environmental writing, first through definition, then by putting it into an Alaskan context. Next, several writers of this research will be briefly analyzed in the context of Alaskan ecofeminism, as defined. The hope for this section is that it will reach beyond the limited labels of masculine and feminine writing by adding another dimension, which in this case may be considered a mindset rather than an explicit gender bias, so that changing roles and mindsets may be explored more precisely in the

35 "Champions & Record Holders." *Iditarod: The Last Great Race.* "http://iditarod.com/about/champions-records/." Viewed on 22/7/18.

context of modern Alaskan, environmental travel writing, increasingly transcending the masculine man and feminine woman stereotype. In making this point, this section also asserts that those like Lord who walk the line between patriarchy and ecofeminist perspective can potentially be more effective in their environmentalist goals in Alaska than those who are entrenched in their own communities and filter bubbles.

In *The Future of Environmental Criticism*, Lawrence Buell starts off a conversation about ecofeminism by calling it a "multiverse," in order to confess outright that its multiple strains make it difficult to pin down to a specific definition, but he goes on to attempt one anyway, dancing around the idea but at least stating confidently that it " … starts from the premise of a correlation between the history of institutionalized patriarchy and human domination of the non-human" (19). He goes on to say, "An ecofeminist might claim that the analogy between 'woman' and 'nature' is inherent, or (increasingly) that it is historically contingent" (20). "Might" is an interesting word choice, as it ultimately points out that, as a way of thinking, not all ecofeminists argue this assertion. Later in the work, Buell points out that many ecofeminists also display many similarities to "deep ecology." Again, not all, which is why employing ecofeminism is so difficult, because the term is so plastic. In *Ecofeminist Literary Criticism*, edited by Greta Gaard and Patrick D. Murphy, various faces of the term present themselves, often leading toward a rethinking of the structural systems that society has placed upon us, in nature, in literature, and in the classroom (Gaard and Murphy, "Introduction," 1–13). This deconstructive conversation is important to a modern mosaic of Alaska that is still searching for an identity beyond that of The Last Frontier to be exploited.

Buell's work was written in 2005, and Gaard and Murphy's work came out in 1998, so neither would have approached what has become known as the "gender spectrum."[36] Gender itself is often less important in Alaska than personal capabilities and one's approach to environmentalism, the latter of which can be discussed very generally in terms of boosterism, conservationism, preservationism, and deep ecology (Coates, *The Trans-Alaska Pipeline Controversy*, 21, 36–49, 157–8, 201). Former Governor Sarah Palin's political capabilities and her boosterish environmental politics, for example, are considered to be much more important than her gender in Alaska. But, ecofeminism can potentially add another dimension to the horizontal gender spectrum, when defining approaches to Alaskan rights and environmentalism, by offering another angle from which to view the argument. I would argue that those writers who

36 Gender Spectrum. "www.genderspectrum.org." Viewed on 5/4/2018.

maintain a balance between an ability to navigate "institutionalized patriarchy" and an ecofeminist perspective[37] are often more effective in their environmentalism than direct, bellicose militarism. This is particularly pertinent when writers are faced with discussing specific issues with the communities who are directly impacted. Not only is approaching a community with an eye to its perspective important but so is employing a respectful knowledge of its umwelt (Vaillant, *The Tiger*, 182–6).[38] To clarify, these writers do not offer entirely different viewpoints to opposing groups. Instead, they arrive at the same point through different means, approaching different communities through the tones and subject matter that may be most relevant to those communities, as we will see with Nancy Lord in the following section, as well as Part 2 of this work.

However, it would be pertinent to begin with the "godfather" of Alaskan environmentalist literature, John Muir. Could Muir be considered either a masculine or a feminine writer? This is an interesting question not only because masculine and feminine as terms are so difficult to pin down but because Muir was an extraordinarily multi-faceted character. One could make the argument that he was entirely masculine for his constant wandering and sleeping outdoors under all weather, on glaciers even, and existing for long periods in deprivation of food without problem, something male nomadic and semi-nomadic hunters have traditionally done for millennia.[39] He also grew up on a strict and structured, patriarchal farm (Muir, *The Story of My Boyhood and Youth*, 241–4) with a father who famously made him spend months digging a well from dawn till dusk, almost ending in Muir's death by carbonic acid (Muir, 274–8). On the other hand, Muir had a romantic, or aesthetic, side. Muir spent long hours talking with friends and writing descriptively about the bounties of nature. Balance is potentially the wrong word because Muir was by all accounts an effusive human being with constant explosive energy. If we are to take into account Buell's assertion that ecofeminism is often akin to deep ecology, and that Muir also fought against what can be seen as "institutionalized patriarchy" in his battle with Gifford Pinchot, which began the long-running feud between

37 In the sense of wanting to do away with traditional thinking of systems put in place by a status quo that is leading toward the degradation of irreplaceable terrain, even rainforest (Kolbert, *The Sixth Extinction*), as well as demographic diversity among Alaska Natives (Haycox, 56–7, 178–80, 195, 208–210).

38 Vaillant, John. *The Tiger: A True Story of Vengeance and Survival*. London: Sceptre, 2010.

39 In *Boundaries and Passages*, Ann Fienup-Riordan limns a detailed portrait of Yup'ik Eskimo life, where women and men have particularly defined roles (18–28). In both Lowenstein's *Ancient Land, Sacred Whale* and Hess' *Gift of the Whale*, men are the hunters portrayed as well.

conservationists and preservationists, Muir can potentially be considered an ecofeminist in hindsight. Many would obviously have a problem with this viewpoint, as retroactively labeling a multi-dimensional and symbolically-important figure with the characteristics of a contemporary movement is easily contestable. Therefore, let me state that this is not an assertion, but rather an exercise in exploring possibilities in order to view the region as a whole from a contemporary angle.

To continue the exploration and finalize a point about gender, the two females prominently discussed in McPhee's *Coming into the Country*, Ginny Gelvin and Donna Kneeland, would not be considered ecofeminists, as, despite the fact that they are undoubtedly authoritative women, and potentially even have strong feelings about low carbon footprints, sustainability, and/or nature, they are both comfortable in their places in "Institutionalized patriarchy." Both authoritative women view their places as inside the home doing the important work of maintaining a household. This is perhaps not unlike the authoritative women often encountered in traditional Alaska Native culture. Though, outside the ecofeminist perspective, the term "matriarchal" may carry a different connotation. Lineages and decision-making in communities may be considered matriarchal even when the women work inside the home while the men are out hunting. This becomes interesting because many men and women in modern Alaska travel writing cannot be easily fit into boxes. They tend to slip from one label to another, and having more dimensions as a way to describe them is ultimately helpful for the expression of the critic.

Both Lois Crisler and Mardy Murie, for instance, were undoubtedly authoritative women who did a lot to gain respect for Alaskan women who came after them. They both broke rules in the structure of patriarchy by going into the bush; caring for a baby in one case and raising wolves in another. Murie, who was content to watch over her infant child and maintain camp while her husband was off researching, could switch roles whenever needed. She was the rare figure who made friends in both the preservationist and conservationist camps, and she instinctively knew when to be a wife and a mother and when to be a strong feminist (in the sense of publicly moving forward her agenda), or ecologist (that agenda included preservation of very large swaths of land). This flexibility would make her one of the more important environmentalist figures of the 20th century, through her literature and through her work protecting Alaskan lands. Susan Kollin goes so far as to suggest that the fieldwork Murie performed with her husband eventually led to the creation of the Arctic National Wildlife Refuge (Kollin, 94).

One could certainly argue against this assertion of approaching different communities on their own terms, employing their vocabulary and their customs, particularly by noting that Muir was extraordinarily successful in his rigid, unswerving environmentalism. And, they would not be wrong. However, Muir was also not careful. He made enemies easily (Pinchot, Burroughs – to an extent –, etc.) and was hot-blooded enough that, as stated earlier, he lived to regret his decision to promote Alaska's glaciers to tourism. Murie, by contrast, is less remembered but was also careful, with no visible regrets and also preservation to show for her work. This leads us to Nancy Lord, who, like Murie, is an ecofeminist and is also not. She is able to walk between two worlds carefully, and she is respected in Alaska by various communities with differing priorities.

5 Nancy Lord

Nancy Lord has been one of the most influential Alaskan environmental writers of the 21st century so far, carving out a niche in the environmental world that is unique to the isolated state, while also exploring multiple literary forms: from the short story (*Survival*),[40] to the essay (*Rock, Water, Wild*), to the novel (*pH*),[41] to journalistic, environmental travel writing (*Beluga Days*), to historic travel writing (*Green Alaska*). This section will analyze how Lord's chosen path in life, mixed with her careful approach to environmentalism, positively affects her relationships with readers of various communities. Excerpts from *Beluga Days* will be concurrently analyzed, with Lord's background kept in mind, demonstrating what is unique about the writer. Finally, this section will use Lord as a lodestar for womens' writing in Alaska, in comparison with previous generations, keeping in mind that she came of age but a few decades after Murie and Crisler opened the region's female writing to the wider cultural turning point of the 1960s. Whereas the act of breaking free was once considered a theme itself, Lord has lived and written about a life that could have been pursued by either man or woman in modern Alaska, and notably, with an environmental approach that could have been employed by either sex. This says a lot about the evolution of The Last Frontier, where lines between traditional masculinity and femininity have begun to blur just as they have in the Lower 48.

Lord has written that she fell in love with Alaska on a college trip to the Brooks Range (*Rock, Water, Wild*, 29–36). After graduation, she and her partner, Ken, moved to the fishing village of Homer, which is situated at the southern tip of the Cook Inlet. As a recap from the introduction to this research, she began working at a canning factory, cleaning rooms in a hotel, working nights

40 Lord, Nancy. *Survival*. Minneapolis: Coffee House Press. 1991.

41 Lord, Nancy. *pH*: A Novel. Berkeley, Ca: West Margin Press, 2017.

at the weather service, and doing so while living out of a truck. She and Ken eventually saved up and bought the sports store where he had been working. Building on to the store and running it six days a week for five years, while working side jobs as well, they finally sold it to finance their commercial setnetting career (*Fishcamp*, 41–2), which in Lord's case would be supplemented by a second career in writing.

The very fact of having started from the bottom, with almost no material possessions, and having worked her way to up to commercial fishing, affords Lord an automatic sliver of credibility when confronting working class and politically conservative communities more in line with what Buell, to repeat, has described as "institutionalized patriarchy" (*Future*, 19). Furthermore, the fact of having hung around as a commercial setnetter for decades has also served her well. She was able to gain seniority in a place with a relatively short history among the white population, who often do not spend an entire lifetime, or even the entire year, in the state of Alaska.[42]

As for her environmentalism, most would likely not call Lord a "deep ecologist,"[43] as she has in fact made her living as a commercial fisherman and has been open to discussions with various communities in various works. In *Beluga Days*, for instance, she notably allows space for communication with

42 "Alaska Population Overview," *Alaska.gov*. "http://live.laborstats.alaska.gov/pop/estimates/pub/02popover.pdf." P. 11–13. Viewed on 31/3/2020.

43 This assertion should be explored in detail. Lawrence Buell links "Norwegian philosopher-outdoorsman-activist Arne Naess'" "deep ecology" to Aldo Leopold's "concept of 'the land ethic,' with its ascriptions of rights to nonhuman life forms," essentially asserting that man treat the environment as if he were a piece of the puzzle and not above it (Buell, *The Future of Environmental Criticism*, 100–1). Or, as Bruno Latour says in *Politiques de la nature*, man should stop treating all that is non-human as an object and start treating it as a subject (113–5). Concurrently, Buell discusses "deep ecology's propensity to think in terms of holisms like 'nature' or 'man' or even 'humankind,' and with its emphasis on identification of self as fulfilled through its identification with nature" (108). Nancy Lord can be viewed as a deep ecologist in the sense that she believes holistically in man as a part of nature and not above it. As a fisherman and writer, she attempts to maintain that connection through hunter-gathering ("I've never taken particular pleasure in nurturing plants or in trying to arrange nature to please a sense of orderly aesthetics, preferring instead to value what exists on its own, in its own time and place" – "Naming the Green," *Fishcamp*, 164), and through fishing as a conscious form of participation in the biotic community. Obversely, the term "deep ecology" carries heavy connotations of often uncompromising preservationism (Buell, 102–3), which is where Lord would break from most modern deep ecologists, as a fisherman. The point Lord would make, however, is that one may kill fish and treat them as a subject at once.

the director of The Alaska Oil and Gas Association (221–3).[44] She also does not harangue her neighbor for having killed belugas decades earlier but shows her disagreement through subtle statements such as, "I was, for a moment, caught off-balance. It hadn't occurred to me that George would ever have wanted to 'get rid of' belugas," and "… because I might have been frowning" (*Beluga Days*, 4). She goes on to recount various stories, pointing out the histories and logic behind why communities have acted against current moral trends, and how the status quo of both the white communities and Alaska Native communities have both become irresponsible in their actions, leading to the degradation of beluga populations in the Cook Inlet. With subject matter, and with her relationship to nature shown in *Fishcamp*, where she spent weeks at a time literally talking with no one and simply enjoying being on her beach and far away from so-called "civilization" ("On Not Being Alone," 203–220), and where she also suggested many ways to the reader of getting by in Alaska on the fruit of its land ("Naming the Green," 163–77), Lord falls somewhere on the spectrum of ecofeminism. But, she indeed moves back and forth in approach, always speaking compassionately and respectfully to different communities, fitting in well with multiple groups.

However, Lord has not always been appreciated by environmentalists from outside Alaska, because people Outside have a different understanding of environmentalism overall from that of Alaskans.[45] This will be discussed in further detail in Part 2, but ultimately, Alaskans tend to understand, as early Russian Americans did, that logistics are difficult (Haycox, 90–1), and that subsistence is important to both Native and white communities (Lord, *Made of Salmon*). By contrast, environmentalists from the Lower 48 much more easily rely on fruits and vegetables, which their lands tend to allow, as well as the fact that their spiritual and cultural traditions are not based on subsistence hunting of Arctic mammals like those of many Alaska Natives (Fienup-Riordan, *Boundaries and Passages*; Hess, *The Gift of the Whale*; Lowenstein, *Ancient Land, Sacred Whale*). Similarly, salmon is the most abundant and important food source for much of the state. It is an essential element of Alaskan culture, which Lord both lauds and takes part in as a commercial fisherman ("You can love what you kill and eat," – "The Conservationist as Woodchipper," 108).

The preface to *Beluga Days* sets up the premise and investigation into disappearing Cook Inlet belugas. Lord is descriptive in her writing about the whales, while also exploring their historical and biological background as a species, as well as her personal experiences with them. She then connects these personal

44 This is not un-reminiscent of McPhee's interview with Robert Atwood in *Coming into the Country* (76–7).

45 unpublished interview with me, 30/3/18

experiences to her investigation into the disappearance. Through this multi-radial process that plays out in the very first few pages, we are introduced to the juxtaposed perspectives of a "matriarchal" environmentalist and a "patriarchal" commercial fisherman who are both part of Nancy Lord. She is able to accomplish the two simultaneously by keeping both in perspective. On the very first page, for instance, she describes "a favorite part of fishing," aside from "hauling in nets full of shining fish," as "the quiet meditative time spent walking the beach or rocking, waiting, in the open skiff." Both sides of her as an Alaskan are present in this single sentence, and she qualifies it by saying, "If my mind had taken a Zen turn, that didn't mean I was oblivious to the world around me" (xv). The word "Zen" carries New Age connotations adapted from Japan, which tend to conflict with traditional Christian values often associated with fishermen in the US. Lord is letting the reader know that she is first and foremost a fisherman, but that she is simultaneously prone to the kind of reveries and meditative moments that clash with the Protestant Christian aphorism "Idle hands are the Devil's playthings."

Toward the end of the preface, Lord allows herself another moment of reflection about the beauty of the cycle of life in fishing for salmon, particularly alongside the beluga whales. She describes herself standing on a rock out on a beach, hearing an eagle squawk, feeling the breeze on her arms, "warmed to tepid by its passage over sun-soaked mud and sand." She then once again removes herself from revery and focuses in on logic:

> I kill fish for a living. I am not generally given to sentiment, don't gush over animals, have no desire to pat whales on their heads. But I did like beluga whales, and I did appreciate living in a time and place that sustained fish, whales, and fishermen – all of us together. (XXI)

The same juxtaposition once again plays out in this section. Lord begins by describing the inlet and its salmon, which leads into an expository analysis of the relation of salmon to the local food system. She also explains how one could easily pass by the waters along the scenic route and not understand the complexities going on just below the surface. She then talks about how the "gifts of the place" make her "ache." Such a description would be in line with many environmental writers, but the term "ache" is a unique choice. The writer does not say "ache" with joy, or transcendence, or some other term potentially expected in ecopoetics.[46] She merely says "ache." In this way, she gets across

46 Ecopoetics – "Similar to ethnopoetics in its emphasis on drawing connections between human activity – specifically the making of poems – and the environment that produces it, ecopoetics rose out of the late 20th-century awareness of ecology and concerns over

her sentiment at the beauty, but she also holds back. She then immediately crosses into the opposite realm, that of the commercial fisherman. She says, "The West Foreland and Cook Inlet are, in general, hard places, and a person has to be tough to live there, on the beach or on water." This is when she moves into the above excerpt, "I kill fish for a living. I am not generally given to sentiment. But ..." This is how Lord continuously walks the line between ecopoetic environmentalism and "systematized patriarchal" pragmatism.

As such, she has the ability to speak to multiple demographics and voices, mediating on a topic so culturally divisive that people tend to turn away from each other before even beginning to listen to what the other has to say. As soon as Lord says "Zen" or "meditation," many people in her local community she is attempting to convince are likely to stop listening. However, on the very same page she goes into the technical details of fishing, so they are likely to stay and keep reading. On the other hand, when certain environmentalists are confronted with her descriptions of unabashedly killing fish for a living, they may also be ready to stop listening, but her "ach[ing]" for the scene that lay before her might convince them to hear out her arguments that follow. And, what follows is the trickiest of moral areas, regarding: Alaska Native rights (which are so often difficult to approach),[47] fishing rights (of the utmost importance to subsistence life in the state), and land conservation (one of the most important and emotionally evocative issues of the past century).

A related technique that Lord employs is to constantly question. For instance, the writer has the natural sentiment that waters should be sustainable, that beluga should have a right to live and survive in the Cook Inlet, as other creatures do. But, early in the preface of *Beluga Days* she asks, "Elsewhere in the world, where you find people in such numbers, whales and other large (and small) creatures generally have a hard time of it, generally don't last. Why should Cook Inlet be different?" (XVIII). On the last three pages of the work, Lord sums up her investigation. She does not directly answer whether or not Cook Inlet belugas deserve to exist, suggesting that she does not have that authority. She plays Devil's advocate by asking the reader: if people knew

---

environmental disaster. A multidisciplinary approach that includes thinking and writing on poetics, science, and theory as well as emphasizing innovative approaches common to conceptual poetry, ecopoetics is not quite nature poetry." ("Glossary of Poetic Terms: Ecopoetics." *Poetry Foundation*. https://www.poetryfoundation.org/learn/glossary-terms/ecopoetics." Viewed on 2/4/2020.)

47 For instance, with regard to an Iñupiat character in *pH*, Lord says in a Q&A at the end, "The biggest challenge was with Helen, because of the cultural aspects – that is, avoiding cultural appropriation while recognizing that Alaska Natives live successfully in 'two worlds'" (250).

they could save the species, would they? Do people care about "... the world's biological health? Anyway, don't you want that gas to run your car, that port that must be dredged, that boat (or Jet-ski) ride, that silver salmon?" (225). On the final page, finishing strong with her true point of view, she suggests that we have a moral obligation "to treat the Earth and its creatures with respect," before also offering the suggestion that we might actually "learn something that may be of great value to us, something that we could not get or learn without them ... " (226). Finally, she suggests that the disappearance of the beluga whale from the Cook Inlet is a harbinger for bigger problems to come, particularly the wiping out of more prized subsistence species such as seals, and especially salmon, and evokes the "necessary social, cultural, and nutritional needs" as well as "spiritual values – for those things that make people feel happy about living in the world ... " We see a mix here of all the juxtapositions suggested earlier, put together to ask important questions and ultimately argue for preservation in a humanist manner, while still not alienating the pragmatic opposition. Again, Lord does not come out and state that we should save the belugas, but the final sentence, vague but evocative, is, "I was getting closer to what I thought were the most essential questions, and I was, still, watching for belugas" (226). The two commas surrounding ", still," are important here, serving as a form of emphasis on the controversial subject-matter explored throughout. The commas are essentially exclamation points. Lord obviously wants to save the beluga whale, though she still does not know exactly why. She knows that her humanistic or sentimental side is in favor of saving the whale, but she also appears to be suggesting that sentimentalism is simply not enough to convince a pragmatic opposition to act. She is also exploring the pragmatic part of herself, as a fisherman and consuming human being, which is why there are still so many questions, why she is "..., still, watching for belugas" at the end of the book.

One could chastise Lord for not "pulling the lever," but she has gotten people's attention in her home state. *Beluga Days* remains a centerpiece of the book store at The Anchorage Museum. Lord was also selected as Alaska State Writer Laureate not long after *Beluga Days*' publication.[48] As for the issue at hand, beluga numbers have since stabilized.[49] In the work, Lord finds out that the numbers of belugas being "taken" are unsustainable and not being considered seriously by either the government or various Alaska Native communities (19–41). Her investigation, despite being virtually unknown outside Alaska,

48 https://www.akarts.org/nancy-lord. Viewed on 8/1/2024.

49 Bissett, Hannah. "Cook Inlet Beluga Whale Population Stabilizes," *KNBA*. https://www.knba.org/news/2023-06-27/cook-inlet-beluga-whale-population-stabilizes. 27/6/2023.

likely played a serious role in the reversal of the mammal's extinction. Therefore, despite the fact that her writing style may not resemble that of many in the activist wing of the environmental movement, it is clear that her care with respect to Alaskan culture, without once giving in to thoughtless boosterism, is effective in the region.

In the process, Lord respectfully covers the multitude of voices of Alaska in a way that is rarely, if almost never, accomplished. Her sincerity in respecting Alaska Native voices is always clear, and this respect conflicts with mainstream understandings of the spiritual and cultural importance of subsistence living in the Lower 48.[50] Furthermore, Lord is always reflective, even in *Fishcamp* when she describes her reverie-filled experiences walking alone for weeks without talking to a soul, and spending pages on end suggesting ways to sustainably benefit from the local vegetation. The point of view of the small-scale setnetter bringing fresh, wild-caught food to a table while thinking reflectively about sustainability and being respectful to neighboring communities is just about all that one could ask for from an environmentalist, except for the fact that many would disagree with taking fish in the first place, particularly for remunerative purposes. However, the perspective of the Lower 48, facing different sets of problems, often neglects the nuances of regional environmentalism, as well as the humanity of small-scale fishing and its cultural purpose. Anthropologist Charles Stépanoff calls this difference of view "exploitection," a portmanteau of "exploitation" and "protection." In *L'animal et la mort*,[51] Stépanoff explores the exploitation that can inadvertently lay within protection when employing the "naturalist" viewpoint of industrial civilization instead of the "animist" perspective of traditional peoples. For Stépanoff, there is often more violence in industrial-scale agriculture that inadvertently kills millions of wild animals and insects than in the subsistence living of traditional cultures. He includes European peasantry in this discussion, which aligns quite well with Lord and other small-scale fishermen in her community. As discussed previously, the feminist perspective can potentially be considered different in

50 In *Beluga Days*, for instance, Lord inverts the negative stereotype of Indian Time, which generally implies laziness. She discusses a different relationship, where "Time moves as it will ... being in the world without being driven by clocks was something I valued about Natives I knew, about village life, even about the life I assumed at my fish camp" (157). Value of time from a different perspective is merely one example of this misconception, but it gets at the heart of misunderstandings between the developed world and traditional Alaska Natives.

51 Stépanoff, Charles. *L'animal et la mort: Chasses, modernité, et crise du sauvage*. Paris: La Découverte, 2021.

Alaska as well. In all of these cases, survival itself outside civilization shifts the modern viewpoint.

Ultimately, to bring this particular section together, Lord fits right in with a long tradition of authoritative women in Alaska who may have a slightly different mindset from women of the Lower 48. This is partially due to the geographic and social barriers required to get to and live in Alaska in the first place, and partially due to the physical lifestyle required outside civilization, or at least at its edges. As discussed, Libby Beaman had to work hard to get to Alaska, an accomplishment in itself due to the barriers set by multiple men along the way. Shortly after, Septima Collis was able to reach Alaska with more ease and comfort, showing her strength through her daring in entering the region without accompaniment of her husband, and with her educated writing. In the 20th century, Margaret Murie and Lois Crisler both showed daring in spending long periods of time not only in cabins or on boats in Alaska but camping with their companions in the bush. The social climate then shifted to the point where by the time Nancy Lord came along, being a woman in the bush or doing manual labor in Alaska, while potentially still slightly controversial among some, was not unheard of. She was free to forge her own path, working where and how she wanted in order to accomplish a dream which would likely not have seemed feasible a generation or two earlier. As stated as well, in the meantime, as Lord has continued writing, a female governor was elected in the state, and multiple women have won the Iditarod. Lord was never a feminist writer in the sense of constantly beating the drum of equal rights. However, short stories like "Survival," (an oddly similar plot to *Into the Wild*, except with a female protagonist, and written before Christopher McCandless ever went missing), "Marks," (bringing light to missing and murdered women) and "The Lady with the Sled Dog" (about a female sled dog racer), combined with the fact of her continued strength in accomplishing her dream as a fisherman and serving as a constant voice in the literary and environmental scene of the state, has built upon the bricks that were laid by women before her, allowing those following her the same, and likely more, opportunities.

There is no single point in time that denotes a sudden shift, but somewhere along the way strong women moved toward the Crisler-like aspect of being more involved in work, politics, and environmentalism in Alaska, blurring lines that defined the traditional "roles" of women. At this point in Alaska, women have the ability to choose their own "roles." If a woman chooses to become a matriarch of a family and live out the traditional role of maintaining the house, the food, the clothes, etc., she may do so. If she chooses to become a commercial fisherman or hunter, she is also free to do so. In the process, men have adapted as well. Examples include much of Jon Krakauer's work (*Into the*

Wild, 145–56), along with that of other modern writers such as Barry Lopez and Peter Matthiessen. Emotion is a much more open theme now in Alaska than it was in previous generations. Chris McCandless, famed protagonist of *Into the Wild*, is not only in touch with his mercurial emotions but flouts the patriarchal responsibilities Krakuauer suggests that McCandless would have seen in McCandless' own father, a successful engineer and entrepreneur (103–26). As will be noted in the "Jon Krakauer, John Vaillant, and Dave Eggers" chapter of Part 2, John Vaillant has followed in this vein, dealing heavily with emotion in his works *The Tiger* and *The Golden Spruce*.[52] Also to be explored, Dave Eggers looks at Alaska from the perspective of a female dentist running away from domesticity in the Lower 48 (*Heroes of the Frontier*), something rarely, if ever, seen from a male writer on Alaska. In this way, Alaska parallels changes occurring in the Lower 48 and across borders in occidental culture. One major difference is that, whereas domestic families may come to Alaska, women, still as influenced by the prevalent frontier talk as men, are unlikely to come into the Country in search of domesticity.

---

52 Vaillant, John. *The Golden Spruce: A True Story of Myth, Madness, and Greed*. New York: W.W. Norton & Company, 2005.

FIGURE 4 Chena river state recreation area
PHOTO BY CHRIS FERGUSON

CHAPTER 3

Native Oral Tradition, Lack of Written Record, and Lack of Respect

This research has discussed women's roles, positions, and voices in Alaskan society in order to better understand the context of their respective circumstances, which helps to put their environmental and literary work in the region into perspective. The reader may also note a lack of Alaska Native literature herein. This is entirely natural, as written literature is not at all part of traditional Alaska Native culture. Rather, it is a recent adaptation to modernity. Like most indigenous peoples of the world, Alaska Native tribes maintain a strong oral tradition instead of writing (Toelken, "Foreword," *Giving Birth to Thunder*, XI–XIV; Haycox, *Alaska: An American Colony*, 14, 98; Hensley, *Fifty Miles from Tomorrow*, 70). Increasingly, university educated Alaska Natives are being published, such as Ernestine Hayes and William Iġġiaġruk Hensley. Hensley, a popular Iñupiat politician, was torn from his home at a young age and sent to the Lower 48 for an education. He ended up in Tennessee, viewing segregation and racism against African Americans firsthand (Hensley, 80–97). His memoir, *Fifty Miles from Tomorrow* (2009), which expertly details this perspective, was boosted in popularity by his success as a politician, taking part in the decision-making process of projects of enormous repercussion such as the formation of Native Corporations and ANCSA (154–60, 168–78). He was also a significant character in Part 2 of John McPhee's work *Coming into the Country*: "In Urban Alaska: What They Were Hunting For" (87, 106–9, 128–33). There have been more examples of successful literary figures coming from the Alaska Native community in recent decades, such as the poetry of Athabascan writer Mary Tallmountain (Kollin, 131), but still few, and almost always by someone with one foot in the modern world and one foot in traditional culture, belonging to both but neither at once. Furthermore, none of these authors have written works of travel literature.

If poetry and memoir are the most common choice of literary genres employed by modern Alaska Native writers, the traditional Native perspective is most often expressed through writing adapted from oral accounts. One example is *Shem Pete's Alaska*, a collection by Dena'ina elder, Shem Pete of the Cook Inlet, in collaboration with linguist and University of Alaska Fairbanks

 | DOI:10.1163/9789004734807_005

professor James Kari.[1] Another example is the work of ethnologists Marie Meade and Ann Fienup-Riordan, who collaborated in their collection of Yup'ik stories (*Agayuliyararput, Our Way of Making Prayer: Yup'ik Masks and the Stories they Tell*).[2] These researchers often attempt to get across traditional Native storytelling as true to natural form as possible. Therein lies the first problem when translating Alaska Native storytelling into written English: not only the words but the spirit and means of storytelling themselves are so divergent that they tend not to be comprehensible to a reader habituated to a western paradigm. Reading Alaska Native stories often comes across as disjointed in direct translations ("[They are] also deadly to read," Lopez says of academic collections, with a "misleading ... pretension to accuracy" – *Giving Birth to Thunder*, xvi), often without a narrative arc to which the occidental world is accustomed. Furthermore, to Lopez' consternation, they tend to be translated with sexual bowdlerization and often additions of pedagogical lessons or biological explanations "perpetuating the notion that Native Americans were anxious to explain everything, which they were not" (Lopez, XVII). Tom Lowenstein, meanwhile, paid careful attention to oral histories and wrote them out in experimental prose/poem form in order to circumvent the jarring lack of Western coherence that one might experience while reading Fienup-Riordan and Meade for pleasure (*Ancient Land, Sacred Whale*, XIII–XV). The other option would be to completely transform the stories, conflating them and adding that narrative arc for Western audiences, which would likely succeed commercially but lose much from the Native standpoint in doing so.

Mary Tallmountain had a foot in both worlds, as part-Koyukon-Athabascan, part-Irish American (Kollin, 148), and she was able to accomplish the spirit of the Alaska Native with the technique of an occidental poet. In a sense, she lost the purist traditional means of storytelling of her community, but she was able to convey something novel, which expresses the true heart of the Native without ringing false. Susan Kollin employs Mary Tallmountain as exemplary of respectful Alaskan writing, particularly with regard to how we view the state as a frontier, in the fact that Tallmountain offers a view of "counterenvironmentalism" that does not present Alaska Natives as proto-ecologists but as a dynamic community evolving via its own epistemological perspective (Kollin, 158). Tallmountain, however, was removed from Alaska at a young age and

1 Fall, James A. and James Kari. *Shem Pete's Alaska: The Territory of the Upper Cook Inlet Dena'ina*. Fairbanks: University of Alaska Press, 2003.

2 Meade, Marie and Ann Fienup-Riordan. *Agayuliyararput: Kegginaqut, Kangiit-llu: Our Way* of Making Prayer: Yup'ik Masks and the Stories They Tell. Seattle: University of Washington Press, 1996.

eventually became a poet in San Francisco in her mid-fifties, contracting cancer three times in what she described as due to "the trauma of being separated from her people and homeland" (Kollin, 149). Yet, she did not choose to go back and live in Alaska, as she occupied a liminal space between two cultures and did not feel that she would fit in with her own people anymore. She eventually reestablished connections with her community later in life by visiting her hometown and writing about the experience in the essay "Koyukons Heading Home" (Kollin, 150). Such is the difficulty of those who know how to convey two vastly different worlds but are nonetheless caught between them, belonging to neither.

This chapter explores the difficulty in crossing such a divide, and how three of the primary travel writers of this oeuvre (Barry Lopez, Jonathan Raban, and Tom Lowenstein) have chosen to go about doing so in their own unique ways, as well as what that essentially means for environmentalism in the vast differences of the lenses through which the world is experienced. With the styles of these three writers approached, the reader should come away with a better understanding of how (travel) writers in Alaska can respectfully tackle the tricky issue of translation of Alaska Native oral stories. The worry is that translations may not only end up sounding cold and academic, or lose their spirit in Westernization, but they may be disrespectful to Alaska Native communities in not properly conveying beliefs and worldviews. In the end, these discrepancies have real world repercussions. Misperceptions of culture, language, and mindset can frame how a people are treated, just as misinterpretations in how the environment is tread upon by respective communities can skew the faith given to their perspectives by broader society.

1 Attention to Language in *Arctic Dreams*

In succeeding in ringing true either in the faithful, ethnological Fienup-Riordan manner or in the creative Tallmountain spirit, there still remains a rather large gap between linguistic worlds being lost in translation. In *Arctic Dreams*, Barry Lopez devotes entire sections to language and the phenomenon of getting lost in translation, particularly in the chapter "Ice and Light" (204–251). The Inuit quite famously have myriad words for different types of snow and ice (Tuan, *Topophilia*, 77).[3] This is important, as knowing whether one should melt a particular type of ice for water or whether one can walk across

3 Tuan, Yi-fu. *Topophilia: A Study of Environmental Perception, Attitudes, and Values*. 1974. New York: Columbia University Press, 1990.

a particular type of ice, maybe on what conditions of ice certain animals tend to congregate, and when particular types of ice have a certain color or texture, as well as what that means for animal migrations or the changing of seasons, these geographic perspicacities are a matter of life and death and of the utmost importance to a subsistence lifestyle (Lopez, 208–14). In fact, the very existence of man in the Arctic requires such knowledge. This was particularly evident in the literature of 18th and 19th century Western explorers who arrived in boats, needing decades of experience in understanding through which ice they were attempting to navigate and how it was affecting their voyage (Lopez, 215–19). Jonathan Raban goes so far as to relate, through a mix of dead-seriousness and comedic effect, that Polynesian sailors would place their testicles lightly on a boat "to pick up slight variations in the rhythm of a swell" (*Passage to Juneau*, 94). Sir John Franklin is a perfect example of the catastrophic effects of European arrogance in the Arctic, as he attempted to navigate its waters employing solely Western traditions (clothes, food, hunting, etc.), and paid for it with his life and the lives of his men (Lopez, 359–63). Likewise, the Russians were never self-sufficient in Alaska, as they attempted to maintain their own culture instead of adapting (Haycox, 110–2). It took men like Knud Rasmussen (*Across Arctic America*)[4] and Roald Amundsen (Roald Amundsen, Bomann-Larsen),[5] as well as Peary and Stefansson (Lopez, 360) to adopt Inuit culture, dogsleds, clothes, food, etc., to finally move through the Arctic with relative facility. Such an explanation is important to this particular research for two reasons: 1) it demonstrates the ignorance and arrogance Westerners so often displayed, for so long, toward the Inuit, who were successfully surviving in places where ostensibly well-prepared Europeans were sometimes dying in droves (Lopez, 360, 379–84), and 2) this success of survival in the Arctic in and of itself serves as a sense of purpose, self-actualization, for the Native peoples of the North, and without this struggle a frigid, barren, tundra of a life can easily disintegrate into drugs, alcohol, and junk food shipped in from the South, as well as dependence upon television from the same place (McGinniss, *Going to Extremes*). For the purposes of this particular chapter, the segue comes when we examine the fact that subsistence, in offering purpose, requires a certain manner of language different from that of the industrialized world.

This section will begin by examining Lopez' rumination on language itself and the effects which languages create on our respective lenses on the world

4 Rasmussen, Knud. *Across Arctic America: Narrative of the Fifth Thule Expedition*. 1927.University of Alaska Press; Reprint edition, 1999.

5 Bomann-Larsen, Tor. *Roald Amundsen*. 1995. Translated by Ingrid Christophersen. First Published in English, 2006. Gloucestershire: The History Press, 2011.

and our relations to our own habitats. We will then take an argumentative leap in expressing the criticism of Lopez that he chose not to concretely discuss the Inuit in *Arctic Dreams*. The criticism will be noted in order to move forward with the supposition put forth that Lopez indeed does not avoid the Inuit, but that the Inuit are omnipresent in his work through their viewpoint. In fact, Lopez moves fluidly between multiple viewpoints, discussing the pertinence of each to the Arctic, the Inuit perspective most idealistically being one of subsistence. Lopez hones in on this importance as an observation most other writers seem not to note, which will be discussed in detail.

Here is the example from Lopez' rumination on language in *Arctic Dreams*, quoting a study by Benjamin Lee Whorf on Hopi in the 1930s:

> Hopi has only limited tenses ... makes no reference to time as an entity distinct from space, and, though relatively poor in nouns, is rich in verbs. It is a language that projects a world of movement and changing relationships, a continuous "fabric" of time and space. It is better suited than the English language to describing quantum mechanics. (Lopez, 274–5)

This analysis then continues over the next few pages, setting up a more thematically pertinent example that sheds yet more light on the importance of language:

> The Eskimo language reaches its apogee in describing the land and man's activity in it. Young people in modern Eskimo villages, particularly in the eastern Arctic, say that when they are out on the land with their parents, they find it much more difficult to speak Inuktitut, though they speak it at home all the time. (Lopez, 277)

The former passage is important in setting up the more general idea of how language can affect thought processes and ways of perceiving the world, and even Jacob von Uexküll's idea of the individual umwelt (Lopez, 268–9, 313), since language physically changes the way we perceive and interact with the world.[6] If the Hopi can better understand the theory of relativity, a notoriously

6 In this way, we could consider language to be an extension of Kant's supposition that our view of the world and its fundamental laws is a construct of our senses and the limits of our mental capacity (Wulf, 34–6). Therefore, our view of the world is also affected by the construction of our language. Jakob von Uexküll's idea of the umwelten came out of the same spirit of philosophy, as von Uexküll essentially posited that each animal has a unique sensorial experience and, in turn, experiences the world in a unique way (Lopez, 268–9, 313).

difficult idea for the Western mind to grasp, then the door to other ostensibly strange aspects of language is left wide open. In the latter passage, Inuktitut is approached directly: essentially, children who remain in the village can lose connection to the nuances of a language based on the land itself. If they know television and Coca-Cola, then the color of the ice that is good for seal hunting and representing luck or danger, whatever the case may be, is more likely to be lost to them. The addition of this passage in *Arctic Dreams* is important. However, anthropologist George Wenzel had a different view, attacking Lopez for not mentioning the actual people of the Arctic enough. In *Animal Rights, Human Rights*, Wenzel says:

> In only one way does *Arctic Dreams* differ from the literary tradition that began with the explorers: until the epilogue there is hardly a reference to Inuit. My surmise is that Lopez, in his travels through North Alaska and the High Arctic, found Inuit too ordinary. (14)

Animal Rights, Human Rights (1991) is a staunch defense of the Inuit against animal rights activists in attacks on fur trade, which affects seal hunting by the Inuit in the Arctic. Entire villages, according to Wenzel, were drastically affected by Brigitte Bardot and Greenpeace's movement to ban fur trade in Europe (48–53). Wenzel was living in Arctic Canada, among the Inuit, for much of this time, and decided to devote himself to the writing of this book (viii–ix, 1–2). Therefore, he had a vested interest in protecting the Inuit and their interests. When Wenzel says that there has always been mention of the Inuit in a "literary tradition that began with the explorers," this is another important point, because he expects Lopez to conform to the others in this fashion. One could, by contrast, take the opposite perspective by saying that Arctic explorers often ignored animals unless they were either a pest, a potential source of food or transport, or a way to make money. Few and far between were the Arctic explorers before the 20th century who focused heavily on animals (Steller being a notable exception – Ford, *Where the Sea Breaks its Back*). Therefore, Lopez was potentially interested in concretely discussing a subject that had not been sufficiently covered. A more important rebuttal in favor of Lopez would be that Lopez approached the Inuit, at length, but from an indirect perspective.

Examining the basic idea, there are absolutely no ethnological sections on the Inuit in *Arctic Dreams* the way that Lopez covers animals in so much minute detail, with multiple specific chapters dedicated to individual species (musk ox, polar bear, narwhal). Yet, as will be argued in depth in a chapter on animals later in Part 1, Lopez often approaches animals and landscapes, not only from a first person viewpoint, and from the viewpoint of various travel writers and explorers before him, but from the viewpoint of the Inuit. The Inuit perspective, though not concretely covered, is spattered throughout the work in a way that offers the reader a constantly shifting viewpoint on the issues the author wants to discuss. Notably, what do the Inuit discuss out on the ice? According to Lopez in the above excerpts on linguistics and the young, landscapes and animals: subsistence. Throughout *Arctic Dreams*, the author is attempting to project himself into the Native viewpoint in order to retain the "Native Eye," which will also be discussed in detail later in this research. Apropos this argument, in a discussion with writer Mike Newell, Lopez relays a story regarding one of his works and Native perspectives:

> A Nez Perce man saw the book and said, "What's this? Another white guy writing about Indian life?" Barre didn't know what to say. The guy walked away with the book, really irritated, a kind of chip on his shoulder. That evening he gave the book back to Barre and said, "This is exactly how my dad used to tell the stories" ... (Newell, *No Bottom: In Conversation with Barry Lopez*, 56)[7]

There is quite literally no better compliment to Lopez than to say that he tells stories like a Native, as the author spent so much of his career attempting to reach that particular place, be it through the "Native Eye" or in respectfully translating their stories. One could potentially criticize Lopez' at-times-saccharine view of the Arctic. Lopez admittedly is searching for the lost mysticism of humanity (*Arctic Dreams*, 250) that he believes still exists in some parts of Native communities more than it exists in the Western world. One clear criticism is that this view risks forgetting the "realism" of his chosen areas, landscapes, and peoples. But Lopez, being a searcher of mysticism (or the "numinous" – a discussion for the following chapter), has priorities that do not always coincide with "realism." He actively avoids the social and cultural issues that go on in many Native communities, the Arctic included. Lopez even states quite clearly in a conversation with Mike Newell, when discussing a criticism of his not having discussed "the drunks" in the Arctic:

> That's the constant refrain. I'm writing the other story, and I expect you to bring that knowledge of darkness with you to the other story. If you don't, what I'm writing could seem sentimental and I'm not writing

7 Newell, Mike. *No Bottom: In Conversation with Barry Lopez*. Gambier, OH: XOXOX Press, 2008.

> sentimentality. I'm offering you, as a writer with his own awareness of darkness, something to hold against the dark, a plausible light. (Newell, *No Bottom*, 35)

To leave this argument at a rebuttal in favor of Lopez' intentions, however, would be incomplete. Lopez does not merely set out to sound like a Native storyteller for vain reasons of technical expertise. He is likewise never merely searching for mysticism, the numinous. His being able to sound like a Native's father's storytelling demonstrates Lopez' search for new ways of narrating stories, conflating two storytelling traditions into one so that his work sounds natural and interesting to both, which is a very difficult task. Take for instance the bilingual work on the Yup'ik put together by ethnologists Marie Meade and Ann Fienup-Riordan, and mentioned above, entitled, *Agayuliyararput, Our Way of Making Prayer: Yup'ik Masks and the Stories They Tell.* For ethnological reasons, the translations were left fairly pure. Here is an example:

> Back then people used to transform into animals. Many things were visible when the earth was thin. Things were distinct when the land was thin, not the way it is today. They predicted that animals and other things that were visible at that time would no longer be visible, and they said there would be many people of different races. (39)

There is then a brief story about a fox that is a human. The fox is hungry and goes looking for food. He finds a needlefish. The needlefish tells the fox that the fox's "anal area" smells bad, which makes the fox turn red with embarrassment. The fox then worries about his reddish color, turns into an animal and paints his nose and ears with charcoal, and his tail with ash, making himself subjectively beautiful. This is an explanatory ending ("So that is why the red foxes are that color today with black on them."), instead of creating a narrative arc with a climax and denouement. As a writer attempting to bridge European and Native storytelling traditions, finding a way to bridge not only the language barrier but the idiomatic and storytelling traditions themselves can be an enormous challenge.[8] From the idiomatic perspective, how does one attempt to translate an idea like, "when the land was thin"? Returning to the two Lopez quotes above, the Hopi language allows for a more natural understanding of concepts moving fluidly through time and space. People and

---

8 Franz Boas' indigenous texts are an example of this. The ethnographer was tenacious in getting his collected stories translated to perfect accuracy, but they are often not easily readable for the Westerner, as his translations favor linguistic perfection and lack in fluid storytelling.

communities, then, are not static in a timeframe the way Western culture views itself. Also pertinent, it is understandable that a young Inuk who understands Inuktitut in a village life dominated by European influence would be lost out on the ice where an extraordinarily different existence itself reigns. It is worth exploring Chief Barre Toelken's "Foreword" to Lopez' *Giving Birth to Thunder*, in order to more fully explore this theme. Toelken devotes multiple paragraphs to the work that Lopez put in toward getting to know the Native perspective. Lopez spent time on the land, talking to people, learning some of the language, etc. Furthermore, in slowing down and observing as opposed to gawking like a tourist and moving on, Lopez was able to capture the little things, such as shades of leaves and behavior or animals. Toelken ends by stating the following about Native translation:

> The trick consists in finding equivalents rather than direct translations of connotative words or rhetorical strategies. In effect, Lopez has done these tales over in such equivalent terms, giving us the story, the style, and Old man Coyote himself, without betraying the magic. (XIV)

This passage explores the authenticity of what Lopez is attempting to accomplish in the face of criticism. As noted with regard to Marie Meade, there is something important in maintaining the ethnological accuracy of stories, but something is also lost. The essence of the stories themselves is sometimes lost. Both Toelken and Lopez are quick to point out that Lopez' work is not academic (XI, XVI–XVIII). Lopez is attempting the trick of maintaining the accuracy of the work, yet also maintaining its essence, the spirit of the stories as told by Native Americans themselves. This is the feat that Lopez accomplished when a Native admitted that the author told stories like the Native's father. And, the author did so in the English language, without sacrificing the essentials for the Native community,[9] while also appealing to European English speakers. Therefore, to once again revert to Wenzel's criticism, even in *Arctic Dreams* where Lopez does not recount Native American stories as he does in *Giving Birth to Thunder*, Lopez indeed approaches life in the Arctic, its animals, its landscapes and encroaching development, from the Alaska Native or First

9 Though there are indeed a few issues in the existence of the book in general: 1) coyote stories are only supposed to be told in winter, and 2) coyote stories tend to be owned by individuals and should not be stolen, a fact around which Lopez tiptoes, according to Toelken by "not pretend[ing] to be an 'Indian Book.' It does not provide the original language, the ritual detail, the full context; in short, it does not give away or betray the magic of the actual storytelling event" ("Foreword," *Giving Birth to Thunder*, XII).

Nations "eye" and observation. This general cautious approach and constant attention to minute detail in achieving the "eye" of another culture should not be under-appreciated, but it should ultimately be lauded as the crowning accomplishment of *Arctic Dreams*, in line with Susan Kollin's suggestion that the importance of subsistence, and the civilization based around subsistence, be given precedence over the draw of the aestheticism of the Last Frontier. The following section will move from Lopez' attention to structure and detail to the importance Raban places on individual names in controlling the narrative in Alaska and the coast of British Columbia.

2 Raban Evokes Wittgenstein, Importance of Names

How important is an individual name: the name of a child to that child's life, the name of a place to the people inhabiting it? Individual words, as well as the ways in which grammar and syntax are constructed in a language, are all important for the identity and culture of a people. For instance, Ben Okri states in a short essay entitled "Writers and Nations" in his work *A Time for New Dreams*[10] that he feels a truly healthy country is one that appreciates its writers, and one can gauge that appreciation by the names countries and cities have given to their streets. Okri mentions Rue Victor Hugo and Boulevard Voltaire in Paris, and he states, "The force that they have incarnated in the imagination is subtly impacted into the people's psyche, as they walk or drive down these roads" (17). This example lies in contrast to Washington D.C., where politicians and men of power, as opposed to the creative class, are almost universally celebrated, "impacted into the people's psyche." Discrepancies could obviously be found within these cities, and each city's priorities may have their own merits, but the connection Okri makes is nonetheless striking. In Alaska, Tom Lowenstein spends considerable effort pointing out that Native place names are almost universally related to subsistence living (*Ancient Land, Sacred Whale*, 19–23). Likewise, in the ethnographic work *Shem Pete's Alaska*, Dena'ina place names are front and center, and they are translated from all around the Cook Inlet. The following passage from a 1985 narrative by Shem Pete demonstrates how tribal band names are employed in speech:

> Iydluytnu Qayeht'ana ghu k'a qeyl dghinih ha t'qighit'a.
>
> They were called 'the plural objects river village people' [Eklutna people].

10 Okri, Ben. *A Time for New Dreams*. 2009. London: Rider, 2011.

> You know from Susitnu Station, Yusdishla Qayeht'ana, Dashq'e Qayeht'ana, ighi Tsuk Qayeh qayeht'ana.
>
> You know from Susitna Station there are the 'little point people' [site on island in Susitna River], then the 'on the shoal people' [Kroto people], and the 'old village people' [lower Yentna Old Village people].
>
> Ndaha village yeh ha'u qayeht'ana hel dghinih.
>
> Wherever there was a village they called them 'the village people.'
>
> That's the way it is from the Kenai to Eklutnu to Matanuska, Niteh, Nitehen, Iydlughet, Iydlughe Ht'ana, K'enakatnu Ht'ana.
>
> 'Among the Islands,' [Matanuska site], 'the person of among the islands' [was the chief there], at 'by the plural objects' were the 'by the plural objects people' [Eklutna people], and the '?-stream people' [Fish Creek-Knik village people]'. (27)

In this passage, Shem Pete demonstrates how bands of people are denoted by the proximity of their physical space to certain rivers, streams, villages, and islands, which themselves are given names such as "plural objects" or "fish." The geography of a place is most prominent, whereas, generally speaking, in Europe and America, "La Seine" or "The Mississippi" have long lost their more specific meanings in the public imagination. Rather, the names have taken on the connotation of the physical rivers themselves. Furthermore, a table entitled "Semantic Analysis of Place Names in *Shem Pete's Alaska*" is presented later in the work, and it is worth replicating here to supplement the argument:

TABLE 1 Semantic analysis of place names in *Shem Pete's Alaska*

A	Natural history		
	Animals	30	5%
	Fish	18	3%
	Birds	8	<1%
	Vegetation	43	7%
	Insects	3	<1%
	Hydrology	123	21%
	Terrain and geology	157	26%
	Weather	12	1%
	Direction	19	2%
B	Human activities		
	Material culture	34	6%

TABLE 1. Semantic analysis of place names in Shem Pete's Alaska (*cont.*)

	Foods	6	<1%
	Trails	22	2%
	Customary activities	54	9%
	Significant human events	15	2.5%
	People	6	<1%
	Historic events or activities	6	1%
	Mythological events	6	1%
C	Others		
	Opaque or unanalyzable -θ- or -?-	11	1%
	+Partially analyzable	7	<1%
	+Ambiguous (two or more meanings)	5	<1%
	Class not clear	9	<1%
	+Substrate O	2	<1%

(*Shem Pete's Alaska*, 40)

It is immediately clear that geographic places named for "People" are extraordinarily rare amongst the Dena'ina. In this analysis, "People" make up less than 1% of all geographic names. "Hydrology" sits at 21%, and "Terrain and geology" at 26%, with no other category reaching into double-digit percentage points. This means that water sources and physical spaces make up 47% of place names, which verifies the importance of subsistence to the language and culture of the Dena'ina people.

Nancy Lord, in fact, references *Shem Pete's Alaska* in *Beluga Days* in order to better understand and live with her local surroundings and truly internalize the places and names she sees in passing and the traditions involved in their histories: " … Pete also reported that a 'long time ago, before Russians or anybody came to this country,' Tyonek [a Dena'ina village] had a reputation for a particular kind of food wealth. 'They call Tyonek "Mother of the Earth" because there are lots of things to eat all the time: clams, fish, beluga, seal, grease, and oil'" (179). She goes on to mention his pointing out that Susitna River was called "Father of the Earth" because it was rich in interior meats and firs such as caribou. These types of knowledge are important to Lord in truly understanding her place, as well as the Native sense of duality in the place, and of the geographical extrapolation of masculinity (the land) and femininity (the sea) by Dena'ina culture. For instance, with regard to her immediate subject, beluga, Shem Pete's stories hint that the whale was once an integral part of the food cycle. This means that it must have been plentiful to the "Mother of the Earth" village, but that it waned in the 20th century, as it has begun to recede from stories (*Beluga Days*, 6, 178–9). To put the matter of names further into perspective, Lord lives in a town called Homer, which immediately evokes the celebrated Greek writer, though it was named for a man named Homer Pennock.[11] Likewise, the primary water source for fishing in the area is known as The Cook Inlet. Therefore, one is confronted diurnally with the fact that James Cook once explored the waters where the local fishermen (the backbone of the economy) are working. The histories being evoked are of personages, of great men. By contrast, when Lord learns the Dena'ina names she is learning stories of places and the relationships of peoples with those places. Yet, these names are not only important for the historical context which could impress important mores on Dena'ina young or impress outsiders with the historic weight of standing upon a spot where Dena'ina have been hunting or fishing for generations. The place names were (and potentially still are in some cases) important for the physical placement of peoples without maps or GPS. In fact, both Barry Lopez and geographer Yi-Fu Tuan point out studies done on the specific geographic knowledge of the Inuit, who have demonstrated the ability to draw fantastically detailed maps of their lands without previously having encountered cartography (Lopez, *Arctic Dreams*, 286–9; Tuan, 34–6). The Inuit traditionally knew their native areas well because they roamed for subsistence and internalized the place names and words for important aspects of their land, which may last as long as the names remain in the local culture.

Yet, modern Alaska continues to be a palimpsest where multiple cultures glare through. Revillagigedo Island, for instance, demonstrates Spain's dalliances in the region. Raban points out that George Vancouver found a rare functioning farm there that was supplying the Spanish who were pushing north from California (Lopez, 152–4). Meanwhile Boston traders (174) and British explorers like Cook and his protégé, Vancouver, were also bouncing around, with the Russians already installed (Haycox, 78–82). In fact, Russian names continue to resonate in Alaska today: Chichagof, Baranof, Wrangell, etc. (Haycox, 37–9). These various European cultures' names are glimpses into Alaska's history of exploration, colonization, exploitation, etc., amongst the spattered Alaska Native names. Interestingly, European explorers also employed names such as Desolation Sound[12] (Raban, *Passage*, 181–4) and Deception Pass (86–9, 154) for the experience of the place, not unlike the Alaska Native naming process.

For the direct purposes of this research, then, I will focus on Lopez' discussions of linguistics compared with Jonathan Raban's discussions of names.

---

11 "Welcome to Homer," City of Homer, Alaska. https://www.cityofhomer-ak.gov/welcome-homer. Viewed on 10/1/2024.

12 An anomaly, according to Raban, as mentally unstable Vancouver was the only one who found Desolation Sound an unhappy place (181–4).

Raban follows threads of both Native cultures and the European explorers who have inhabited or traveled through the region he is exploring In Passage, noting accents and slight differences in culture and geography, treatment of the land, subtleties in the water, etc., as he moves up along the coast (86–9, 119–21, 222–3). He focuses heavily on water throughout, citing names that had been influenced by various phenomena:

> The names of passes – Arran, Yuculta, Dent, Hole-in-the-Wall, Greene Point, Whirlpool – were for me the most evocative by far of all the names on the coast. Each set of rapids had its own haunting character. If I went through a pass once, the whole face of the water left its features printed on my memory, from where it was likely to sneak into my dreams. (154)

Raban buys wholeheartedly into the idea that language and names influence us. Furthermore, he states that they represented the power and influence of colonization to the European powers. Raban even goes so far as to evoke Wittgenstein:

> The world we live in is the words we use, wrote Wittgenstein – and the world through which the cutter now moved under oars was one freshly recast in English words, and a Who's Who of the 1790s King's navy. The 'extensive sound' with its 'rugged snow barrier' to the north was now Burrard's Channel, named for Sir Harry Burrard … (151)

Raban puts the reader into the mindset of the British navy in this passage. He explains how the British navy and nobility were coming to physically dominate the spaces of the world, impacting places as far off as Alaska. On the following page, Raban evokes the emotional impact of such importance to the British. In one example, George Vancouver finds out that a place he named Point Grey has already been found by the Spanish and is named Punta Langara:

> Point – Punta. Grey – Langara. Vancouver's names were suddenly robbed of their authority by their Spanish alternates. Where they had sounded natural and inevitable, now they seemed pert and capricious. In his Voyage, he confessed humiliation caused by the first true discoverers of the Strait of Georgia."[13] (152)

---

13 The importance of official names also reared its head again in 21st century Alaska when President Barack Obama changed the official name of Mt. McKinley, for once-presidential -candidate William McKinley, to the Athabaskan name, Denali: "The Great One, The Mighty One, The High One" (McPhee, *Coming into the Country*, 95; Gregory Korte, "Obama

Wittgenstein's own teacher, Bertrand Russell, says this in his "Introduction" to Wittgenstein's notoriously difficult *Tractatus Logico-Philosophicus*[14] in attempt at explaining part of his student's early work: "What we cannot think we cannot think, therefore we also cannot say what we cannot think" (16). Shortly thereafter, Russel quite poignantly states, "That the world is my world appears in the fact that the boundaries of language (the only language I understand) indicate the boundaries of my world" (16). This idea is pertinent to this research as it demonstrates the sheer force and importance of language to experiences of reality. For instance, in the Lopez example from *Arctic Dreams* discussed in the previous section, certain young Inuit quite literally lost the ability to express themselves out on the ice despite speaking Inuktitut fluently at home (277). Likely much of the issue with the young Inuit in this case is that they are entirely out of their element. Therefore, they cannot think what they cannot think. They cannot speak what they cannot think, do not understand, when it comes to the boundaries of their reality. Being on the ice, then, would expand the boundaries of both their reality and their language, together, potentially making both richer upon returning to the village. This aspect of Wittgenstein is much more in line with Lopez' work, as Lopez so often focuses on the importance of landscapes. Raban, by contrast, was speaking about the importance of words, names in particular, to our view of reality. He who controls the names controls the narrative. Vancouver understood well that ownership, not only physical ownership, but ownership of the historical narrative of a physical area, was wrapped up in language. By conquering the words, the names themselves, the physical places became part of the empire. If a place then became associated with empire or frontier it was therefore a place to be conquered, made British, Russian, American, etc. This is obviously an issue in that Native peoples were consequently erased from consideration of their own personal narrative in what could be argued a civilization-level form of solipsism and part of a cultural genocide.

With this in mind, Raban's voyage from Seattle to Juneau is set up as a technique to explore the past and the past's tread on the land and people themselves. Therefore, if it is important to note facts such as British Columbia continuing to be deforested at a rapid rate (Passage, 175), or the British Columbian need to separate themselves culturally from Americans (113–4), or even the contemporary ethnological rejuvenation of tribal languages and dances

administration renames Mt. McKinley to Denali," *USA Today*. "https://eu.usatoday.com/story/news/politics/2015/08/30/obama-rename-nations-tallest-mountain/71426656/." 31/8/2015).

14 Wittgenstein, Ludwig. *Tractatus Logico-Philosophicus*. Introduction by Bertrand Russel. 1922. Sweden: Chiron Academic Press, 2016.

(71–7), then the names spattered throughout the work and discussed in detail are just as important in demonstrating this imprint. Names even give scope and power to a multitude of voices in the modern argument surrounding environmental issues, as they broaden a solipsistic colonial narrative based on Anglo-dominance that claims a right to natural resource extraction.

However, neither Lopez nor Raban is attempting to focus in on one single theme, as both have done with other respective works. In *Passage to Juneau* and *Arctic Dreams* they both employ the same technique of constantly changing perspective. Both writers employ 1st person travel writing at times, Raban more than Lopez, but both also reach deeply into stories of exploration, colonialism, and tribal history. Raban focuses a bit more on people themselves, on tribes, communities, and even fishing culture from Seattle to Alaska in the summers (3–37), whereas Lopez focuses more on landscape, with animals often at the forefront. The important point here is that both writers consistently cross perspectives, chronologically as well as in personage and umwelt, in order to convey a more complete and inclusive view of the region, exposing both the glories and the scars. And, for both writers, language is an important facet in understanding these glories and scars. Yet, the differences between the two foci are indeed real. The idea of verbal structure, syntax, etc., interest Lopez more than names, as they evoke the native perspective. Raban, by contrast, is not so much looking for a bygone perspective from another umwelt as he is interested in exploring mankind's intentions through its history, and names. For example, when Lopez deals with exploration, he discusses the human element in terror and mysticism and experiences with something beyond oneself (*Arctic Dreams*, 240–51). Raban is more interested in how man reacts with other men in such circumstances. He is fascinated with the downfall of George Vancouver on the explorer's voyage in the footsteps of his former captain, James Cook, at the hands of Britain's young noblemen. Similarly, watching a tribal dance, Raban can only think of the concept of colonialism and its effects on what he considers to be a false tradition resuscitated by ethnologists (*Passage*, 76–7). He later feels more interested when he comes across a group of First Nations people dancing to an imported rock band, calling it "more real" (318–21). Both Raban and Lopez could be considered humanist in their approach, then, in the sense of moral responsibility to fellow man, yet they offer decidedly different perspectives of humanity and its ethics.

Ultimately, the evocation of Wittgenstein by Raban is important in understanding Raban's humanist attitude toward the Arctic. His explorations are in a way based on Wittgenstein's supposition that "The world we live in is the words we use." Raban discusses place names throughout *Passage to Juneau* and their historical relations as well as his own impressions of them, which makes for an interesting technique in a layer of Raban's work. He draws the reader's attention toward the colonialist history of the region through the names themselves while simultaneously offering his own first-person impressions of place. Lopez could be said to follow Wittgenstein as well, but in ethnographic linguistics, mostly outside a historical context. Tom Lowenstein also fits into this mold, particularly with regard to names. Lowenstein does not exactly examine linguistics as a technique, as Lopez does. However, Lowenstein learns the Iñupiaq language of the Tikigagmiut (*Ancient Land: Sacred Whale*, 185), and he directly approaches how the Tikigagmiut are affected by names from the very first pages of his work. The next section will discuss Lowenstein's perspective on the importance of language to Alaska Native empowerment.

3 Tom Lowenstein and the Experiment of Translation

Language is more than just a mere form of communication. The way we communicate is quite literally related to how we experience the world. Many Outside writers have been unsuccessful in their translation attempts of oral stories told by Alaska Natives and the Inuit in general because they tend to translate directly, often underneath or side-by-side with a version of the Native language. Even if this is very helpful to the ethnologist, it often falls flat for the general reader, or even bilingual Native peoples, as the story cannot truly be translated into English or other Western languages. What Lowenstein does in *Ancient Land, Sacred Whale* is to work with the nuances of languages and cultures, trying to represent the realities of the people reflected in the stories. He subsequently makes the stories work in English according to the ethnographically-reflexive, poetic rules he explains at length in his commentary. If he were to have translated the story directly, the spirit of the work would have been lost in the clunky way the story was told in a second language. His work is in this way distinguished from that of most academic writers.

Tom Lowenstein acknowledges at the beginning of *Ancient Land, Sacred Whale* that he pulled much of his research from the anthropological work *The Whale Hunters of Tigara*[15] by Froelich Rainey, as well as Rainey's unpublished field notes dating from the 1940s (*Ancient Land*, VII). In the "Conclusion" to Rainey's work, that anthropologist states that he was told by the chief of Tikigaq, Old Peter Kuniaq, that white men tend to come into the Arctic and be genial at first, even openly interested in Inuit culture, before isolating themselves off

15 Rainey, Froelich G. *The Whale Hunters of Tigara*. Volume 41: Part 2. New York, Anthropological Papers of the American Museum of Natural History, 1947.

from the Native community, as the gulf between Inuit and Western cultures is so large (280). Rainey's point was primarily to discuss the relevant adaptation of Western culture, technology, and the Christian religion by the Tikigagmiut and how these adaptations are almost never clearly understood by Westerners. Nor are concepts such as Christianity truly well-understood by the traditional Tikigagmiut, who employ what they find useful through their own prism of the world. Yet, roughly 50 years later Lowenstein attempts the ostensibly unachievable in associating two realities that so deeply tend to misunderstand one another. In the process, Lowenstein decided to employ a plethora of techniques, which may seem inextricable, and even lacking in cohesion at times, but the final product, taken as a whole, comes together quite clearly.

First, the author spends roughly 5 pages explaining the process of what he is attempting to accomplish with the work, which sets the stage in an admittedly scattershot approach. For instance, in explaining the gulf between Inuit and Western culture, Lowenstein has this to say:

> Tikigaq people walk in and out of company without feeling it necessary to explain themselves. The dialogue likewise comes and goes. The language of the narrators attempts to reproduce the plain, often repetitive, slightly didactic, sometimes a little sinister … storytelling medium. The stories they tell are direct translations from Inupiaq. (XIV–XV)

There is so much to explain that Lowenstein approaches wider topics, but he brings them back together through particular examples. In this particular case, the author chooses to discuss the abrupt nature of pivoting in storytelling. He says that the Tikigagmiut "walk in and out of company without feeling it necessary to explain themselves." Likewise, their stories may be abrupt, with pauses, starts, endings, etc., effected in places and ways that the Western reader may find grating. He also warns that other techniques such as tone, repetition, plain speech, didacticism, etc., may be outside the realm of the Western reader's scope of acceptance. This idea is given much space because it is so important to approach another culture's technique with an open mind. Cultures become so accustomed to a way of doing something that the strange is often not accepted (Boas, *Anthropology and Modern Life*, 66, 140–3).[16] Part of why Lowenstein's approach appears convincing is his candor in explaining the

16 Boas, Franz. *Anthropology and Modern Life*. 1928. New York: Dover Publications, 1986.

narrative and prosodic choices he has made, while also demonstrating the differences between his poetry and the stories in their original forms.

He then gives a basic background of bowhead whales, followed by that of the Tikigaq whale hunt itself. This is a technique that separately limns the biological and the cultural backgrounds of the tradition before unifying them in a duality, creating one of the backbones of Tikigaq culture: land/sea, sun/moon, brother/sister, etc. This explanation of the whale hunt leads into two important Tikigaq origin stories that serve as a prologue and an introduction into Tikigaq culture. In the "Introduction" just before the stories, Lowenstein quotes a Tikigagmiut proverb: "Never tell one story. Always add a second. That way the first one won't fall over" (XIII). The two stories therefore support one another. In recounting these stories translated into English, Lowenstein attempts to maintain what he can of the essence of the Tikigaq language (in the same sense as Lopez; with a focus on how a language itself affects a society's worldview, even potentially its umwelt). Lowenstein likewise attempts to maintain a physical resonance:

> All the lines written for the two narrators have short beats and are justified to a left-hand margin. The idiom of their commentaries is modeled on the language of the stories themselves, and this in turn approximates the dialect that Tikigaq people use today when they speak English. (XIV)

Lowenstein maintains the flow of language that the Tikigagmiut employ in English and makes a convincing literary story. Franz Boas attempted to maintain a similar absolute veracity in translation and failed, from a literary perspective. This section ultimately asks why this is the case.

The difference between Lowenstein's and Boas' styles is primarily that Lowenstein has taken literary liberties with his own stories, maintaining what he deemed essential to the plot, the lessons, the humor, etc., from the Tikigagmiut standpoint. Lowenstein also maintained the flow that the Tikigagmiut actually employed in their cadence. Meanwhile, Boas wanted the translated words to remain exactly as said by the individual people. Boas, as a renowned anthropologist, curator at the Smithsonian American Museum of Natural History, and professor at Columbia University, most certainly would have been attempting precision over spirit. However, it is important to remember that Boas was writing for an academic audience.

Lowenstein, in maintaining a literary stance rather than academic, is further able to maintain Native rhythm and cadence by stripping the language down to a bare form influenced by Zen poetry, as Buddhism is the other subject on which he has written extensively. Here is an example from *Ancient Land*:

> Asatchaq:
> The old woman did this.
> She chewed burnt lamp oil [puiya].
> It was black.
> It turned white when she chewed it. (3)

Employing a very concise, Zen-like, poetic style allows the reader to take in the story with a natural flow. It also remains more familiar, as concise, Asian poetry has been integrated into American poetry by the likes of Gary Snyder (*The Practice of the Wild*, 130–3)[17,18] and Ezra Pound (*Cathay*).[19] Importantly, Asian poetry appears both familiar and "other" at once. Therefore, the exotic nature of the work remains, as well as its informative nature, while also remaining oddly familiar to the average reader and not off-putting enough to turn one's attention away in frustration.

Following the two origin stories, "Land Whale Story" and "Sun and Moon Story," Lowenstein moves into a discussion of names, which eventually reaches into the mythology of the people themselves. It also explains how "the land-whale myth was understood in terms of local geography, and how the Tikigaq whale hunt echoed the creation story" (*Ancient Land*, XIII). Essentially, Lowenstein is attempting to explain the importance of whale mythology to Tikigaq culture. Of "Part II," Lowenstein says this:

> "Part II is largely about the whale cult and the hunt itself shown in the context of the annual cycle, its hunting patterns, journeys and thought systems. Within this sketch of the annual cycle the book's focus is Tikigaq's shamanistic religion, in which two great events dominate. First, the rituals of autumn; second, the spring whale hunt, which was itself meticulously ceremonialised." (XIII)

The entirety of "Part II" is broken up into four seasons, which is also how Froelich Rainey organized a chapter of *The Whale Hunters of Tigara* (244–68). Tikigagmiut stories discussed in "Part II" are employed in the same poetic form as the two origin stories of the "Introduction."

---

17 Snyder, Gary. *The Practice of the Wild: Essays*. Berkeley: Counterpoint Press, 1990.

18 Snyder actually likens the Japanese poet Dōgen to a Coyote story in *The Practice of the Wild* (120).

19 Pound, Ezra. *Cathay*. London: Elkin Mathews, 1915.

Another technique which Lowenstein employs is to speak in the voice of "two fictional narrators," whose names and likenesses are taken from "my [Lowenstein's] own storyteller and teacher Asatchaq and his uncle Samaruna (c. 1870–1935)" (XIV). What Lowenstein has done, in other words, is to employ the literary voices of actual elders from whom he has gathered stories, but he makes their voices his own by telling the stories his way, while still offering the elders credit for their influences. This decision gives the author free rein to make changes while also not culturally appropriating from the storytellers. Lopez has discussed the importance of a "moral contract" between the writer and the reader (Tydeman, *Conversations with Barry Lopez*, 38),[20] and Lowenstein accomplishes this by being up front about his process.

One further dilemma facing Lowenstein was the fact that Westerners are accustomed to a narrative arc and often some sort of subjective point or moral to be expressed. Even if that subjective point or moral is that there is no subjective point or moral, this is a conscious decision made to prove a subjective point. Alaska Native culture sometimes has subjective points to make and moral commentary as well, but sometimes the point of a story is to be entertaining (XVI–XVIII). Different cultures, broadly speaking, also have different senses of humor. A typical direct translation of an Alaska Native story can make the Westerner feel awkward and confused, waiting for the punchline of a joke or the conclusion that never comes. Lowenstein tasked himself with maintaining this mixture of occasional morality, entertainment, "other" humor, etc., while also accepting the world of the shaman, where shape-shifters exist (Lowenstein, *Ancient Land: Sacred Whale*, 3–15, 43–50) and people become seals and move to houses at the bottom of the sea (Fienup-Riordan, *Boundaries and Passages*, 3–45), while also making his work presentable to outsiders. The writer could have easily turned his poetry into a sci-fi or fantasy story, but he did not. He maintained the general spirit of the people with whom he had been living and experiencing the world.

This is an important form of respect that could aid the reader from outside the community in better understanding local traditions from a foreign mindset, while also potentially helping the bilingual community from within understand itself better from a Western perspective (i.e. how Lowenstein chose to present Tikigaq to the world outside). Lowenstein could in this way give his chosen community pride in themselves and their traditions. Such an accomplishment is not always a given, considering the hostility that so often faces ethnographers in the field, who at times wear out their welcome and make Natives feel like specimens

20 Tydeman, William E. *Conversations with Barry Lopez: Walking the Path of Imagination*. Norman: University of Oklahoma Press, 2013.

instead of real people (Malaurie, *Hummocks 2: Alaska*, 291).[21] Lowenstein's work is an example of someone really attempting to learn the language and culture, and using it not only to go back and write to academics and the curious Outside, but to truly give the chosen community a solidified understanding of itself, its struggle, and its place in the larger world. For a young person from Point Hope or Utqiaġvik, standing over a precipice in life and looking for something worthwhile before jumping in, seeing no future worth living beyond instant, dopamine-inducing gratification, this type of work can potentially be life-changing, on both the individual and larger social level. It is a powerful form of mobilization, particularly considering that Jean Malaurie went into these same communities not long before and was chastised by the young, who were depressed, killing themselves at enormous rates, and placing blame on anyone with white skin, telling them to get out (*Hummocks 2: Alaska*, 291–4). Joe McGinniss reported on a similar phenomenon when he went to Utqiaġvik and witnessed alcoholism and violence (64–6), with white teachers brought in and constantly harassed to the point of developing alcohol problems as well (66, 81–3): "The bar had been shut down since August, closed by the state police. Too many fistfights and knifings. The Eskimos of Barrow, apparently, did not handle their liquor very well" (59); "In Barrow, when John Anttonen's children walked the streets, they were called 'honkies' [derogatory for a white person] and cursed at, and spit at … The feeling of racial hatred in Barrow was deep and strong. This was not country where the white man belonged … " (63). There is always a story behind a story like the one McGinniss tells, and in this case, that story might just be Lowenstein's *Ancient Land, Sacred Whale*, a careful, experimental collection of stories that brings pride to the traditions of the Tikigaq community.

Examining Lopez and Lowenstein's perspectives on language, and how they conveyed their messages, and, notably, the subjects and themes they chose to cover, allows us to better understand traditional Alaska Native perspectives, which risk being lost in translation. Without this analysis, just as we may visit Anchorage and see the visible social issues magnified by alcoholism in city parks and on the streets (*Hummocks 2: Alaska*, 291), we may read Alaska Native stories and completely miss their essence. When truly understood from multiple angles, Alaska Native writing allows us to see, if only slightly more clearly, with the "Native Eye," as Lopez suggests. Though, as will be examined later, Alaska Natives are beginning to reclaim their own narratives, these travel writers of the late 20th century who respectively conveyed oral traditions played an important role in the aftermath of a cultural genocide at the hands of multiple colonial powers.

---

21 Malaurie, Jean. *Hummocks 2 : Alaska, avec les chasseurs de baleine, mer de Bering: Tome 2 – Livre I*. 2nd ed. Paris: Éditions Plon, 1999.

FIGURE 5 Yukon River
PHOTO BY CHRIS FERGUSON

CHAPTER 4

Depictions of Alaskan Natives

The rise of New Age spirituality over the past half century has led to a shift in respect toward indigenous cultures around the world. During the second half of the 20th century, Native Americans developed a reputation of relative peacefulness, respectfulness, and environmental stewardship, and this has tended to be a boon. Communities who were once maligned as heathens suddenly had powerful allies in the likes of intellectuals such as Gary Snyder and Barry Lopez, who began to teach the industrial world that there was something to be learned from Native culture. Yet, as will be demonstrated, this shift has created a whole new stereotype to which Native peoples are expected to conform, and the result often impedes self-determination.

Pigeonholed by stereotypes of ecological stewardship that are actively cultivated to promote subsistence rights (Lord, *Beluga Days*, 29), Alaska Natives can find themselves adversely affected due to diverging understandings of what sustainability actually entails. In fact, Natives have not infrequently clashed with environmental groups in recent decades (Wenzel, *Animal Rights, Human Rights*, 36–7, 41; Lord, *Beluga Days*, 37–8). Outside environmental groups have been open to an aesthetic and emotional idea of Arctic cultures, but often not in practical terms (Wenzel, 41). For example, as Native spirituality has been artistically cultivated in broader society as a form of mysticism with a close relationship to the Earth, and with an ancient wisdom (Raban, *Passage*, 348), it tends to spark the "exotic" in the Western imagination. Traditional Native spirituality is also inherently unlike Christianity, a fact that appeals to many New Age practitioners the way that Buddhism does (Kerouac, *Dharma Bums*; Matthiessen, *The Snow Leopard*; Gary Snyder, *The Practice of the Wild*). In effect, one of the primary goals of the New Age is to overturn Christianity's patriarchal and colonial history. The practical aspect, however, is that Native peoples are often hunters, Alaska Natives in particular "taking" whales and seals. This act tends to be viewed as taboo in the New Age paradigm, as it rests at odds with Buddhist opposition to the killing of all living things due to a respect for the cycle of reincarnation (Brinkley, *The Quiet World*, 192, 392–5). Furthermore, Native peoples are often not against adapting Western technology, an extension of colonial patriarchy that interrupts a truly spiritual relationship with the land under this philosophy. The Native preference for the use

 | DOI:10.1163/9789004734807_006

of whale guns with exploding harpoons (Hess, 203–4), then, can be shocking for some Outside, as can be persistent use of snow machines during a hunt. What is sustainable, environmental, and careful are constantly scrutinized by activists and researchers, but they themselves must be careful not to impose an industrial society perspective on Arctic communities.

However, careful depictions of Native peoples allowing for self -determination involves a multiradiality that reaches far beyond environmental and spiritual stereotypes developed in the 20th century. For example, Susan Kollin is mostly complementary toward John McPhee for not falling into stereotypes and not extrapolating modern beliefs onto traditional values, but she is nonetheless a bit negative toward his depictions of Natives (168–9). Picking on McPhee the way a professor tends to challenge their best student, Kollin expects more from him. She begins a paragraph with:

> Nature writer John McPhee, for instance, who otherwise tries to dismantle mythic images of a wilderness Alaska in his book Coming into the Country, addresses in a less than complimentary manner what he refers to as the new class of 'business-smart' natives. (168)

She goes on to say on the following page that creating a "shareholder" out of the Native "disturb[s]" McPhee, "as they [the changes brought by the pipeline] signal to him the profound transformations that capitalist development and modernization have brought to the nation's Last Frontier and its indigenous inhabitants." Like so many others from Outside, McPhee does not allow for self-determination. Being worried about the forcing of capitalism onto traditional Alaska Native peoples is arguably justifiable, but the question is whether or not McPhee is more uncomfortable with the fact that Native peoples are stepping outside their position as "representative men" and becoming "business-smart" or whether he is really worried about their futures as peoples and cultures as they continue in their self-determined evolution. Kollin seems to be arguing the former, and there is some evidence for this sort of predilection in McPhee's work as discussed in the McPhee chapter herein.

Kollin also criticizes Joe McGinniss for his judgment of Natives in *Going to Extremes*, saying:

> Ultimately, McGinniss's comments here say less about Alaska Native life in the post-pipeline era than they do about his own desires for a stable, pre-modern primitive, a figure that supposedly embodies elements of the national past which he and other white Americans have somehow lost. (169)

In this work, the journalist goes to visit a friend near Bethel, way out in rural Alaska, where a family is found in a small home. McGinniss is impressed by the stolid, concentrated gaze of the father who is carving a stick. This man apparently does so for hours, with McGinniss finally, ostensibly sensing the authentic experience of Alaska Native culture, when all of a sudden the son rushes into the room screaming that *The Six Million Dollar Man* is on television. The father emphatically throws the stick and knife aside and hurries away to be enraptured by an hour of this program. McGinniss is absolutely dismayed. He feels betrayed, that the things he has come to Alaska to escape are suddenly front and center in the home of this authentic Alaska Native (108–12). For Kollin, this is an example of McGinniss' expectation of the "stable, pre-modern primitive," and his lack of allowance for self-determination, just like McPhee. From the Native perspective, adoption of television seemed as obvious as it did for everyone else. Again, as with McPhee who was worried about the encroachment of industrial society into traditional Arctic communities, the question is whether McGinniss was worried about a representation of Native peoples in his own mind or for the peoples themselves. As Natives adopted snow machines after being forcibly moved into villages by Alaskan and Canadian governments and suddenly needed to travel farther, faster, to get to their hunting grounds (Wenzel, 66–74, 94–6; McPhee, *Coming*, 29–31), they also adopted the metal knives Robert Peary offered at the turn of the 20th century (Harper, *Give Me My Father's Body*, 1–10), and they have adapted and lost various technologies for millennia. Yet, McGinniss, as a New Journalist, does not attempt a multiradial perspective, and the result is the weakness of his limited perspective combined with the strengths of firsthand experiences he is able to uncover that John McPhee's journalistic style never would: the prevalence of casual drug use by politicians in Juneau (16–8), a British man with an infected eye trapped in the Arctic after being scammed into a high-level position at the Iñupiat University of the Arctic (73–9), etc.

Overall, despite not particularly grasping or prioritizing the depths of the plight of the Alaska Native, McPhee is efficacious at describing the moment of the pipeline's arrival and mixed public sentiment in Alaska toward conservation. In his own way, McGinniss is likewise efficacious. Yet, both have their flaws. This can partly be explained by the fact that it is enormously difficult to cover a subject from a completely multiradial perspective. However, considering colonial histories, and how Alaska Natives have so often been portrayed, extra care should be taken into account whenever a writer moves through their traditional lands. This chapter will explore how Jonathan Raban, Nancy Lord, and Tom Lowenstein each navigated complex Native issues in their respective works. All three of these authors offer extraordinarily different approaches to

Alaska Native depictions, and all of them are remarkably respectful, keen to overturn stereotypes in exploring nuance.

1 Raban's Conflicted Contribution to Native Peoples

On the one hand, Raban is extraordinarily complementary toward Native peoples and has gone out of his way to study them at some length. On the other hand, he is also an avid reader of Western explorers linked with colonialism, such as James Cook and George Vancouver, making sure to research his chosen geographic space as extensively as he reasonably can before traveling there. In doing so, Raban makes comparisons between the reflections of different explorers on certain tribes in the places he is visiting, as well as that of ethnological accounts (Boas and Lévi-Strauss, for instance, and the way to view coastal Indian art – Raban, 21–33), and he even alludes to guide books and their misinformation with regard to the unfriendly demeanor of certain tribal groups (323–7). This thorough understanding of history engrains Raban with a certain amount of cultural sensitivity. He is respectful and admiring of the complexity of totems, of Native art and culture more broadly, and he even explores Native navigational skills that are generally disregarded by the West. The issue with Raban is that he can be rather brash, or ostensibly insensitive, in displaying his discomfort with experiences he has had with Natives themselves. This section will explore Raban's somewhat iconoclastic methods, pointing out that they can ultimately be helpful for depictions of Native communities. This is because Raban's educated humaneness remains in line with current cultural trends, which lean toward Environmental and Social Justice. His work, however, does not fall in line with mainstream narratives. Rather, it often highlights the exclusionary narrative of mainstream environmentalism and social liberalism (in many ways this line of criticism is Raban's "bread and butter" in the Pacific Northwest American phase of his writing).

To begin, when Raban is invited to attend a traditional Native dance in Washington state, he describes feeling uncomfortable and unimpressed, daydreaming of hitting the casino and desperately wanting to leave the local hall that resembles nothing of an indigenous abode, and serves Kool-Aid (76–8). This section is worth hovering over because it is the kind of passage that is hard-hitting, a passage few writers would dare put to paper due to fear of reprisal. When this point is made, the nuanced argument of the author nonetheless remains easily misconstrued. The simple message, a privileged outsider "punching downward" and judging a long-repressed minority group, may be the takeaway if not read carefully. Raban is critical in this scene, however, because he feels the dances can no longer be seen as an authentic expression of a living culture, but that they have essentially become a watered-down version of traditional practice. To him, the dances are a static and sentimental recreation promulgated by ethnologists who do not truly understand the minds of deceased people. Later in *Passage*, as briefly mentioned above, when Raban comes across a group of First Nations people in British Columbia dancing to an imported rock band he tells someone that this experience is "more real" than the traditional dance he had seen (318–21). If the writer is likely as presumptive as the ethnologists, he almost never leaves his impressions vague. Far from it, he delves into ethnological and exploratory accounts on his own. For instance, he examines how the records of meetings with Natives in older travel writing can be tainted by meetings and tradings those Natives have already had with Westerners, which may have changed their entire cultures before supposed first contact (29). Raban employs totem poles as an example, implying that there is good reason to believe the oversized iterations of carved wooden statues only appeared along the British Columbian and Alaskan coast once wealth from trading with sailors appeared, and were noted shortly thereafter by official expeditions (29). The point Raban is attempting to make is that our understanding of the authenticity of Native peoples is almost always skewed, and so is our understanding of their cultures and desires. The author, in this way, ultimately criticizes this idea of what he calls "John Muir Indians and Fenimore Cooper Indians …" (348), because Natives in these depictions become stereotypes of themselves. Native American writer Tommy Orange calls this "the copy of a copy of an image of an Indian in a textbook" (7) in his celebrated recent novel *There There*.[1] Raban suggests that such views are "insulting to the Indians" (348). In citing this passage, it is important to restate that Raban is searching for a respectful and authentic version of Native peoples, and that he views them with incredible dynamism and complexity. Natives, in his eye, were always likely as varied and nuanced in their ethics as Westerners. In other words, neither are Natives "Noble Savages," nor are they a lower race.

Concerning Raban's portrayals in *Passage*, the writer comes across drunken Natives at times on his way from Seattle to Juneau (318–21). He does not avoid the issue, at least being honest in his portrayal (in Raban's defense, he also points out white drunkenness quite frequently, actively seeking out seedy bars whenever possible – 364–70). Raban also narrows certain situations down to the personal level, discussing how the Native dance he thought distasteful in Washington turns out to be an exercise in rehabilitation from drug use and a

1 Orange, Tommy. *There There*. London: Harvill Secker, 2018.

boost of self-esteem for the young and jobless (76). A participant, interviewed during a car ride home that Raban offers, does not sentimentalize the dance but feels that it is indeed empowering. The technique of this interview taking place in a car ride home is important because of the first-person narrative of the individual affected and the safe space as the two of them smoke cigarettes. Again, Raban gives his opinion, which is backed up by historical and cultural context (70–3),[2] before he allows his subject a personal say, which is a thorough and inclusive process while also allowing for skepticism. Interestingly, Raban expels a sigh of relief when he sees the teenager involved smoking a cigarette, as the cigarette is a symbol for the writer that there is dynamism to the boy in not taking his own potentially misinterpreted traditions too seriously, again, like "good school children of the 1990s, following Mother Nature's rulebook" (348), especially after Raban felt so uncomfortable during the dance in the hours previous. This notion can easily be employed to dismiss Raban, but the writer is describing nuance and humanity in people who deserve respect. His methods are merely unconventional.

It is ultimately the writer's narrative technique reflecting his ambition of debunking stereotypes and convention that allows him to approach such controversial subjects as the criticism of ethnological re-creation, particularly when considering the plight almost all Native communities have suffered in the past few centuries in managing their engagement with European communities. There is no doubt that Raban is a privileged person, and many would say that he has no right at all to criticize an ethnic/cultural group of which his own nation and people had spent hundreds of years taking advantage, particularly from the perspective of Social Justice. Raban published *Passage to Juneau* in 1999, however, in another socio-political era. This must be kept in mind in criticism of his work. Yet, notwithstanding the socio-political pendulum, Raban is nonetheless thorough in his technique, attempting to approach the argument from multiple angles, both physically and academically, as we have seen. Of course, Raban loves to "walk down Pound Lane" (283), a technique the writer

2 One important point to add here is that on page 70, Raban, always the skeptic, notes explicitly that the "Indians had lived on tiptoe, in small numbers, on the extreme fringe of the forest, barely grazing the surface of the water," before "the great self-important juggernaut of American capitalism rumbled through the Pacific Northwest, clearing it with chainsaws, paving it with asphalt ..." Making this strong, environmentalist statement in favor of the local Native community around Seattle, Raban goes on to explain how "The reservation economy had turned away from the sea to the six-lane highway, where the Tulalips ran a round-the-clock casino. He then calls it, shortly thereafter, "The rest: half tax haven, half open prison" (71). This explanation is meant to imply that the cultural degradation of Native communities is entirely the responsibility of Western society.

takes from his father, an Anglican priest, who would begin sermons by discussing a walk he would have taken down "Pound Lane" (a random street) when something would happen all of a sudden that would lead to a religious life lesson. Raban employs this technique in his travel writing to expound on multiple ideas he considers pertinent, but he is also quite physically "walk[ing] down Pound Lane," which in this case is the physical space being discussed in his travels. So, on the one hand, Joe McGinniss is the oft-criticized travel writer of this research who physically travels into the Arctic and discusses what he sees, exposing important aspects of the political and social system never mentioned in other literary sources but leaving out much nuance and historical context. On the other hand, many careful, well-intentioned anthropologists have traveled extensively through the Arctic, collecting considerable amounts of data, but, as noted in the Toelken's introduction to *Giving Birth to Thunder*, some may "gawk" but not "see" (xi–xiv). Or, very often, travelers come to a region with an agenda and blind themselves to other aspects, a problem to which almost no one is immune.[3]

Raban, like Lopez, comes into the Arctic with an educated and informed eye about Arctic peoples. The difference in technique arises as Lopez often maintains an authorial distance from his story in *Arctic Dreams*. By contrast, Raban takes the direct travel writing route that New Journalist Joe McGinniss takes, except Raban is informed by ethnology, exploration, literary and artistic history, etc., displaying that knowledge throughout, in pertinent moments particularly in order to create a nuanced, patchwork picture. As a metaphor, Joe McGinniss walks straight down Pound Lane and tells the reader what he sees; Jonathan Raban walks down Pound Lane, gives his personal situation as to why he is walking down Pound Lane and how it may affect his view, and explains the background about why Pound Lane and its people are the way they are; and, Barry Lopez mostly avoids Pound Lane altogether, discussing reflections and giving occasional glimpses into Pound Lane but mostly describing its minutia, history, and culture. Of course, all three techniques may be seen to have their respective virtues.

Apropos the techniques of Raban and Lopez, there is another interesting dichotomy between *Arctic Dreams* and *Passage to Juneau*. One is a meditation on land, and the other a meditation on water. Both stage an interplay of views, examining the world from the "Native Eye" in their own unique ways.

---

3 As briefly alluded to, one aspect which separates Raban from so many others is that he is adept in academia, and equally adept at grabbing a cowboy hat and a cigar or a Seattle Mariners baseball cap and marching into a dive bar, and fitting in enough to get interviews (412–3).

Having grown up in Britain, Raban jokes that every man, even one who has never set foot on a boat, feels like a swaggering sailor in his soul due to the nation's history of navigation (226–8). Viewing the Northwest coastal peoples from this perspective, Raban attempts to invert the normal view by suggesting that anthropologists such as Boas were unable to truly understand certain peoples due to their lack of ability to view the water as the primary world and the land as secondary, as he insinuates the British have more of a penchant for doing (201–7, 226–8). Whether there is hard truth in his assertions or not, his viewpoint is an interesting and evocative one in that it truly does invert the general narrative, and inversions of narratives can often offer heretofore unexplored insights. For instance, Raban notes that one would not truly understand the repeating bubbles of art by Pacific Northwest Natives if one has not paid attention to the bubbles rising out of the waters that he has repeatedly seen while sailing. Raban, in fact, cannot not find a reference to this view anywhere in his studies on the region's art (201–7). He also notes that belief in the god, Komogwa, and of whirlpools as truly living entities, was often derided by ethnographers and early preachers, yet Raban expresses the attitude that spending time on the water, in the presence of entirely unpredictable abysses, one could completely understand the rationale behind this belief (218–21, 404). Lopez would likely not disagree, as Lopez has always been a proponent of maintaining the complexity of Native beliefs. For instance, stories about Coyote and Raven are legitimized in Lopez' mind when viewing the complexity of the animals he limns at length in *Arctic Dreams*.

What these insights by both writers do is reinforce Native viewpoints as dynamic. Essentially, this is a similar notion to the dangers of considering Alaska as the Last Frontier. If Alaska Natives and First Nations peoples are taken at face value with a 2-dimensional viewpoint, which fits a certain mold that Raban derides, it is not to the benefit of the affected peoples. The danger is in delegitimizing peoples through symbolism, which means they are not dynamic. They become static. In this way, when anthropologist George Wenzel discusses the Inuit of Northern Canada, he says that many people view them as no longer Inuit, having outgrown their customs in adapting modern technology (*Animal Rights, Human Rights*, 5–6, 34). Wenzel argues that this is a wrong-headed argument, as the Inuit are not a static people, but that their technologies have always been in flux, with peoples employing different techniques that serve them well (34). Therefore, when Raban says that a 100-year-old tribal dance rings false and that a modern rock dance by tribal peoples rings true, the point is that culture is not static, as it evolves organically, in stark contradiction with awkwardly-imposed anachronisms presented as respecting/continuing tradition.

Raban's criticism, then, can ultimately be viewed as freeing for Native peoples and more in line with modern Native literature, such as Tommy Orange's 2018 novel, *There There*, mentioned earlier in this section. Orange's novel presents Native American kids in Oakland learning traditional dances from *YouTube* videos, alongside native hip hop music infused with traditional elements (118–32), and the broken, repressed families with histories of spousal and substance abuse (15–26, 197–207), and even a deadly shootout with 3-D printed guns at a local pow wow (269–90), all while individuals and families are attempting to come to terms with their objectified past and move forward into an uncertain future, unsure of the importance of Native American tradition in moving forward but knowing they are different from other ethnic groups all the same. Tommy Orange's work seems to be the personification of the kind of self-reflectiveness and dynamism Raban expects and is attempting to get across about the pitfalls of exploring Native culture in *Passage to Juneau*. And like *Passage*, *There There* often comes across as iconoclastic. Yet, it is also respectful to traditions in many ways, and it even begins with 9 pages noting historical context of American subjugation/objectification of Native American peoples. Interestingly, there is absolutely no Native Eye in *There There*. The people are without exception creatures of a concrete jungle. The "rez" or reservation in New Mexico is only briefly alluded to (136, 197–207, 217–8, 236–8), and as a place where a woman is beaten by a manipulative partner before she flees back to Oakland. Tommy Orange, then, does not focus on the natural world the way the writers discussed in this research all do. He, like Raban, is interested primarily in people, in humanity. Raban, though not often connected to environmentalism, is taken enough by the sea that *Passage to Juneau* is still truly about the sea in many ways. This, and Raban's overtly academic bent, are where he and Orange diverge. Though, it must be noted that notwithstanding a lack of environmental curiosity by Orange, the two nonetheless share a similar worldview, separated by generations and backgrounds. Raban comes across as caring in a skeptical, scrutinizing manner. Orange is hard-boiled and wants to explain to the wider world his reasoning: cultural objectification, cultural subjugation, cultural appropriation, racism, and, increasingly, gentrification of physical spaces in his native Oakland (3–11, 216).

In the following section, Nancy Lord will also be shown to attempt to dig into tribal histories, often with an interest in anthropology. However, *Beluga Days* is almost never based around her personal story, as opposed to Raban who is right at home in bars, and at dances, and includes his own memoir and personal history in the work. Lord is much more driven by agenda in *Beluga Days*, which is a purposeful deviation from two of her works around the same era: *Rock, Water, Wild* and *Fishcamp*.

2 Nancy Lord's Sensible Objectivity with Regard to Natives

As has been stated herein and elsewhere, one of the problems Westerners run into with regard to Native peoples is that Natives tend to be viewed through a preconceived lens according to the industrial world's principles. This lens can be condescending, or it can be apotheosized as a kind of Rousseauian model of The Noble Savage in order to extrapolate entire nontransferable existences upon the Occident (Joan Pau Rubiés, "Travel Writing and Ethnography," *The Cambridge Companion to Travel Writing*, 256). Nancy Lord succeeds in avoiding both damning judgment of and hyperbolic appraisal for the Dena'ina Athabaskans of the Cook Inlet where she has spent her entire adult life. Instead, she focuses on nuance. This section will explore Lord's techniques of endearing herself with local Dena'ina culture through her respectful approach of striving for objectivity in representing the issues she discusses, as well as benefitting from the general resemblances of her own life as a setnetter to that of Native subsistence culture. Likewise, she employs a similar means (fishing, subsistence) to create solidarity between Native communities and sourdough culture in Alaska, something not always easily accomplished as there is much historical mistrust between the two.[4] Lord's embrace of Traditional Ecological Knowledge will be discussed at the end of the section as well, pointing out that such a technique does indeed allow her to maintain sensible, journalistic skepticism, but it simultaneously invites what the author has previously described as "felt life."

Though not a journalist by trade, Lord takes on the task for *Beluga Days*. She travels all around the inlet, and beyond, working tirelessly to obtain an informed view of humanity's relationship with the endangered whale. Along with a few others, Lord raises the alarm bell of wildlife mismanagement. As a result, the Alaska Native community around Anchorage, which is collectively hunting too many whales without realizing it, self-imposes a hiatus on "taking" Cook Inlet belugas. Not yet knowing for sure whether they themselves are to blame, the Native communities nonetheless care about the whales, and furthermore, are hypersensitive to cultivating their reputations for sustainability in subsistence (29, 40). They are conscientious of the "Noble Savage" stereotype. They understand quite well that a binary exists for them, and that the other end of the imposed spectrum of public opinion available to their communities is childish heathenism.

4 This is a common thread throughout much of McGinniss' *Going to Extremes* (57–85, 103–18) and Lord's *Beluga Days* (35–9).

However, a Native hiatus on subsistence hunting is only part of the solution. For example, multiple individuals state in a Nancy Lord-edited collection *Made of Salmon* that king salmon are likewise on the decline, and smaller than before (113–8, 157–63). As salmon are the beluga's principle food source, salmon decreases in size and numbers obviously affect the beluga's chances of survival. Also, commercial fishermen once considered belugas a rival, a "predatory species," and consequently hunted them for sport (*Beluga Days*, 4). Furthermore, many surviving whales are reputedly ill due to industrial waste and likely even noise pollution (91, 104–9, 118). Ultimately, then, the reasons for the disappearance of the Cook Inlet beluga whale are multifold, with all parties of humans in Alaska to blame. Nancy Lord exposes this fact and makes sure that everyone is well aware of the situation so that no one party can indiscriminately blame the other, becoming heated to the point where nothing gets done. If everyone accepts responsibility, theoretically they can all move forward with protecting the species.

In *Beluga Days*, Lord journeys to the tribal village Tyonek, located on the hard-to-reach, far side of the inlet in order to get a feel for the local community. She also talks with her neighbors, some who are with and some who are without biases against the Native community (1–18), and she speaks with government officials, an oil company representative, and multiple scientists (19–63). She also dives into Dena'ina ethnographical works to better understand the culture, discussing *Shem Pete's Alaska,* Peter Kalifornsky's writings (177–9). Frederica de Laguna's anthropology is also mentioned for its lack of focus on the beluga, as well as Captain James Cook's journals (160). Lord pours through any source she can think of to find historical context on beluga in the region, especially the Dena'ina beluga whale hunt (5–6). The overall result of *Beluga Days* is that Lord portrays political and social context with methodical, journalistic objectivity. She rarely interjects her own emotions, and in this way her writing resembles that of John McPhee. In refusing to become emotionally tangled in the subject, Lord maintains enough respect for Alaska Natives to question the lack of restraint in their practices with a cool head and no preconceived notions (or, at least remaining open to her notions being overturned). Her objectivity also enables her to surmise that the young of Tyonek are hurting partly because they no longer prioritize the subsistence of their ancestors. This technique of journalistic travel writing indeed risks criticism, as does that of McPhee, but it is also effective and endearing to multiple perspectives: that of the Native community, that of the average Alaskan even vaguely interested in the political situation, and even that of those looking in from Outside. In fact, the average reader, likely only having glimpsed Alaska from the point of view of the saccharine commercialism offered on mainstream television, is offered a nuanced glimpse into the Cook Inlet, which the Dena'ina call home and where the city of Anchorage and the village of Homer are located.

The success of reaching across perspectives so successfully allows Lord certain advantages. For instance, she must convince the local community elders of the village of Tyonek to allow her journalistic access, as they are wary of outsiders, and journalists in particular, misconstruing their practices and culture (164–5). Enmity and distrust between the Native and White communities still run deep (157–68), and Lord charges herself with the task of threading this proverbial needle by once again appealing to all parties in an effective manner. The fact that she has worked at a canning factory, saved money for years to buy a boat and spend much of her adult life as a commercial setnetter, and is generally respectful, allows the author the role of the fellow traveler among both Alaska Natives and Sourdoughs. Even if many Native communities are in the process of losing touch with subsistence traditions at this time (181), the village chief in Tyonek is desperate to keep the younger generations connected to their subsistence roots (182–5). Anything that Nancy Lord can do to help that connection is ultimately supported.

However, unlike so many of the anthropologists that Lord mentions,[5] she avoids the topic of Native spirituality almost altogether in *Beluga Days*, focusing on the subsistence perspective instead. She is pragmatic, humanist, and unreligious herself, and even if the Natives with whom she is discoursing or to whom she is breaking through with her writing are highly spiritual or religious, she is often able to reach out to them with a respect for their subsistence traditions. Here is an excerpt of her recounting of the conversation she had with the Tyonek village elders:

> A couple of them spoke about subsistence generally—how subsistence rights had been eroded ... They asked about where I fished and who had fished there before me, and who I knew along that beach, and about the new oil development at Kustatan. They were, I thought, trying to judge my character. (165)

In the complete recounting, Lord notes that the elders are attempting to judge her character, and the concerns are not her connection with animals but her

5 Ann Fienup-Riordan stresses all throughout *Boundaries and Passages* how the Yup'ik traditionally filled every aspect of their lives and thought with animals: subsistence, dance, the spirit world, cleanliness, home, rules, etc. Such is the case as well, at least to some extent, with the work of so many other Alaskan anthropologists (Boas – *Primitive Art* –, de Laguna – *The Tlingit Indians* –, Meade – *Agayuliyararput: Our Way of Making Prayer* –, etc.).

respect for animals. This is much in line with the author's thinking. When the elders speak of subsistence rights being eroded, they mention that they want to be able to hunt at the time most respectful to the community traditions of the hunt and not when modern scientists tell them it is most advantageous, knowing that traditional patterns have their own respective and respectful reasons. They also test Lord's motives. Notably, the elders question not only where Lord fishes, which shows an interest in her credentials as a true fisherman likely to be understanding toward their subsistence culture, but who fished her spot before her. Understanding this shows a respect for the historical tradition of the region. If she were to have come from Outside and shown no interest in the history of her fishing spot, but only in the profession, and not uncommonly only seasonally, they might view her as someone not wholly in line with their own interests. But who fished before her is exactly the kind of question that interests Lord. In fact, she had already written about the history of her fishing spot in an earlier work, *Fishcamp*, which also details the surrounding land and water (25–38, 43), and focuses on issues such as others moving away when profitability diminished (22–4). Her interests in this way endear her to the elders and others still committed to subsistence living, and who are influential in their communities. Furthermore, by avoiding discussions of "orientalist"[6] exoticism, and Alaska Native aesthetics, Lord also avoids falling into the ever-appealing trap of the Last Frontier. As a result, however, *Beluga Days* remains a cult classic in Alaska but is almost unheard of Outside.

Lastly, as with other issues, Lord approaches Traditional Ecological Knowledge (TEK) with respect, albeit infused with (respectful) skepticism. The decision to discuss TEK at all, and particularly in place of spiritual tradition, is interesting. Lord wants to make a connection between the hunters, whom she believes truly understand what they are doing (deeply understanding the patterns of the beluga, the waters, and the fish when they are in touch with their own traditions), and the scientists, whom she also knows have a wealth of knowledge. The consulting of the two by one another can only be a win-win scenario, and in fact, sometimes the National Marine Fisheries Service (NMFS – traditionally charged with supervising the hunt, and who brings in

6 Billie Melman notes in the essay "The Middle East/Arabia: 'the cradle of Islam'" in *The Cambridge Companion to Travel Writing* that 'Orientalism' "draws on a binary epistemology and an imaginary geography that divides the world into two unequal and hierarchically positioned parts: the West and the East, the Occident and the Orient, Christianity and Islam, rationalism and its absence, progress and stagnation" (107). This idea is extrapolated, here, onto the Alaska Native community, for which the same binary could be detailed, with the exception of Islam. The West is dynamic and allowed to progress while the stereotype of the traditional Native must remain stagnant, part of the landscape itself.

TEK studies) is chastised (91–7), while at others it is discussed with approval (the chief of Tyonek notes that the NMFS was entirely respectful during the yearly beluga hunt, and several people, despite being frustrated with the NMFS's lack of action in the declining Cook Inlet beluga population, note that the multiple parties involved make it difficult for NMFS to work with everyone simultaneously – 184–5).

Regarding TEK, the accusation could be made that it has led to Native hunters taking too many whales, which is what caused much of the current crisis. But even if that is the case, the fact that Anchorage has legally been labeled a "village," consequently giving hunting rights to multiple Native groups not traditionally from the area, means that Native groups were not necessarily communicating their numbers with each other at the time and did not know the sustainable numbers to be killed. In Point Lay, high in the Arctic, TEK would suggest that thirty to forty whales be taken each year (188), but when these Iñupiat brought their TEK into Anchorage, they may have been unaware that such a number would rapidly deplete the Cook Inlet's much smaller population. Therefore, TEK and the scientific community work best when they work together in modern, industrialized Alaska. This is what Lord appears to believe, and she attempts to explain the concept of TEK and its background over multiple pages, consulting both scientists and hunters and asking them if they agree with the suggestions made by TEK-influenced scientific papers. One hunter tells her to remain skeptical but not to dismiss the idea. For instance, the months when beluga pass through the inlet are disputed (177), but the paper is more or less in the right frame, as the quantitative shift may be due to climate change. Lord notes that warming waters have meant changes in vegetation and predatory hunting patterns, as well as animals being chased into new territories (28, 173–6). The overall discussion of TEK, then, remains an important tool for pushing forward her agenda for respectfully managed subsistence. In Lord's world, subsistence is incredibly important for maintaining a connection with, and understanding of, the habitable Earth. Moving forward, large-scale industrial society is obviously here to stay, if sustainability of human civilization is supposed, so she is suggesting that communities must work together in managing sustainability while concomitantly pushing for said connection through subsistence. In these ways, mostly involving a mere show of respect, Lord is able to remain connected with the Alaska Native community, honestly portraying it.[7]

---

7 Lord does indeed bring up the well-known issues of Alaska Native problems with alcohol, drug abuse, and waste of millions of dollars on frivolous purchases such as Hawaiian vacations, but she does so respectfully, discussing the background (the cultural norm, deriving

In the following section, Matthiessen, in contrast to Lord, mostly ignores the natives of Nunivak in his work, *Oomingmak*. Nonetheless, he reaches out to the reader with a few very short sections that, while not necessarily negating criticism of a lack of discussion of Natives, demonstrate his support for them. These sections also offer a nuanced view of the plight of Alaska Natives in the mid-20th century.

3 Matthiessen's Sharp, Concise Reflections on Alaska Natives of Nunivak in *Oomingmak*

The criticism levied upon Barry Lopez of not discussing Arctic Peoples enough in Arctic Dreams could easily be transferred to Peter Matthiessen's *Oomingmak*. Throughout *Oomingmak*, there is only passing discussion of the Cup'ig people of Mekoryuk on Nunivak Island. To the author's defense, the work itself is short, even comparable to the length of a short novella, so it must, by its very nature, be concise with its ideas in focusing primarily on the musk ox. Yet, there is no getting around the fact that there is still very little about people. This section will attempt to explain how Matthiessen employs this very little as a concise means of making a brief but powerful statement regarding the social context of Alaska Natives.

First, Matthiessen describes how Bethel, "a sprawl of shacks, mud and rusting oil drums, dumped from the sky onto the barrens, has its admirers, but neither Teal nor myself, who had been here before, could be counted among them" (13). He also describes the civilizational surroundings on Nunivak, with drying racks for salmon, parkas and mukluks still being hand-sewn (15), the adaptation of the outboard motor and frustration with "the mission ban on dancing, drink and smoking" (16). However, these descriptions are few and far between, and one never gets a strong sense of this group of Alaska Natives. This is partly because, as Matthiessen notes, the Cup'ig are afraid of "the Bearded One" (the musk ox) and "leave him alone," finding it funny that these foreigners would want to trap him (25). Matthiessen and his partner, Teal, are there for exactly this scientific purpose, which drives a wedge between his work and the people who are purposely staying out of his way.

---

from potlatch society, of sharing wealth among the community and spending it, enjoying it, upon reception – 167), and she also keeps this discussion to a minimum, preferring to push for both an increase in beluga numbers and an increase in Native subsistence participation (157–85).

Yet, Peter Matthiessen was not a decades-long writer for *The New Yorker*, with prolific publication, by pure chance. In one moment of sharp, concise criticism, Matthiessen reaches into the cultural situation of mid-century Alaska and produces an insightful commentary, while ostensibly throwing out an offhand anecdote.[8] This occurs in "Chapter 6," toward the end of *Oomingmak*. Matthiessen describes his partner, Teal, calling in a helicopter to help round up the musk ox. The pilot tells an Inuit man to "… keep his cotton-picking hands off …" the helicopter, "… suggest[ing] that the man might steal some parts for his own boat" (73). The young child of the insulted Inuit man suggests to the pilot that the father is trustworthy, and the pilot retorts that the child probably is too, if he is kept in view. The child then responds, "Not all natives are bad." This very brief exchange then evokes a concise, yet cutting, two-paragraph commentary by Matthiessen. He first writes that the pilot notices no one is backing up the racist comment and walks it back. The boy remains visibly confrontational. Matthiessen suggests that the younger generations of Alaska Natives are different from the older generations. The young have often served in the US military and are proud Americans. They have begun to assimilate and are less likely to stand for open racism. Matthiessen continues:

> Yet the older Eskimos, who were born in a much older time, know something that their sons may never know. They know what matters. They do not fight against new ways but neither have they lost touch with the old ones. At the loud words of the pilot, the old man had merely raised his brows and smiled. (73)

Perhaps most powerfully from a technical standpoint, Matthiessen does not hover over this conflict, but instead pivots immediately to the situation directly pertinent to the book: the helicopter coming for the musk ox. The two sentences following the passage above are, "Teal flew out immediately in the helicopter and returned delighted. The musk ox encountered had been wounded and put to rout, less by the racket than by the rush of air." This is important, because not only did Matthiessen pivot, he truly pivoted hard. There is absolutely no transition at all from social commentary on Alaska Native relations generationally to the corralling of the musk ox. Matthiessen simply moves on. It is conceivable that one could rush over this passage and never notice it, but it is likely that this passage would also resonate in the mind of the reader, as the scene continues to unfold, at least slightly. Had Matthiessen chosen to lean

8 It is to be kept in mind that desegregation had only recently begun (Brogan, 626).

more heavily on the passage, and pivot slowly, with more commentary, in a book which had nothing else to do with social commentary, the passage could have come across as overly didactic. Instead, as with the sharp, concise commentary on the city of Anchorage (11–2), we are left with a small-but-indelible imprint of an important aspect of the social fabric of the state, and we pivot back to the musk ox themselves.

As for the particulars of the two-paragraph passage, Matthiessen's technique in criticism is subtle yet clear. He does not outright criticize the pilot, but he conspicuously employs a child's morality and innocence to contrast against the pilot's inconsiderate remarks. It is likely that Matthiessen witnessed racism in many forms in 1960s Alaska, but this one scenario was particularly powerful, as it involved a child and that child's faith in his own father. Any culture would be struck by that relationship, as no parent wants to be derided in front of the glowing eyes of their child. Furthermore, the next idea after the child's interaction with the pilot is to point out that young Alaska Natives are "proud Americans" who go away to school and serve in the Army. This leans heavily on American patriotism. It also displays the ambition of young men going away and getting educated, helping their future socioeconomic position, which has been a theme in the American consciousness from the very beginning (Brogan, 48–50, 93, 241, 258, 447, 516). This is a demonstration that America was just as important to young Alaska Natives as it was to anyone in Alaska, which was the way to reach into the assimilation needed to be accepted, sympathized with by larger communities when the book was published in the mid-1960s. Unfortunately, racism is still rife, and the fact that Alaska Natives traditionally identify with subsistence rather than a career makes them seem unfocused and lazy to many Americans (Lopez, *Arctic Dreams*, 401).[9] Matthiessen continues to push all the right buttons for the time period, adding that the young "do not like the taste of seal meat," which is another aspect of Alaska Natives that make them seem less civilized to broader society. Also, the fact that they do not "give the white man the benefit of the doubt," which is described as "sensible and just," shows that they are less gullible.

Moving on to the next short paragraph, Matthiessen decides to discuss older Natives in comparison to the young. He essentially admits that the older generation is quicker to submit to this type of racism. "The old man had merely raised his brows and smiled." But, they have not lost the Native Eye. "They know what matters." They have not lost "the old ways." This statement is left

9 Raban notes that early missionaries could not understand why Natives would not go from one place to another directly, but meandered, ostensibly aimlessly; what was in fact close observation of their lands (*Passage to Juneau*, 99, 129).

vague, but one can assume that Matthiessen means that they have maintained their community traditions, which even those with racist tendencies from a small town of their own may nonetheless understand and value.

In conclusion, Matthiessen could justifiably be criticized for his lack of attention to the local Cup'ig of "Musk Ox Island," Nunivak. Yet, for a work of writing specifically based on the musk ox, Matthiessen has found a way to throw in piercing cultural observations of Anchorage and Bethel, and of Native/White relations, while still maintaining focus on his chosen topic. Not only that, but the writer has done so in a format of writing that was short enough to be serially published in *The New Yorker,* at roughly 75 pages. There is, then, no throw away sentence in *Oomingmak.* This little-discussed book of Matthiessen's carries no flab in the sense of extra stories, descriptions, etc. Everything that Matthiessen chose to say is concisely placed in the book, unlike some of his other, more ample works such as *Tigers in the Snow* and *End of the Earth.* Discussed previously, *Oomingmak* does not engage in technical descriptions. It moves along swiftly as a narrative, becoming descriptive in short spurts, which make this work very readable, and, ultimately, while his cultural commentary can be considered a bit inconsequential, the entire context of the work should be put into perspective, understanding that concision is one of the chief narrative strategies employed.

4 Conclusion

The three works focused on in this chapter present direct relations with Natives in unique and interesting ways. One should come away with a more nuanced view of the techniques employed by modern travel writers to both directly and indirectly discuss Alaska Natives and their viewpoints with regard to the biome in which they live, offering the kind of inclusivity and respect for alterity that Part 2 of this research will discuss as being so ineluctably important to environmentalism in the 21st century. Yet, pertinence was not at all a requisite in deciding the contents of this chapter. The pertinent works of Barry Lopez and Tom Lowenstein have both already been discussed and/or will be discussed in subsequent chapters. However, there is indeed a fortuitous link between this chapter on Alaska Native depictions and the following chapters on spirituality and animals, with these themes so often being inextricably enmeshed. At this point, many of the ideas surrounding Native oral tradition and Native relations have been quite extensively covered, but given the importance of spirituality and animals to animistic Alaska Native umwelten, individual chapters are fully merited.

FIGURE 6 White mountains national recreation area
PHOTO BY CHRIS FERGUSON

CHAPTER 5

Religion and Spirituality

What place does religion have in environmental writing? Many non-Western peoples of this world believe that religion and government, religion and food, religion and life, etc., should either not be separated, or even that there is purely no distinction between religion and other aspects of life. Jean Malaurie, for instance, comments on the Inuit:

> "Lesdites 'superstitions et sorcelleries' sont l'expression même d'une règle de vie, d'une religion très élaborée qui inspire tous les actes de la naissance à la mort. Religare, relier La religion esquimaude, par définition, relie l'homme aux forces naturelles ; un sens est accordé à l'existence : être en accord avec l'ensemble de ces forces'. (*Les derniers rois de Thulé*, 395)[1]

From this point of view, man's connection with all aspects of existence creates (an) all-encompassing link(s) with religion. Western civilization, by contrast, has tended to push religion away from secular life since the Enlightenment, creating clear-cut distinctions between religion, politics, and science. A question continuously arises, however: is Western civilization losing something by pushing religion away from other aspects of life? This is a particularly pertinent question when considering that the struggle for purpose in life itself involves such an inextricable connection with both food and spirituality, which are themselves closely tied (not to mention entire local biomes). Throughout Ann Fienup-Riordan's *Boundaries and Passages*, for instance, there is not even a discussion of Yup'ik "religion," as beliefs, taboo, etc., are considered to be a connected way of life, not separate from any other aspect, "… reli(é) aux forces naturelles …" ("… linked to natural forces …"). In her work, as will be discussed shortly, the Yup'ik have traditionally believed that seals and whales give themselves to the hunter, but only if the hunter is a generous person, and if the hunter is respectful to the seals and to other humans. Instead of focusing on religious piety, Fienup-Riordan is adamant that the traditional Yup'ik

1 "Said 'superstitions and sorceries' could even be thought of as the expression of a life rule, of a very elaborate religion that inspires all acts, from birth to death. Religare, to link … The Eskimo religion, by definition, links man to natural forces; a sense accorded to existence: to be in line with the totality of forces." Translation mine.

 | DOI:10.1163/9789004734807_007

focus is heavily linked with cleanliness and taboo (XIII; 7). This case illustrates the primary point being promoted here because Christian, Jewish, and Muslim religions tend to pray thanks to a monotheistic caretaker for their meals, and in doing so the holistic connection with one's surroundings and one's food itself can easily be taken for granted. Emerson, by contrast, began to view the world in terms of a complicated web of existence very much in line with Native thought, as he backed away from a patriarchal, care-taking vision of God (Buell, *Emerson*, 158–67). This is why both religion and spirituality are to be explored in this chapter; Gary Snyder, who could be considered a 20th century descendant of Emerson's Transcendentalist movement, alongside Barry Lopez, insinuates a difference between what he calls "shared spiritual perspectives" (presumably: spirituality) and "ceremonial institutions" (presumably: religion) before calling "the Judeo/Christian/Islamic faith" both "exclusive" and "divisive" (*Practice of the Wild*, 45–6).

Barry Lopez notes in both *About This Life*[2] and *Arctic Dreams*, that academics once could not consider anthropomorphism in animals without being scrutinized, so they consequently did not approach the taboo subject for decades. On the other hand, Native peoples who live side-by-side with animals have a much stronger understanding of multi-faceted animal intelligence, going against the prevailing view of academia since the French Enlightenment of animals as machine-like (Lopez, *About This Life*, 120–2; Lopez, *Arctic Dreams*, 250; Wulf, *The Invention of Nature*, 32). The scientific Western mind, therefore, historically and culturally linked with the traditions of Judeo-Christianity, immediately tends to believe that the seal has no desire to be "taken," but rather to survive in a mechanical manner without the advanced cognition required to make complex existential decisions.

It is important to note here that Non-Western peoples, the Chinese included, are most definitely not disciples of Western logic. Arguments and discussions between disparate cultures often fall short, not merely due to cultural differences, but because of a different way of looking at and experiencing the world altogether. The discussion of language in the previous chapter is one part of this barrier, but it remains just a small part. For example, Westerners have become so technocratic and so heavily invested in the scientific method, in logic, that they quite often miss what they do not see, what they cannot experimentally verify through the scientific method, that of which their descriptive minds cannot conceive. From an intuitive level, Kant, von Humboldt, and von Uexküll have made forays into a way of thinking about life beyond the basic,

2 Lopez, Barry. *About This Life: Journeys on the Threshold of Memory*. New York: Knopf Doubleday Publishing Group, 1998.

human, sensorial experience (Wulf, 34–6, 88–9; Lopez, *Arctic Dreams*, 268–9, 313), and modern physics has possibly more than any other field understood this, beginning with Einstein and Bohr (Lopez uses this example as the only form of spiritual elegance in Western civilization after the conception of the cathedral, in *Arctic Dreams* – 248–50). However, the Western mind does not broadly register the world beyond its five senses, nor is it broadly capable of reflecting up what lies beyond its language structures and ways of thinking. An interesting example of this evolution is the technocratic expansion into a virtual reality "metaverse," which comes across as more real to the Western mind than the reality of existential decision-making among "Animal People" communities where our reality is not stable but thickens and thins in relation to a spirit world. This is where ethnologists and geographers such as Jared Diamond[3] have been able to come in and study non-Western peoples and note that there actually is something that is important and pertinent to modern cultures, which is being missed when we denounce beliefs of Native peoples offhand.

Therefore, when an Alaska Native suggests that a seal wants to be taken by a respectful person, it is not logic that should be considered. We should recognize that there is something in what is being said that is a bit esoteric to us, misunderstood. For instance, Lowenstein's consistent focus on duality in Tikigaq culture, discussed in the previous chapter, is very difficult for the average Western mind to comprehend. Alaska Natives have thousands of years of experience in their territories, however, observing their prey and other animals and plants and are not merely the youthful bystanders or uncivilized barbarians/heathens ("in need of everything from a bath to baptism") some have attempted to portray them as (Rubiés, "Travel writing and ethnography," *Cambridge Companion to Travel Writing*, 245; Fienup-Riordan, 46–7), which has led to so much racism ever since first contact with Native Americans, by all European powers (Malaurie, *Les derniers rois de Thulé*, 602–3; Haycox, 53–8). We may enter the Arctic and view what we deem backward practices like infanticide, lack of medical expertise, and primitive hunting techniques condescendingly, but once Western civilization (sedentary village lifestyles, television, store-bought food, etc.) replaced primitive subsistence, Arctic peoples all over the world lost their own form of daily struggle, became depressed, overpopulated in unsustainable environments[4] and overweight. Alaska Native traditions, wrapped up

3 Diamond, Jared. *The World Until Yesterday: What Can We Learn from Traditional Societies?* New York: Viking, 2012.

4 Northern Alaska is essentially unsustainable under current conditions. It is very expensive to maintain Western civilization at the ends of the Earth, in places like Utqiaġvik, where there

in spirituality and what we might call "religion," have served them well since before the Common Era (Borneman, 18–28); maybe not each individual but society as a whole. When one says the seal likes to be taken by a respectful person, maybe part of the truth is that a strong hunter gets up early, takes care of his equipment, is meticulous in preparation and paying attention to elders regarding suggestions such as remaining upwind of prey and taking care of one's body. According to Fienup-Riordan, keeping a clean path to a house was also important for the traditional seal hunt (88–91), and the manner with which a woman served food was as important as her appearance, as she and her husband would have to depend on each other in life and death situations, the more meticulous spouse likely being a better partner in survival (172–3). The discussion of keeping a clean path, then, could have become tradition in the same way that the belief in seals being attracted to a respectful hunter potentially did. One may, for instance, reduce the human smell by maintaining cleanliness, keeping a smell free of the path. This understanding, combined with sophisticated animal agency and years of patient, almost zen-like watching by Native peoples, must create a vastly different experience of the world from our own. With such a consideration in mind, there is much importance in studying Native spirituality and knowledge, for clarification between cultures and for deepening understandings of what it means to be human. This chapter will explore the spiritual perspectives of travel writers as they come into contact with Native cultures and/or the lands where Native peoples have themselves moved for thousands of years.

1 Lord's Aesthetic Atheism

John Muir famously climbed to the tops of mountains and across glaciers, in snow storms, exalting in pouring rain to the point of frightening Alaska Natives who needed to be calmed by the writer/environmentalist's good friend, Samuel Hall Young (Brinkley, *The Quiet World*, 6–7). Muir saw God in these aspects of nature the way most people did not at the time (9–10). He experienced an earthquake in the Yosemite and ran outside giving thanks and praise to God for offering him such a beautiful experience in life. Many of his contemporaries fled the region soon thereafter, but Muir stayed (WIlliam O. Douglas,

are no roads in, basic infrastructure is lacking, and technology must keep people from freezing in houses not made from traditional means (McGinniss, *Going to Extremes*, 58–62). If oil money runs out in the state, these Western-style communities may be abandoned due to lack of funds.

Muir of the Mountains, 93–7).[5] Nancy Lord may not be such an outlier as John Muir, but she did come to the Brooks Range and find something beyond herself there, discovering something in Alaska that was simply greater than her and drew her back, while her partners on the trip merely enjoyed the unique experience as a vacation (*Rock, Water, Wild*, 31–2). We can only speculate as to whether that something was a religious or spiritual experience, or even an overwhelming confrontation with mortality, but she doubtlessly saw a beauty outside her experience of the world that spoke to her and in which she felt she belonged. Lord does not often actually approach Alaska from a religious or spiritual perspective, but she does at times fall into the kind of reveries only one who spends a lot of time outdoors alone can express. In fact, in *Fishcamp* she writes about going days, sometimes weeks, without speaking to anyone at all, and appreciating the experience (203–4). In the same book, she expresses her satisfaction for the time she has and making the most of it, living it how she wants to live it, on her own terms and with a sort of aesthetic priority that feels right to her. She has what she calls "beach time": when she's most at liberty, at her fishcamp. She aligns the passing of this "beach time" with kairós, or sacred time, instead of kronos, or chronological time (14–5). There is a soft atheism in this passage, and a soft spirituality, or at least humanism, which is not at all antagonistic[6] and even zen-like in constant reflection (76). Later in *Fishcamp*, the author heads back to the Greek language to reflect deeply upon her life and beliefs, mixing her Greco, or European, intellectual heritage with that of the Dena'ina who are native to Kachemak Bay:

> There's a lovely, well-grounded word – anchorite – that comes from the Greek word for "retire" or "retreat" and refers to a person who withdraws from the world for religious reasons. Although my belief system is informed by a variety of traditions … I would settle for calling myself, if anything, a sporadic and atheistic anchorite. (210)

In this same passage, Lord also discusses her mixture of beliefs, which include her experiences with Dena'ina culture. The Greco-Dena'ina mixture is evident not only in name but in context in this passage, mixing spiritual traditions of the West with what is seen in the lines subsequent to this excerpt, which explain a Dena'ina appreciation for the intrinsic value of all things. Furthermore, this

5 Douglas, William O. *Muir of the Mountains*. Boston: Houghton Mifflin Co., 1961.

6 Antagonistic here because declarations of direct "Atheism" not infrequently have a tendency in Anglophone literature to be declarations of antagonism toward "Theism" (Christopher Hitchens and Richard Dawkins, for instance).

idea of "retreat" shows a deep appreciation of observation and reflection. As will be discussed frequently in this chapter, Barry Lopez calls this observation of nature "The Native Eye" (*Arctic Dreams*, 96–7), insinuating that it can be achieved by anyone but is most closely connected with Native peoples through proximity to nature and observational genius, which will be shown to subsequently be reflected in spiritual belief systems. As Lord notes in this passage, the result of adopting, or at least understanding, Native spirituality allows one to reconnect with the wider environment she later insinuates has been lost by Western civilization, which affects how one sees and interacts with the wider world, "separat[ing] herself to some degree from the din and domination of her own kind." This idea could easily be passed over for the myriad layers the passage attempts to convey, but for the purposes of this research, and connecting this wider chapter of spirituality and religion to environmentalism, it is interesting to note how the author is suggesting that reconnecting oneself to what Lopez calls The Native Eye can impact environmental decisions on a pragmatic level, and, potentially, on a daily basis, with decisions people make.

Interestingly, instead of either praising God for the beauty she experiences around her in Alaska, as Muir so often did, or on the other hand going out of her way to denounce religious piety found in nature, Lord writes about taking advantage of her "sacred time," appreciating observation and reflection, which is rife in *Fishcamp*, and powerful in her written word. This aspect of Lord is rarely approached in *Beluga Days*, as that book is an attempt to reach across a spectrum of people with differing spiritual beliefs who may be interested in saving the Cook Inlet beluga but not interested in a writer's personal spirituality. Therefore, *Beluga Days* smartly remains journalistic, and much less emotional, in order to achieve a specific goal. But, understanding the spiritual beliefs expressed in *Fishcamp* goes a long way in seizing the motives behind *Beluga Days*.

Concomitantly, it could be said that there is openness in the writer's work. As previously noted, she has the adaptability to Western patriarchal masculinity to get along with aging, conservative-minded fishermen living alone in cabins of rural Alaska (*Beluga Days*) while simultaneously maintaining relations with the Native communities, remaining open to the disparate traditions of the two groups. She talks about the history of the Dena'ina found in *Shem Pete's Alaska* and in the work of Peter Kalifornsky (*Beluga Days*, 177–9, 223). She is drawn to explorations of Alaska Native histories, their spirituality, their geographic roots of language. This is not about religion, exactly, but rather being a balanced humanist. Though, she does attempt to employ Western logic in understanding Dena'ina concepts of spirituality in order to understand the local culture's broader world view. In *Fishcamp*, Lord says that she goes as far as her Western mind will take her, finally reaching a point at which she admits that the whole concept breaks down and will go no further with said logic (70). Western logic, become almost a deity in European culture, is therefore shown as incomplete. Many writers would likely come to the conclusion that any school of thought not solvable by Western logic that is not a paradox is therefore below Western civilization, but Lord comes to the opposite conclusion, the same conclusion that Lopez finds: spirituality in Western civilization has lost its way by eschewing the natural world in lieu of larger civilization and deeper technocracy and logic. Lord employs a simile of her "ancestors let[ting] their spiritual lives wither like vestigial tails and develop[ing] instead mental lives focused on rational thought, involved not with felt life but with ideas" (70). This is highly pertinent and visually effective in expounding upon what is lost between understandings of her two cultures. Lord never fully expounds upon this notion, but it is a pertinent and perhaps important one: feeling life is perhaps as important a genius as intellectually understanding life. This is why the memoir *Fishcamp* is a strong means of discussing such ideas, as Lord is given free rein to meander. For instance: "air, tree, hole in the ground, myself standing still" (69). In this free association that begins the discussion a page earlier, the author is able to bring together the surreal, or descriptions lacking in logic, poetically and with the felt (the smells, the temperature, the air in her chest, etc.), before moving on to a specific, logical discussion.

Yet, in exploring these topics, Lord does not fully accept Native spirituality as an explicit genius. She spends much time attempting to rationalize Native beliefs before admitting defeat and her own culture's shortcomings, but she finishes essentially feeling uncomfortable with Dena'ina spiritual life as well. Essentially, she knows the Dena'ina have greater spiritual advances over Western civilization in many ways, but that still does not mean they have more advanced thought. In other words, there is doubt sewn, the idea that maybe what seems childish and lacking in logic actually is childish and lacking in logic and not just a genius European culture cannot understand. The grey area in the short paragraph at the end of her passage is important. It is not accusatory, but it leaves open the idea that both cultures have something to offer which the other does not possess.

As noted in the previous chapter, language as a whole is intimately connected with geographies and peoples, and Lord is no stranger to this fact. Barry Lopez mentions to Mark Tredinnick in *This Land's Wild Music* that certain geographies are more prone to synchronicities and spiritual experiences than others (42). This aspect tends to find itself in languages, even if idiomatically, and Lord finds importance and meaning in language, particularly place names. Studying place names in *Shem Pete's Alaska* allows her to have a more complex understanding of her surroundings through the names written in history, and she even reflects upon Shem Pete's discussion of Ridge Where We Cry, which seems to agree with Lopez' assertion of synchronicity and spirituality connected to place (*Fishcamp*, 63, 67, 70–1; *Beluga Days*, 177–8). If Lord is not religious, this is an example of how she maintains her own spiritual connection with the land. She does not tend toward the overzealous or emotional, as Lopez does when he is rendered speechless during a perceived synchronicity with Tredinnick (42). She does, however, display her spirituality through constant contact with the land. If she is not a cultivator, a gardener, as she states in *Fishcamp*, she is a hunter-gatherer (164), maintaining constant physical contact with the geographies around her and keeping herself attuned to the space she inhabits. When she does this, with the knowledge that the Dena'ina for hundreds or thousands of years before her named a place for a certain salmon fishing spot or a ritualistic site, she is in fact maintaining her own form of spiritual connection with the local biome. *Fishcamp*, then, succeeds in conveying how spending one's time alone, in physical contact with the Earth, can aid in maintaining a broad spiritual perspective. This approach, subtly employed in *Beluga Days*, makes that work more effective than it would be if it were to solely lean upon investigative journalism.

It is perhaps equally notable that Lord seems to have mastered the technique of knowing when to insert emotion and spirituality, and when to back away and present her findings. Often in her work, cold hard facts are juxtaposed against a tendency toward emotion in order to demonstrate the dynamic between the two. Her spiritual and emotional side breeds curiosity and humanism, while her journalistic side steps back from emotion, which can blind one from important truths when moving forward in an investigation. In this way, if Lord were too militant in either her atheism or spirituality, the effect of her work would be less powerful. The following section will likewise examine Peter Matthiessen's penchant for Zen Buddhism and his decisions on when to discuss it in his work and when to leave it out completely, as well as an interesting lack of discussion regarding Alaska Native spiritual beliefs.

2 Peter Matthiessen, the Missing Zen Master

Matthiessen's National Book Award-winning work of travel writing, *The Snow Leopard*, is celebrated not only for its depiction of Buddhist and B'on culture in Inner Dolpo, and the elusive snow leopard itself, but of Matthiessen's personal relation with Zen Buddhism and its intertwining into the larger narrative. The work was published in 1978, roughly a decade after *Oomingmak*.

This short-but-imperative section will attempt to dig into the deep spirituality revealed in *The Snow Leopard*, comparing it with the surprising lack of spirituality in *Oomingmak*, which took place on Nunivak Island. The local Cup'ig were still heavily traditional at the time and likely maintained remnants of shamanism potentially interesting to a curious and open-minded travel writer.

To begin, in *The Snow Leopard*, Matthiessen indeed mentions the Inuit on occasion. At one point, he refers to their appearance as having the same mongoloid features as the Tibetans, adding that their cultural symbols and "animistic kinship with the world around" is similar throughout Asia and the Americas (50–1). At other times, he goes beyond appearance and basic cultural outlook, into spirituality. For example, the writer compares multiple cultures' forms of trance for spiritual purposes, including the Bushmen and the Pueblo Indians, but also the Eskimos, who "use a sharp rock to draw an ever-deepening circle into the flat surface of a stone," which "achieves the same obliteration of the ego" (86). A subsequent exploration into the writer's personal experiences in Zen meditation is then explored. Perhaps more interestingly, Matthiessen later digs into the teachings of the Buddha, comparing man as a form of the primordial cosmos, or the cosmos "contemplating itself," to the "inner light" experienced by Eskimo shaman (62). But, Matthiessen does not go into much depth with Alaska Natives in *Oomingmak*, which is much more educational, or informative, in tone.

With one's obvious spiritual proclivities all throughout *The Snow Leopard*, a pertinent question would be: why would Matthiessen choose not to approach the exotic spirituality of the Cup'ig of Nunivak Island? The answer is unclear. What the reader is made aware of in *The Snow Leopard* is that Peter Matthiessen is on a spiritual journey in search of relief from the face of mortality in the form of his recently deceased wife with whom Matthiessen was having marital problems (72–4, 100–4). The couple had gone on a spiritual journey together, described briefly in *The Snow Leopard*, in the form of Zen Buddhist retreats of strict mediation (100–4), and Matthiessen further notes in *The Snow Leopard* that he was introduced to Gurdjieff's principle of maintaining awareness in the present moment while living in Paris, as early as 1948 (40), well before *Oomingmak* was written. Therefore, combined with the fact that this was the center of his conversation with Mark Tredinnick in *This Land's Wild Music* (140–5), Matthiessen was obviously more interested in the spirituality of Zen Buddhism brought over to the Pacific Northwest from Japan and popularized in the U.S. in the 1960s and 1970s (44; Brinkley, *The Quiet World*, 392–5, 41–21) than in the spirituality he must have experienced during his visit to Nunivak Island in Alaska in the mid-1960s. One might think that someone so moved by the exoticism of Buddhism would be struck in some way by the ancient shaman rituals of the Yup'ik the way Tom Lowenstein was fascinated with the shamanistic Tikigaq tales told to him. One could also understand how the potential Christianization, or Westernization, of Inuit culture in Alaska could potentially drive such a spiritually curious person[7] away, as the shaman rituals had often been corrupted or forgotten by the mid-20th century. Tom Lowenstein, in fact, had to search out an old man known as the village scholar, Asatchaq (*Ancient Land*, 92), who had been somewhat shunned by many in the Tikigaq community, even at a younger age, for not converting to Christianity (92). Therefore, when Matthiessen traveled through Mekoryuk on Nunivak Island, it is not inconceivable that Christianity was the dominant, if not only, form of religion followed (even if adopted in a bastardized form, as will be discussed later in this chapter). But, even with traditions likely merged or ostensibly gone there likely would have been remnants interesting enough to explore. Those remnants of Inuit culture were just as ancient and full of prospects as the Buddhist, B'on, and Tantric traditions of Nepal, especially for someone so interested in learning from travel and spirituality and world religion in general. Yet, in *Oomingmak* there is not even half a page. Whether the answer lies in a cultural zeitgeist or a personal, spiritual preference, the technique of keeping spirituality and religion out of *Oomingmak* entirely was a conscious choice.

Yet, Tom Lowenstein, who has also long maintained an interest in Buddhism, chose to extensively involve Native spirituality in *Ancient Land, Sacred Whale*. For Lowenstein, animistic spirituality was inextricable with the relationships between the people and the animals they hunted.

3 Tom Lowenstein, or, How Does Buddhism Relate to the Inuit?

Tom Lowenstein is a British writer who has focused extensively on two primary topics: Buddhism and the Tikigagmiut of Point Hope, Alaska. One question briefly approached above is whether or not there is any relation between the two. And also, how writing about one can affect the writing about the other.

First off, Buddhism is sometimes described as not so much a religion[8] as a philosophy: a way of viewing the world and one's grip on mortality rather than

7 The term "spiritually curious person" is employed here because Matthiessen expressly notes in *The Snow Leopard*, when discussing his paths into religion and spirituality with his wife, "Not that D and I considered ourselves 'seekers': we were embarrassed by such terms, and shied from people who employed them. We read and talked and read again, but what we needed was a teacher and a discipline" (42).

8 Religion in the sense of Merriam-Webster's definition 1b: "the service and worship of God or the supernatural"
"https://www.merriam-webster.com/dictionary/religion," viewed 12/18.

devotion to a God or gods. In the "Introduction" to Tom Lowenstein's *Buddhist Inspirations*,[9] in a broad explanation of Buddhism, there is no discussion of a God or gods beyond the mortal realm (8–13), and enlightenment is discussed as achievable by all men (13). As opposed to the Christian Heaven, or an afterlife in general, creditable to a Creator(s), whether fallible or infallible (theism, in other words), Buddhism distinguishes itself from many other major religions in its historical Hindu association with reincarnation (33, 40–1), as well as the teachings of "crossing the ocean of samsara" to essentially become one with existence beyond an individuated ego (10, 40–1). Buddhism did not initially survive in the Hindu stronghold of India, however, moving eastward toward Tibet, China, and Japan (12–3) and mixing with religious traditions along the way.

How do the various forms of Buddhism, then, compare with the spiritual beliefs of the Inuit? Just like the teachings of Buddhism in different regions, animistic beliefs of the Inuit vary, often due to mixing with religions. Froelich Rainey says in *The Whale Hunters of Tigara* (1947), "The present tendency among the Tikerarmiut to associate Alignuk and God, or, at least, to compare the two as Supreme Beings, may have elevated the position of Alignuk beyond his normal status in aboriginal theory" (271). This example demonstrates the fluid shift in religious understanding by the Tikigagmiut over time with the arrival of new influences. Rainey covers another, similar, aspect of this religious fusion a few pages later:

> The absolute faith of the Tikerarmiut in angatkoks [shaman] who had proved their power paralleled that so often described for other groups of Arctic Eskimo. In spite of the teachings of Christianity, there was and is now [1947] no doubt among them that men can fly to the moon, that they can restore themselves and others to life within a four-or-five-day period, heal the sick, eliminate a mortal wound in an instant, pluck out their own eyes and replace them, kill men and animals at any distance, control the weather, halt the movements of ice, bring or send away the animals (famine was always due to an angry angatkok), and perform endless miracles at any time. (276–7)

In explaining how the Tikigagmiut beliefs have merged into hybridity with Christianity, just like Buddhism has with so many other world religions, Rainey then likens spiritual belief systems to technology. The Tikigagmiut accepted "darting irons" and "exploding bombs" for hunting whale but continued to

9 Lowenstein, Tom. *Buddhist Inspirations*. New Jersey: Chartwell Books, 2013.

employ "charm songs and return the skull to the sea" (280). Regarding foodstuffs, Rainey says that the Tikigagmiut had no use for vegetables, but eagerly accepted "coffee, tea, sugar, and tobacco" (280). The Tikigagmiut have also taken strongly to modern machinery, particularly airplanes (280–1). Like technology, then, the plasticity of spiritual beliefs should be accepted and self-determination respected.

As Sheldon Jackson famously lobbied multiple branches of the Christian Church to send missionaries to disparate parts of Alaska at the turn of the 20th century (Haycox, 184–7), the Episcopal Church gained a foothold in Tikigaq. According to Rainey, many villagers went to Episcopal services, as they were "the chief social event each week" (283), and prayer meetings outside the Church "are probably a half-conscious protest against the more formal group services of the Episcopal church. The interest in the individual experiences probably harks back to the aboriginal pattern of religious theory" (283). Rainey's ethnological accounts, dating, again, from the 1940s, are pertinent to this research in understanding the fluid nature of spiritual belief as it moves from "aboriginal patter of religious theory" to Christianity, and what is maintained in the process of integration between the two. Such is also the case with Buddhism and likely most other of the world's religions.

The conversion away from animistic beliefs systems toward Christianity via proselytism is often considered problematic in Postcolonial studies. However, with self-determination such an important concept, there is much room for nuance in discussion of hybridity in religion and spirituality. The travel writer, then, has a heavy multiradial responsibility when it comes to the belief systems of others.

Lowenstein chose not to focus so much on the fusion with which Rainey poignantly concluded *The Whale Hunters of Tigara*. As the purpose of Lowenstein's work is more in line with expressing traditional Tikigaq stories in modern English, the integration of the two religious views could potentially have been interesting, yet distracting as well, to the writer's purpose, blurring lines in an already inextricable situation that required much subtlety of planning to pull off in a comprehensible manner, described in the "Introduction" (XIII–XV). Perhaps just as pertinent to this research, what Lowenstein does choose to focus on with regard to spiritual belief is an attempted explanation of the soul from the Tikigaq perspective, as well as its individualist construction of the spiritual world. Each person, and group, traditionally had fluctuating beliefs based on the mythology of the culture, which was never systematized into an organized religion. There are indeed basic ideas which Lowenstein approaches as standard, however. Here is his description of the idea of a soul in Tikigaq culture:

> Tikigaq people believed there were three human souls. One soul, the 'life force', was mortal and vanished at death. The 'soul of personality', on the other hand, moved 'somewhere to the east, where the light comes from', or hung around the village, perhaps to become a shaman's helping spirit. The third soul lay in name. (*Ancient Land, Sacred Whale*, 22)

Alongside this dynamic spiritual understanding is also belief in shapeshifting, in the transformation of human into animal and vice versa, particularly in the case of the shaman, who may transform into a raven or a seal.[10] In Buddhism, likewise, one may become a flower, insect, or animal in the next life, and therefore, all things should be respected (Lowenstein, *Buddhist Inspirations*, 142). Most traditional Buddhist sects are either vegetarian or do not kill living things themselves, even if they do eat meat gotten from others (Matthiessen, *The Snow Leopard*, 139). The Inuit have a parallel here, but it is slightly different. They both hunt and partake of meat, but they also typically find a use for every body part, including the fat and skin. They also respect animals tremendously, calling the process of hunting "taking," as has been noted. Meanwhile, Inuit hunting of whales has always been sustainable and particularly respectful and thankful. A whale or a seal must be thanked for allowing the hunter to "take" it. As a result, those who maintain reverence toward their traditional Inuit traditions tend to be much better stewards of their environment as a whole. If they pollute the streams, if they overfish or -hunt, if they strip the earth bear of its nutrients, they know the animals will "go away."

Several authors of this research wrestle with this idea in different ways, as it is so important to Alaska's view of itself. Whereas Lopez attempts to find the positive in the Native Eye, and Lowenstein attempts to find the genuine in traditional culture, Jonathan Raban essentially argues that a traditional portrayal of Native peoples is a saccharine view which limits the dynamism of peoples to "New Age" storybook versions (*Passage*, 76–7). Raban makes an important point, but a robust tradition of Alaska Natives nonetheless exists. In evoking this argument, however, it is also important to understand that morality and ethics, often considered important aspects in religion, are largely subjective. If Raban does not approach the subjectivity of traditional beliefs, both Lopez and Lowenstein indeed do. Alaska Native and Native American mythology is full of erotic storytelling that was often bowdlerized by late 19th- and early 20th-century ethnology for its crudity (Lopez, *Giving Birth*, XVII), and along

10 One example would be Sedna, a girl who drowned and became the goddess or spirit of the sea, giving or withholding food sources in many northern cultures. She often appears animal-like in Alaska Native art (Brinkley, *The Quiet World*, 421).

with the sexual indiscretions in Lopez' *Giving Birth to Thunder* (155–6), Tom Lowenstein's *Ancient Land, Sacred Whale* must deal directly with rape. At times that rape is incestuous, and at times it is even performed in a gang ordered by the leaders of a qalgi (Tikigaq group or clan) (Lowenstein, 10–5, 113). Both Lopez and Lowenstein decided to confront rape in traditional storytelling in a forthright manner, in the way that it would be approached in a classic Alaska Native or Native American narrative. Lopez, who was sexually abused himself when he was young, has discussed how the experience has affected his own writing (Tredinnick, 49–51). He has also discussed how modern cinema increasingly does not allow the watcher to avoid high levels of violence, unannounced, short of fleeing the theater (Newell, *No Bottom*, 31). Lopez has, then, addressed the issue of violence and sexual exploitation on multiple occasions. Lowenstein approaches the issue, but he never refers it back upon his own life and perspective, as Lopez does, preferring to leave it within the primary text itself.[11]

In Lowenstein, the text approaches a bygone era anyway, with subjective positives and negatives of the culture portrayed as raw civilization without an ethical filter imposed upon it by the West. Often, there is retribution for rape and murder, and Lowenstein repeatedly presents females as peers with men.[12] In the story of the sun and the moon (10–5), for example, the sister oddly gains an amount of self-respect by publicly admonishing her brother for incestuous rape, employing self-immolation. In another story, a woman who does not want to marry is gang-raped, and a man who makes her bleed is eventually killed by a whale at the behest of the girl's father (113). These approaches can be viewed as being oddly shaped by Lowenstein's spiritual affiliations. In Buddhism, unlike in Christianity, there is not so much importance placed on sin as much as there is upon reflection and enlightenment.[13] Lowenstein's approach, then, may be viewed as an objective and reflective Buddhist perspective on Tikigaq

11 Rainey also approaches the issue in his work. He mentions that most women did not in fact appreciate wife swapping. The loose morals discussed by explorers were, then, likely misinterpreted (242).

12 Ann Fienup-Riordan comes to a similar conclusion. Though traditional Yup'ik wives had ostensibly subservient lives with too much taboo involved, the author devotes an entire section to "The Power of Women" (*Boundaries*, 163–9).

13 This idea is also the focus of multiple pages in Pico Iyer's "Introduction" to Matthiessen's *The Snow Leopard* in the form of Matthiessen's reverence for the porter, Tukten, who is oddly monkish but also untrustworthy and potentially a thief. "Tukten appeals to Matthiessen, even perhaps teaches him (more than does the obviously wise but matter-of-fact lama of Shey) by taking everything in his stride, as the way things are; he will look unmoved, Matthiessen says, on "rape or resurrection" (xxii–xxv).

culture, releasing an egoistic need for agenda as much as possible for a writer who necessarily has an ego due to individuated consciousness.

To further illustrate, Tom Lowenstein spent time at the University of Washington learning to speak Inuktitut before moving to Point Hope, where he lived with the people and studied them in a non-obtrusive way. He brought with him the objective respectfulness of the Buddhist religion, along with a general open curiosity. The success he had in translating the culture's stories into English language poetry was due to this mindset. It invited a sort of unbiased compassion not common in travel writers moving through the Arctic. Even Bishop Veniamov, revered among the Russian Orthodox Church, never relinquished the superiority of his religion over that of Alaska Natives (Garrett, *Saint Innocent*, 105–11).[14] Lowenstein does not judge the traditions of this people in his work. Instead, as mentioned, there is pure story-telling and emotion. He opens all aspects of Tikigaq life, showing both the beautiful and the ugly, stressing that the life of this village is beautiful when taken as a whole. Not everything is perfect. People die. People struggle. But, at the end of the day, these traditions, the intimacy with the land, the communal whale hunt as intense human and spiritual experience, offer a way of life with purpose, one worth living, which is the overarching message Lowenstein is able to convey in his work *Ancient Land, Sacred Whale*.

In the following section, Lopez and Raban's spiritual views, informed by their respective youths in differing situations, will be compared and contrasted. As religion and spirituality are such slippery concepts, different aspects will be approached spottily, not unlike Lowenstein's scattershot approach, in the hopes of pulling together a broader vision of the diversity of thought regarding the subject by the various authors presented in this oeuvre, as well as what unites them.

4 Wordsworth and Emerson, and Their Descendants

In this section, the primary line of thought to be explored is Jonathan Raban as descendent of Wordsworth and Barry Lopez as descendent of Emerson, a canny comparison which Raban makes in criticism of Lopez and other Northwest US writers in order to defend a broader view of wilderness and its relationship to humanity. This comparison sets up a spiritual chasm between the two writers in that both Raban and Lopez view wilderness as symbolic of

14 Garrett, Paul D. *Saint Innocent: Apostle to America*. Crestwood New York: Saint Vladimir's Seminary Press, 1979.

their disparate spiritual beliefs: Raban, a whirling abyss of existence; Lopez, a beam of light or a corolla.

Raban makes the important point that he "[came] from a country logged out completely by the mid-sixteenth century … By the new standards [he] was trying to master, England had no nature at all … [his] childhood – bird's-nesting, fishing, butterfly hunting, pressing wildflowers between the pages of old books – had been … a case of error." This is to say that wilderness is viewed differently in Seattle from how it is viewed in Europe. Raban, however, does not dig much further into the environmentalist question, preferring to stick with the literary comparison between his home culture and his adopted culture and his guttural response to America. He states openly that he struggles with what he calls the "Northwest nature writers," who include Gary Snyder, Barry Lopez, and Richard Nelson. He talks about how their paternal line stems from Emerson and his own from Wordsworth. He calls Lopez' view of the world too "tidy." For Raban, the world is chaos, and there is no logic to it. Furthermore, he struggles with their overall style:

> … I found myself an agnostic in their church … I wished there were more jokes. But humour was not their line. I liked the microscopic particularity of much of their writing, their intent and well-informed gaze, as they tried to penetrate the veil of the natural world. I thought individual passages were beautiful. But I couldn't join in their hymns … (*Passage*, 190–1)

From the standpoint of religion and spirituality, Raban is highlighting his discomfort with the Northwest US mental construction of the word "wilderness" so often connected to spirituality and the numinous. There is even a brief taste of sarcasm, as Raban jokes that he was under the misapprehension as a child that he was playing in nature. However, what does not quite come across, yet is nonetheless a major theme of Raban's work, is that he finds this school of writing to be part of an urban society that likes to play at utopia but ultimately creates dichotomy between itself and its rural counterparts. Raban has spent much time at the docks, and traveling, driving, outside the city, back east to places like Montana where he has often identified with the cultural zeitgeist under attack by the city of Seattle (*Bad Land, Driving Home, Passage to Juneau*). The value system Raban finds in these working class milieus is more religious culture than spiritual or mysticism, something that suits his Anglican roots (and at least borderline Atheism or Agnosticism, as noted in the passage) just fine.

Interestingly, Raban employs a comparison between Wordsworth and Emerson in the above passage, noting not the difference between the two writers, exactly, but that of their descendants. If the Northwest US writers are descendants of Emerson "at his most vatic," then what does that mean for Raban's own tradition of Wordsworth, and of England? The British have for centuries meditated on Wordsworth's descriptions of daffodils in the Lake District of England in "I Wandered Lonely as a Cloud." But, as Raban insinuates, to Northwest Americans these daffodils are part of the problem, the encroachment of man into the wilderness, a viewpoint which Raban wants to put into broader perspective. Let me be clear here: Raban's weakness in environmentalism is his myopic viewpoint with regard to looming climate change and species devastation under the Anthropocene; yet his strength is a broad, global view of the world, spanning centuries of exploration and across the gamut of ethnological knowledge, which allows the writer the vision to see what is happening to multiple communities, in multiple layers, and how the effects of those relationships affect the environmental argument through human emotions, tribalism, egoism, etc. In this respect, Raban is every bit the peer of Lopez.

As for Raban's Wordsworth-inspired view of nature, until urban communities destroy their own wilderness they typically have no major appreciation for it. Protestant American preachers like Cotton Mather even went so far as to insinuate that wilderness was Hell, full of demons, and paradise needed to be carved out of it. (Nash, *Wilderness*, 28–9, 36–7). Raban understands and respects the ferocity of nature in this regard, without going so far as to call it a literal religious Hell. Keeping in mind that much of *Passage to Juneau* maintains the perspective of the roiling waters outside Seattle, where the Puget Sound drops off a subaqueous cliff almost immediately after leaving shore, and "five-hundred-pound halibut," "… slimy creatures with tentacles and fleshy suckers, of the kind that surface in exceptionally bad dreams," lurk about beneath the boards of his sailboat (Passage, 43–6; *Driving Home*, "Julia's City," 179), Raban states in the same passage as his discussion of the Northwest US school, "Turner's whirling abyss[15] seemed to me a true picture of reality as I generally experienced it; Lopez's version struck me as improbably tidy and benign." All of this maintains a connection to spirituality and religion because

15 Referring to the painting "The Pass of St. Gothard," by J.M.W. Turner (Birmingham Museums and Art Gallery), painted in 1803–4. "The painting … is a work of scrupulous realism, but it also represents that mystical Abyss, the great deep, the primal chaos into which reason itself might topple. It's a painting designed to make you lose your balance and fall into it, through the circling rapids" (Raban, *Passage*, 186–8).

the Northwest US school is almost by definition inherently spiritual, mystical, with wilderness interconnected in the ideology, as they are the descendants of the spiritual Transcendentalists, and of John Muir, of whom the same could be said.

Except, Raban views the Northwest US as sometimes hypocritical for its gentrification into a soulless, capitalist space formed by companies like Starbucks, Amazon, and Microsoft, and where diversity is pushed away to the fringes. It can be seen in this view as having no true soul or character, as benign and somehow void of the reality of the dark sides of nature and civilization, which discussions of true wilderness require. This is a legit criticism with which to be reckoned, but we must keep in mind that Raban finishes the passage with, "after a few pages I grew restless and began to ache for more profane company" (191). This negates any serious attempts at putting the writer into the camp of either environmentalism or the British Romantics of Wordsworth. Raban, rather, is the writer of the "Soft City" of developing one's own identity according to whim (*Soft City*, 1–2),[16] who at times aligns himself with working class America, and at times with the punk fringes of London, but no matter with whom he chooses to align at a given moment he remains highly informed.

As for Lopez' "benign" view, that author interestingly cites the notoriously dark Cormac McCarthy as being a writer whom he admires (Newell, 33, 52), and in any in-depth reading of Lopez the reader will note tinges of darkness and the power of nature which linger just beneath the "benign" light he is limning.[17] For instance, Lopez' interest in the centuries of deathly stories of Arctic exploration take up the bulk of the second half of *Arctic Dreams* (particularly the chapters "The Intent of Monks" and "A Northern Passage"), along with a fascination with the bizarre nature of the lives of oil workers at Prudhoe Bay (393–401). The same goes for an interest in the darkness of Dorset art (182–3). To conclude his chapter "The Intent of Monks," Lopez indeed employs a chiaroscuro of light and darkness, as he explores the suffering, sickness, and loneliness of those early Westerner explorers. Many of them died painful, extended deaths of scurvy and starvation (354).

In a nutshell, Lopez is one of the few modern writers who would find the light of spirituality in the dark suffering of the men who explored the Arctic. Raban, by contrast, writes as a humanist who sees the random nature of life, of terror in Arctic waters. Lopez indeed evokes some of the most terrifying scenes humanly imaginable, in starvation, Arctic darkness, loneliness, and extreme

16 Raban, Jonathan. *Soft City*. London: The Harvill Press, 1974.

17 Mark Tredinnick agrees and devoted multiple pages to the idea in *This Land's Wild Music* (67–73).

discomfort, but he evokes these scenes and thinks not of the horror, but of the spiritual reaching for something greater, for spiritual transcendence even, in the face of human constraint.

Likewise, Lopez finds deep spirituality both in landscapes and the architecture of cathedrals, where he is affected by a form of transcendence. For Raban, by contrast, if the cathedrals are gentrified, left cold and out of touch with a mixed and "whirling abyss" of a city, they have neither importance, nor, in a sense, reality, as they can become caricatured versions, or false versions, of their true selves, and therefore, false spirituality. Though Raban is not even remotely spiritual in his work (in either the religious or mystical sense), he is most definitely humanist, and the spirit of place is important to him in its humanism. At the end of the essay "Boomtown USA" from Driving Home, the writer discusses searching out a cigar from one of the last remaining cigar shops in the city (411), which he employs as a symbol for both Seattle's moral austerity/health conscious nature and its becoming a false city lacking in identity. Here is the final paragraph of the essay:

> The cigar butt is flipped into the dock, making circles in the water like a rising fish. For me, it's back to Seattle's present, to the screen blossoming to Windows XP, the black Thermos mug emblazoned with the green Starbucks logo, and the peculiar silence of a twenty-first-century boom city minting money … (412)

Raban ends this passage by suggesting that he could be anywhere, which highlights that lack of identity and how it leaves people behind in its push forward.

But, what does that mean for Raban's view of Alaska and the waters off the coast of British Columbia? As McPhee has noted, and as is a common saying regionally, Anchorage "might be a sorry town, but it has the greatest out-of-town any town has ever had" (*Coming into the Country*, 116). Raban would likely be unimpressed, and, in fact, does not discuss Anchorage at all, stopping his journey completely at Juneau. But, he is also talking about cities here, and not about rural areas, which he views differently. What is clear in *Passage* is Raban's disdain for the "false." He views the tourism of yachts in Desolation Bay as distasteful and searches out First Nations representatives and dive bars when he can, instead, as they somehow ring true to him. As the traditional, re-enacted dance rings false to him, he seems on a constant search for some kind of anchor in his drifting life as things around him become caricatures of themselves. This leads to his being wary of inauthentic or misguided environmentalism, appropriation of inauthentic Native morality, false architecture, false views of ethnology, other yachtsmen playing at explorations of the sea, etc. In these worries, Raban often sees symbolism: in his car, in his house, in his habits, and in his company. These symbols are employed to make suggestions in his work that often subvert ideologies with which he disagrees.

Raban is a close observer of people, cultures and even water, focusing time and again on the dichotomy between the urban and rural, whereas Lopez has been drawn in by the contemplation of nature and the fascination with the wild, in the footsteps of an Emerson "at his most vatic," attempting to purvey a broader view of human endeavor and sense of achievement. Yet, Raban, through his European heritage, understands political dichotomy and the pragmatism in Wordsworth's relatively small and domesticated fields of flowers. In doing so, he often comes across as somewhat prescient. Indeed, *Passage*, *Bad Land*, and much of *Driving Home* can be considered a bellwether for many social problems that have since made their way into global politics, including politics in Alaska, where various marginalized demographics have begun to display their frustration. In the end, both Raban and Lopez have their fingers on important perspectives with regard to Alaska that are rarely discussed, and both express these perspectives indirectly.

5 "The Native Eye" in Lopez and Lowenstein

This section is perhaps the linchpin to this entire chapter. Having previously referred to the Native Eye on multiple occasions, the concept merits a more detailed discussion. Therefore, it seems pertinent to begin with an analysis of Barry Lopez' text on the subject from *Arctic Dreams*. The discussion will then move into the various ways in which the Native Eye is portrayed by other authors within the œuvre.

In the "Tôrnârssuk" (polar bear) chapter of *Arctic Dreams*, Lopez and a friend spotted a bear from a helicopter and discussed the desire to follow it out of curiosity. The author says this of the experience:

> The Arctic reminds one of the desert ... because it puts a like strain on human life. It favors tough and practical people, people aware of the vaguest flutter of life in an environment that seems featureless and interminable to the untrained eye. People with a predator's alertness for minutiae, for revealing detail. (96)

The first paragraph breaks down the exigencies of living in the desert, as "predator," which is an impactful word choice for Lopez who tends to search for reflection and beauty in the world. This leads to the hint of a suggestion of nobility in the predator, the subsistence hunter. In this view, the hunter is not a prowling killer but a respectful watcher, part of the chain of nature. The words "alertness for minutiae" are employed in conjunction, and they compare well with Ann Fienup-Riordan's analyses of Yup'ik respect toward prey that have already been discussed. Immediately following this passage, Lopez says that "the loss of a 'native eye'" has been noted by Vladimir Arsenyev who lived in the Russian Taiga and by Laurens van der Post who lived in the Kalahari Desert, among so many others. What he is suggesting is that people who spend extended periods of time in the "wilderness," in places like the Arctic, are able to regain some semblance of the "native eye," a way of seeing that has been lost to the civilized world. Lopez goes on:

> It not only takes a long time of watching the animal before you can say what it is doing; it takes a long time to learn how to watch. This point is raised, deferentially but repeatedly, in encounters with Eskimos. They are uneasy ... about the irrevocability of decisions made by people who are not sensually perceptive ... in these northern landscapes ... (96–7)

This passage continues with a brief discussion about the idea that Western Civilization bases its understanding of such places on field biologists who have not spent their lives observing to the same extent as those hunters with a Native Eye. Consequently, Lopez states that those who have made such lifelong observations feel uncomfortable with decisions made based on short-term observations of field biologists who are suggested to have less complex understandings of the holistic nature of respective biomes. The suggestion here is that true observing requires more than sets of data. Along similar lines, Bill McKibbon notes this about Thoreau: "[Thoreau] was struck by the idea that the lives Americans had begun to lead were desperately divorced from reality, a reality that could be found in nature. 'In wildness is the preservation of the world,' he wrote famously—one of the great koans of American literature" (*American Earth*, XXIII).[18] McKibbon is not only suggesting a relationship with nature in order to understand it, but he is suggesting that not understanding nature creates a divorce from reality itself. These observations indeed have concrete repercussions. Aside from epistemological discussions on our understandings of reality based on experience, language, and thought, not listening to nature can lead to grave consequences. Natives understand this well, as Alaska Natives with a Native Eye understand how much pressure

18 McKibben, Bill. *American Earth: Environmental Writing Since Thoreau.* New York: Library of America, 2008.

can be put on their own land, and how their biomes work, based on the reality they experience merely by existing in it.

Furthermore, there is an insinuation in Lopez' passage that observation is actually a genius for Native peoples. The author even suggests that scientists look to Natives and their methods with respect moving forward. The previous European propensity had been to look down upon indigenous culture in order to subjugate it and to propagate empire through colonialism (Carr, "Modernism and Travel (1880–1940), *Cambridge Guide to Travel Writing*, 73). Even if this is no longer a priority, the tendency remains engrained like an appendage no longer in use. Interestingly, not only does Lopez push for this updated view, but he also goes so far as to flip the scenario altogether. When discussing the discomfort the Inuit feel with societies disconnected from the Native Eye, Native peoples are vaulted to a perspective of supremacy, at least in this field of understanding the natural world. Lopez, interestingly, moves these arguments forward without actually presenting a single, direct perspective. Rather, hints and shades of perspective are thrown in, orienting the readers' understandings of the issues Lopez hopes to present. As a result, The Native Eye argument comes across clearly, as do its repercussions and suggestions of implementation, while maintaining an opaque, almost prophetic, style working between the spaces of thought enough to avoid the literary pitfall of didacticism.

Moving on to comparison, what Lopez and Lowenstein have in common is an attempt at conveying the spirituality of Alaska Natives (and broader Native American communities in Lopez' case) in the Western literary tradition. They both also demonstrate the connection between said spirituality and a holistic view of regional biomes (animals, landscapes, food, subsistence, etc.), for the benefit of both the Western reader and Native peoples themselves. Interestingly, Lopez begins gaining steam as a feted writer around the time that Matthiessen's *The Snow Leopard* is published,[19] with the acclaimed *Of Wolves and Men*, before attempting to offer a positive light to the Arctic in the very middle of the Reagan years with *Arctic Dreams*. The Reagan years being rife with religious revival in American culture, after the dissolute 1970s (*Consider the Lobster*, "Certainly the End of Something or Other, One Would Sort of Have to Think," 51–9[20]), one can begin to better understand the world that Lopez wanted to convey at that moment. Had Lopez been writing *Arctic Dreams* in the late 1990s, when the Northwest technological boom was more important than the oil boom in Alaska, and contemporary to Raban's writing of *Passage to Juneau*,[21] *Arctic Dreams* would likely have carried a different tone. For instance,

19 As well as McPhee's *Coming into the Country*, perhaps not coincidentally.

20 Wallace, David Foster. *Consider the Lobster: And Other Essays*. London: Abacus, 2005.

21 And documented throughout Raban's *Driving Home*.

Lopez published *Light Action in the Caribbean*[22] in 2000, a year after *Passage to Juneau* was published, and the same year the 1990s tech bubble burst.[23] Mike Newell describes the stories of *Light Action* as "appear[ing] post-apocalyptic in their savagery and portray[ing] a series of plots which, on the surface, seem inimical to Lopez's usual venues" (110). The following book Lopez published, in 2004, was *Apologia*,[24] about the nature of overlooked roadkill. So, we see that broader political moods reach as far as isolated areas in the Arctic. Yet, Lopez, despite any shift between fiction and non-fiction, or melancholic moods and optimistic ones, has never swayed from his interest in the Native Eye. The following chapter on animals will discuss this idea in more detail: explaining how Lopez shifts points of view constantly in *Arctic Dreams*, creating a mosaic of perspectives as a means to inclusion (and ultimately, to spirituality, which trickles through as though through osmosis; that is to say, mostly indirectly), instead of in-depth studies of individual cultures and communities the way we see with McPhee.

Lowenstein, by contrast, digs into the actual storytelling process itself, much more along the lines of what Lopez does with his fictional *Giving Birth to Thunder* than in *Arctic Dreams*.[25] Lowenstein also breaks up Part II of his book into seasons (51–172), employing a mixture of stories and poetry to cross this difficult divide. Poetry, for Lowenstein, serves as a technique for expressing the inexpressible, the ineffable, the cultural differences of spirituality in storytelling. He attempts to explain his method in the "Sources" section at the end of *Ancient Land, Sacred Whale*. He then copied a pertinent part of the explanation in his "Introduction" to the "Ritual Year" section of *Ancient Land* in its entirely reprinted version in the compilation, *Ancestors and Species*.[26]

> My effort has been to reconstruct an old [sacred] drama and to show how people lived within the structure of its poetic idea ... Local stories embody their own poetry. My own contribution is simply an attempt to encapsulate or interpret in a Western idiom experience which may lie under or parallel to the narrative genre. (*Ancient Land*, 185; *Ancestors*, 36)

---

22 Lopez, Barry. *Light Action in the Caribbean*. New York: Knopf Doubleday Publishing Group, 2000.

23 Geier, Ben. "What Did We Learn From the Dotcom Stock Bubble of 2000?" *Time*. "http://time.com/3741681/2000-dotcom-stock-bust/." 12/3/2015.

24 Lopez, Barry. *Apologia*. Athens: University of Georgia Press, 1998.

25 Interestingly, Lopez notes that his own early work blurred the lines between fiction and non-fiction, but never lied about whether or not his work was fact-based, which would be a "dishonest relationship with the reader" (Newell, 20, 36)

26 Lowenstein, Tom. *Ancestors and Species*. Bristol: Shearsman Books, 2005.

Essentially, what Lowenstein is saying is that communities and lives can be displayed in a poetic form, or narrative, and controlling the narrative is a very important aspect for peoples, and what the author has done is attempt to carefully relay the Tikigaq narrative to a Western audience without losing its power, its essential nature.

In reading Raban's *Passage*, it is clear that British explorers benefited from viewing Native peoples as childlike and immature. Yet, Raban also discusses more directly how the Northwest School has typically had the opposite agenda from that of the colonialists, iconoclastically stating, as previously noted, that the Northwest School thought of Native peoples as "... people who apologized to salmon before killing them, who hugged trees before turning them into war canoes, like good school children of the 1990s, following Mother Nature's rulebook" (348). Lowenstein's poetic idea, then, can be interpreted as the controlling of a narrative combined with a holistic view of the traditional lifestyles of the Tikigaq people in a complex form that finds its greatest truth in the nuances of its stories more so than in collections of structuralist data, for instance, or directly translated stories lacking the spirit of the people. The author should be lauded for accomplishing this careful task without succumbing to bowdlerizations or the naive, folklorish stereotyping against which Raban warns (though Raban may be exaggerating in his description). In summation, conveying this sense of cultural nuance to a Western audience, what Lowenstein calls "a Western idiom experience," the writer ultimately hopes to bridge the rather pronounced cultural divide through poetry.

Lopez, by contrast, is no poet. Asked by Mike Newell about his opinions on poetry as a form for him, as his prose is descriptive, expressly beautiful, thoughtful, therefore poetic for many people, the author stated that he preferred to stick with prose as a storytelling mechanism (34). Lopez preferred to find the Native Eye and convey it to readers through his own experimentation with the art of storytelling: a hybrid Native/Western mode with compelling narrative capacity (*Giving Birth to Thunder*), and fluidly switching between Native, historical, and first-person accounts (*Arctic Dreams*). Lopez has also said in interview that the narrators of his first-person accounts in certain works, such as *Desert Notes* and *River Notes*, are not supposed to be viewed as the writer, but as a "non-autobiographical, 19th century, first-person" narration (Newell, 20), presumably representing the Western viewpoint. *Arctic Dreams*, then, follows the history of European exploration in the Arctic, while simultaneously viewing the landscapes, animals, etc., from the point of the view of its native peoples.

As far as his mysticism is concerned, Lopez is without a doubt a naturally sensitive, reflective person, which comes through in his writing style. In Mark Tredinnick's time with Lopez in *This Land's Wild Music*, Lopez is moved to silence "at the sight of a halo of ice encircling the sun" at a place of personal spiritual importance as the two drove through the forest around Lopez' house (42). One could never imagine this happening to Jonathan Raban, the stoic cigar smoker in his Dodge Daytona muscle car. In fact, unlike his sanctuary of run-down bars on archipelago islands of British Columbia and Alaska (320–1, 349, 364–66, 412), mingling among Seattle fishermen (3–37), the only time Raban comes across as truly uncomfortable in *Passage to Juneau* is during the Native, spiritual dance ceremony he has been invited to witness around Seattle that rings "false" (71–7). The sensitivity witnessed in Lopez through his being moved to silence, also coming through unequivocally in his style on the page, inclines the writer to be categorized in ecopoetics (essentially, descriptive writing about personal experience in nature – Buell, *The Future of Environmental Criticism*, 104), whether he likes it or not (Newell, 76), or at least landscape writing. When confronted by Mike Newell, who tells the writer, "There's a divinity in your early writing. Not dignity, it's beyond dignity," Lopez begins his response by saying, "Right, the numinous, I guess. The landscape of the numinous" (Newell, 30). This response speaks volumes, because, in discussing Lopez' style, we so often end up discussing landscapes and perspectives of landscapes from myriad angles. Lopez was once even a photographer focusing on the same aspect, but he felt his personal gift was in words, where his explorations of landscapes are so often connected with the numinous, wherever they may be (Tydeman, 44–6, 106–9).

If Mark Tredinnick's *This Land's Wild Music* has any continual purpose or theme, or any lasting effect on the greater world of ecocriticism, it is to truly show his chosen writers in their habitats, where they write. Lopez spent his youth in California, New York, Indiana, and traveling all over the US, including Alaska, but he made his home in a conspicuously numinous space "forty miles upstream of Eugene, Oregon," in a "clapboard cottage," in the midst of a rainforest rich in wildlife (Tredinnick, 45–7), apparently not too far off from the Burroughs model of minimalism in that author's celebrated "Slabsides" (Brinkley, *Wilderness Warrior*, 368, 617, 768). According to Tredinnick, "Nothing anonymous ornaments his [Lopez'] cottage. On its wooden walls hang paintings, drawings, and tapestries, made by friends or by the peoples he has traveled among. He tells me the stories of their making … We eat at a table made by a friend from a tree Barry felled, a slab he cut" (53). Tredinnick goes on to say that the stories of Lopez' personal history objectified in the house make the place "light, thus, in the dark forest." Like Tredinnick's brief story of Lopez teary-eyed in an old pickup truck, the landscape Tredinnick describes, in which Lopez surrounds himself, and even inside his house, are a constant search for the numinous.

It is worth restating that when Newell said "divinity," Lopez responded with "numinous" (30). "Divinity" is deeply attached to the Christian religion, as Jesus Christ was "Divine."[27] Lopez did not disown the religion of his youth,[28] his culture, but he was searching for mysticism, for the numinous, in landscapes (Tydeman, 132), much like John Muir did (Brinkley, *The Quiet World*, 1–21). Lacking true knowledge in the Asian religions, Lopez looked to Native Americans to find the numinous in their perspective. Lopez said that Christianity is religion codified. It also, like the other world religions, supports colonialism (Tydeman, 132). He lambasts these aspects of Christianity for their stifling of imagination (133).

One could suggest that there is likewise the potential for ruthlessness and stifled lives in Native cultures and spiritual beliefs that can rival that of Christianity (Bartram,[29] 527–67; Fienup-Riordan, 162–9; Lowenstein, 10–5, 113; Rainey, 242; Cournut, *De Pierre et d'os*[30]). Yet, as with Jean Malaurie (*Les derniers rois de Thulé*) who spent extended periods within Inuit culture, the exoticism of Native spirituality to an outsider can be much more exhilarating and evocative for the imagination than it is for one who is born into a culture. Both Malaurie and Lopez do indeed seem keenly aware that Native culture has a potential to stifle in different ways. Throughout their observations, they both constantly show compassion and empathy, while also attempting to remain scientifically objective. However, Lopez, as seen in *Giving Birth to Thunder* and in the multiple perspectives of *Arctic Dreams*, has always been an open and fierce advocate for Native spirituality. Whether those Outside judge it to be a frivolous form of spirituality or not, whether it is effective or childish or whatever pejorative adjective we may decide to use or not, it is part of an ancient culture of accumulated wisdom and TEK that is connected to the "numinous" landscape that Lopez explores. Furthermore, connection through one's historical, familial, and spiritual roots in this way is a strong means to healing communities that are struggling with drugs, alcohol, and ennui.

---

27 Merriam-Webster defines "divine" as, "of, relating to, or proceeding directly from God or a god," with secondary definitions all relating to deities and God, with the words "being" or "directed to" involved. "Divine," then, relates much more directly to the organization of religion than the view Lopez tends to hold.

28 Of such an importance that Lopez considered becoming a monk at Gethsemane, a Trappist Abbey in Kentucky, just after graduating from the University of Notre Dame (Newell, 28, 70).

29 Bartram, William. *Bartram: Travels and Others Writings*. Edited by Thomas P. Slaughter. New York: Library of America, 1996.

30 Cournut, Bérengère. *De pierre et d'os*. Paris: Le Tripode, 2019.

One way that Lowenstein is unique in his technical approach to finding the Native Eye is that he attempts to explain the worldview of the Tikigagmiut in a spiritual and linguistic manner. As noted in a section above, the soul can be viewed as a life force, a personality, and a name, all taken seriously as the individual person. Likewise, the stories are told in the Tikigaq manner of storytelling and fuse meanings in a way that is wrapped up in ethnology, but that have potential spiritual and cultural repercussions. For instance, Lowenstein goes out of his way on multiple occasions to attempt to explain to the Western reader that in Tikigaq mythology the land is quite literally a whale ripped from the ocean bottom with a harpoon. From the scientific perspective, the earth quite literally trembles due to the pressure of the ice against Point Hope, Tikigaq, and this is taken by the Tikigagmiut to be a living aspect of the breath of the harpooned whale (5–9, 21–3, 28–32). Therefore, viewing the earth and landscape with a Native Eye quite literally means understanding it as a living creature. And, not only that but animals are animal people, who can often shape-shift and were once different than they currently are. Understanding the multiple layers of lessons, entertainment, and spirituality within the story, then, involves a complex scenario of which the Tikigagmiut were likely aware in explaining their worldview to themselves.

Lowenstein also dives into other difficult forms of symbolism. For instance, there is a constant theme of dualism in Tikigaq mythology, with male shaman negating the power of female shaman, or their working together as couples, strengthened. Furthermore, there was a "dual character" understood in many symbols, as noted with the land itself, which was also a whale. As Lowenstein points out: "The dual nature of the mythic couple and the doubleness of the earth they 'created' extended further than myth. As we shall see the 'opposites' of earth/whale, land/sea, male/female were vital and almost endlessly evoked symbolic forces to Tikigaq people" (38). In other words, duality, or a plurality of symbolism and meaning, was a constant in Tikigaq society. Lowenstein adds:

> But Tikigaq hunters took the coincidence of myth and subsistence a large step further. Where the central acts in the myth and in whaling coincided, Tikigaq elaborated on the mythic structure of their whale hunt by re-enacting Tulunigraq's narrative and adopting the creation story's myth roles and characters. (38–9)

Jonathan Raban repeatedly made the point in *Passage to Juneau* that the narrative of Native Americans as somehow at one with nature and pure was disrespectful because it was too restrictive of complex cultures. One only has to look as far as Rainey and Lowenstein's suggestions of the normality of rape to understand that the Iñupiat community is not somehow sublimated in comparison to that of the West. Yet, Lowenstein does not necessarily fall into this characterization of Native peoples as somehow nobler or "at one." Rather, he makes the point, as Lopez does, that subsistence and myth are wrapped up in an inextricable worldview. He does so by explicitly saying as much, before pointing out examples in story form (Lowenstein, *Ancient Land, Sacred Whale*, 35–50). Furthermore, whether the Tikigagmiut were respectful to the whale out of fear, or appreciation, or as a deity, one may not be sure, but if the worldview is nonetheless intertwined with the animal being hunted itself, a sustained relationship of respect is much more likely, even if some of the meaning has been transformed or even lost. This supposition is only strengthened by the understanding of dichotomy in the symbolism that penetrated Tikigaq mythology: a raven is not merely a raven, but a raven and Raven Man, The Sun and the Moon chase each other as woman and man, the land is also whale harpooned, etc. This duality goes a long way in Lowenstein's taking in the Native Eye, which Lopez appears to view as a somewhat utopian, pre-industrial, holistic world (though admitting an Inuit "sober capacity for violence") (*Arctic Dreams*, 242–51). In such a view, it is likely true that more attention is paid to a holistic relationship with the earth and one's food. Native peoples, however, potentially played a part in killing off entire species of megafauna when they came across the Bering Strait.[31] Also, it would not be a stretch to suggest that many women would find a modern, urban city a more amenable place than traditional communities in touch with the Native Eye of holistic, subsistence living. Despite often playing extraordinarily important roles in the community, as noted previously, Native woman can be susceptible to rape, unwanted marriage, and unwanted wife exchange. Lowenstein does not attempt to gloss over the negative aspects. As has been mentioned multiple times in this research already, these aspects are a part of traditional Tikigaq life and displayed as such in his work. In this way, Lowenstein's technique is highly informative and true to its goal of explaining the traditional culture, the traditional people of Tikigaq, and their whale hunt.

The trope of the Native Eye has been thrown around quite liberally in Part 1 of this research, but, like so many other tropes, it constantly drives home an important point. One of Lopez' most important contributions to Western literature is to peek beyond the modern, industrial world, where food itself has become overly sanitized, processed in multiple stages in factories, even bowdlerized away from its own face and form. This aspect of Western culture

---

31 Groves, Pamela. "Pleistocene Megafauna in Beringia." *National Park Service*. "https://www.nps.gov/articles/aps-17-1-4.htm." Viewed on 6/2/2019.

is a tragedy for Lopez, who does not suggest that society revert to small tribes, but more likely that we continue to learn from their worldview, keeping a life beyond the industrial world in perspective. Likewise, this is what Lowenstein accomplishes in *Ancient Land, Sacred Whale*. There is a reason that Lowenstein does not approach modern issues of living in Inuit society, nor does he approach many modern aspects of Inuit life, such as snow machines, Coca-Cola, Christianity, and airplanes much at all. He interviews older people in the community in an attempt to recreate the Native Eye that the community itself had once held, and, in doing so, he was attempting to give it back to them, along with a greater understanding to his own, larger community, the Western world. In portraying this people in an artistic, poetic sense, attempting to create something different from the academic work of ethnologists, Lowenstein had to believe that everyday Western people have something to learn from the Tikigagmiut.

6 Conclusion

As an aside that will also form a cohesive conclusion to this section, splitting up much of Part 1 of this research has been a difficult undertaking. On the one hand, in the section above there is an emphasis on language; in this section, there is an examination of religion and spirituality; and in a section to follow, on animals, this research purports to focus on symbolism; but simultaneously, in Alaska, language, animals, religion, and spirituality are all interconnected. Animals are part of the spirituality, the community of Alaska Natives, who have traditionally lived and breathed and even spoken in terms of the numinous landscape and subsistence living. This has long been the purpose of life in a region where survival is so difficult, as Barry Lopez eloquently discussed. The local landscape provides or hordes everything, is literally everything to the traditional culture of Arctic peoples, including religion.

In *The Invention of Nature*, Andrea Wulf discusses a conversation between Thomas Jefferson and Alexander von Humboldt, in which the two agreed that "only an agrarian republic brought happiness and independence" (103). The insinuation is that a connection to the land is almost spiritual. Geographer Yi-Fu Tuan goes into this idea for multiple pages in his work *Topophilia*, discussing the idea that farmers naturally maintain a deep intimacy with their land, as it is almost an extension of the self through its provision and withholding, serving as both friend and enemy, affecting mood and health accordingly, and ultimately offering self-worth and a point to life to many (96–8). Wulf goes on to say that Jefferson and von Humboldt saw colonialism as bringing destruction, pointing out how the Spanish had "annihilated ancient civilizations, native tribes and stately forests" (103). Likewise, this view of yeoman agrarian life ("As long as a man had his own piece of land, Jefferson believed, he was independent." – 106) measured against the unsustainable resource extraction of colonialism (104–6) could be extrapolated tenfold to the Arctic: Alaska Natives do not lose their land and integrate into a new form of life, the modern city, but they lose their land and waste away in an uninhabitable climate with long, dark winters. Throughout Wulf's work, she latches onto von Humboldt's idea of the interconnectedness of all things, what he called Naturgemälde (88–9), which is much the same assertion being made in this section, regarding the numinous and interconnected repercussions of nature and how to view it. Kollin would suggest that we pay attention to the subsistence on display in the pages of the *Tundra Times*. Lopez' work may lack the graphic depictions of subsistence we see at times in the pages of Lowenstein and Lord, but in his search for the numinous, he also supports Kollin's assertion for viewing the Arctic, from Native and subsistence perspectives. Founding president of the ASLE,[32] Scott Slovic, begins his chapter on Barry Lopez in *Seeking Awareness in American Nature Writing*[33] by saying, "To overcome the absurdity – the meaninglessness and destructiveness – of man's estrangement from the natural world is precisely the goal of Barry Lopez's *Arctic Dreams*" (137). This is a pithy statement that summarizes Lopez' intentions well. Despite the fact that the author has said that he does not consider himself a nature writer (Tydeman, 33–4, 55), Slovic's remark gets at something important, as it characterizes the ultimate pragmatic goal of finding the numinous and the Native Eye in *Arctic Dreams*. If Lopez is searching for that connection lost around the time of the construction of Europe's cathedrals (250),[34] nowhere is the negative resulting contrast more poignant than in Lopez' recounting of his visit to the oil extraction communities of Prudhoe Bay. In that place of oil extraction, men live inside with no relationship to the natural world around them, speaking to the author almost exclusively of "seduction, domestication, domination, control" (398). Frustration and mistrust are rampant, and Lopez is even explicit in extrapolating this experience to the wider national paradigm, "This is part of factory life in America, an ugly way the country has

32 Association for the Study of Literature and Environment.

33 Slovic, Scott. *Seeking Awareness in American Nature Writing: Henry Thoreau, Annie Dillard, Edward Abbey, Wendell Berry, Barry Lopez*. Salt Lake City: University of Utah Press, 1992.

34 "Not only was God light but the relationship between God and man was light. The cathedrals, by the very way they snared the sun's energy, were an expression of God and of the human connection with God as well" (Lopez, 248).

arranged itself …" (398) (397–400). Prudhoe Bay, then, is the embodiment of that "estrangement from the natural world," with the further example that men look outside and see "a great wasteland 'with a few stupid birds,' too vast to be hurt" (398). These men are falling for a trap against which Lopez warns: viewing nature as beautiful or ugly according to a paradigm set by society, and pushed by "photo pornography" (Tydeman, 108–9). Because the landscape seems empty, the oil workers do not see any reason to be gentle with it, or protect it. Lopez, then, as Slovic notes, is attempting to reconnect people with the numinous in the ostensibly barren landscape, which "can open suddenly, like the corolla of a flower, when any intimacy with it is sought" (Lopez, XXIV). Lopez aids the reader in achieving these glimpses of reconnection by searching for the Native Eye, by describing the lives of animals on the landscape, and viewing the Arctic through this holistic perspective of subsistence the way Susan Kollin suggests we do. Such a perspective can ultimately be viewed as a form of spirituality, even codified into a loose set of religious beliefs forming the backbone of much of Alaska Native tradition.

FIGURE 7 Port Heiden
PHOTO BY CHRIS FERGUSON

CHAPTER 6

Animals

The Cartesian view[1] of animals being considered robotic, not cognitively dynamic, during the Enlightenment stems from a desire to systematically replace fable and myth with truth, with observable fact (Wulf, *The Brother Gardeners*, 46). This is understandable, as much erroneous information had theretofore been spread (think of the search for paradisiacal Thule – Malaurie, *Ultima Thulé*, 9–10; Mund-Dopchie, *Ultima Thulé*[2]), and a standardized, experimentally proven understanding of the world was a laudable goal. Yet, a consequence of such a shift was that observing anthropomorphism in animals became unacceptable in academia, likely as it had been associated with myth and was therefore considered unscientific, unempirical (Lopez, *Arctic Dreams*, 176–8, 199–203, 268–71). Not only has this shift objectified the natural world, animals in general, so that they lost their complexity and dynamism along the way, but this objectification became wrapped up in colonialism, which was itself consciously intertwined with travel writing from at least the time of Richard Hakluyt. What this means is that animals have lost what is now being called "personhood," or seeing animals and humans as part of a community together. Animals are now commodified, treated as numbers and part of not just a binomial, but an economic system (industrial scale chicken farms, for instance).[3] The animal perspective, then, in maintaining personhood, may offer us something similar to that which the ecofeminist movement offers. As ecofeminism hopes to break away from patriarchal domination, so too, perhaps should the perspective of animals. After discussing Farley Mowat's *Never Cry Wolf*[4] in Ecology and Literatures in English, Françoise Besson says: "We should listen to those wild voices. Nonhuman animals, who only take what is necessary for them to live, show us the way to the preservation of the planet and of those

1 Descartes, René. "Lettre à Morus du 5 février 1649." AT, V, p. 277–278. *Œuvres de Descartes*, éd. Adam et Tannery [AT], Paris, rééd. CNRS-Vrin, 1964–1974,11 vols.

2 Mund-Dopchie, Monique. *Ultima Thulé : histoire d'un monde et genèse d'un mythe*. Paris : Droz, 2009.

3 Buller, H., Roe, E. "Modifying and commodifying farm animal welfare: The economisation of layer chickens." *Journal of Rural Studies*, 2013. "http://dx.doi.org/10.1016/j.jrurstud.2013.01.005."

4 Mowat, Farley. *Never Cry Wolf*. Toronto: McClelland and Stewart, 1963.

 | DOI:10.1163/9789004734807_008

who live on it, including human beings" (367).[5] This trend of thinking otherwise, again, with travel writing historically complicit, has led to bloated fur trade (Lopez, *Horizon*, 113)[6] and exhaustion of animal communities to near extinction in certain cases, such as the buffalo (Brinkley, *The Wilderness Warrior*, 154–5) and the sea otter (Brinkley, *The Quiet World*, 426–7). Aside from arguments of a lack of empathy and compassion, the general absence of holistic embrace in Western culture is increasingly leading toward unsustainability of ocean life, of undomesticated, unindustrialized animal and plant life everywhere. Yet, there remains a reluctance among the scientific community toward anthropomorphism, a word which itself is a symptom of scientific prejudice, as it connotes humanity risen above the natural world, allowing sublimation of certain aspects or personality traits via comparison, instead of placing the human species back into the natural world, as part of a community, which, as Barry Lopez says in his latest work, Horizon, "is a sign of hubris" (297). Lopez also notes the argument that ceasing "to exaggerate our own importance as a species … might eventually lead to better politics and to the development of more equitable social and economic systems worldwide," before observing that man "is exceptional," but is, in Deep Time, part of a Darwinian spectrum and not a fixed, apotheosized and omnipotent creature, is not infallible and is, indeed, capable of extinction. So Lopez finishes the subject by asking of H. sapiens, "Where will his exceptionalism take him?" (297).

Packaging the enduring legacy of empirical Cartesian thought together with empirical colonialism and moving forward into the modern world, we are once again drawn to Susan Kollin's warning against Frontier as Empire: "The frontier therefore functions as a complex rhetorical strategy that simultaneously advances the United States' desires for and denial of empire" (20), and "Precisely because Alaska is still regarded as a Last Frontier in the dominant American imagination, the region has not remained immune to the problems of development" (56). If, then, the scientific community in Alaska refuses to see animals in a holistic, dynamic sense, the latter become static objects caught up in the idea of the Last Frontier, something to be controlled and conquered through human action, which has not gone well for animals in the past few hundred years.

This chapter will examine five of the primary authors of this research, as all but Raban employ unique methods of discussing animals. Interestingly, while we are moving forward in the discussion of animal subjectification, we are also

5 Besson, Françoise. *Ecology and Literatures in English: Writing to Save the Planet*. Newcastle upon Tyne: Cambridge Scholars Publishing, 2019.

6 Lopez, Barry. *Horizon*. London: The Bodley Head, 2019.

moving backward chronologically, as this technique for personhood, for animal respect, seems less common in literary works than it was earlier in the 20th century. If there is one aspect, however, that brings together the entirety of this diverse array of approaches to representing animals and their relationships with humans, as well as with their own environments, it is that the animals almost always represent some purpose beyond themselves.

1 Animals Speak in the Work of Barry Lopez

It is quite notable to this research that Barry Lopez decided to focus much of *Arctic Dreams* on descriptions of animals in their habitats. Much time is spent describing the lives of polar bears ("Tôrnârssuk: Ursus maritimus"), musk oxen ("Banks Island: Ovibos moschatus"), and narwhal (Lancaster Sound: Monodon monoceros"), respectively, with an entire chapter devoted to the migration of multiple other species as well ("Migration: The Corridors of Breath"). The result is a rare work of biological animal study gaining commercial success. Yet, Lopez does not focus on one specific animal as a beacon of some particular situation, as noted in the examples above and as many environmentalists have done with the polar bear, for instance, in the fight against climate change.[7] The overall view, rather, explores the depth and complexity of life in the Arctic, often confused as a lifeless barren (XXIV, 397–8) at a time when oil exploration and oil money are both flowing freely (393–400). However, not only does Lopez make the reader rethink the complexity of life in the Arctic, he also accomplishes what Bruno Latour suggests in *Politiques de la nature*:[8] treat what is non-human as a subject and not an object (Latour, 109–117), as objectification implicitly invites exploitation.

Toward the end of Chapter 6 of Arctic Dreams, "Ice and Light," Lopez discusses man's relationship to the world and man's advancements as a species. He ponders whether the advancements since the brilliance of cathedrals have been solely technical, and whether we have lost our mysticism as a species,

7 "Polar bear population decline a wake up call for climate action," *World Wildlife Fund*, "https://www.worldwildlife.org/stories/polar-bear-population-decline-a-wake-up-call-for-climate-change-action" viewed 6/10/18

"Polar Bears Really Are Starving Because of Global Warming, Study Shows," by Stephen Leahy, 1/2/18, *National Geographic*, "https://news.nationalgeographic.com/2018/02/polar-bears-starve-melting-sea-ice-global-warming-study-beaufort-sea-environment/"

8 Latour, Bruno. *Politiques de la nature : Comment faire entrer les sciences en démocratie*. 1999. Paris: Éditions La Découverte, 2004.

becoming specialists, or "singular adepts ... performers" (250). He then justifies himself by saying:

> Of the sciences today, quantum physics alone seems to have found its way back to an equitable relationship with metaphors, those fundamental tools of the imagination. The other sciences are occasionally so bound by rational analysis ... that they recognize and denounce anthropomorphism as a kind of intellectual cancer. (250)

This passage is, in a nutshell, the crux of Lopez' overall message: mankind has lost its edge of mysticism and, thereby, its connection to the natural world. Lopez spent essentially his entire career attempting to reconnect the developed world with Native cultures and viewpoints, which he views as maintaining a closer relationship with landscapes, biomes, and "the passionate and spiritual realms of life" (*Crossing Open Ground*; *Arctic Dreams*; *Of Wolves and Men*; *About this Life*, 12, 130–43; *Giving Birth to Thunder*, XVI–XVII; etc.). The chapter of Arctic Dreams on the polar bear, "Tôrnârssuk: Ursus maritimus," is where Lopez delves into his idea that Western Culture has lost its Native Eye. To revisit part of Lopez' quote on the Native Eye, the author says, "It not only takes a long time of watching the animal before you can say what it is doing; it takes a long time to learn how to watch" (96). Over the next roughly page and a half, Lopez discusses how the Inuit tend to be uneasy with examinations by "people who are not sensually perceptive"; in other words, observers on the look-out for hard facts who come to collect data without truly "seeing" behavior, treating animals as statistics and not individuals. As the passage states, the developed world and its science has long been uncomfortable with anthropomorphism. Since *Arctic Dreams* was published in 1986, however, there has been a gradual and uneasy shift toward acceptance of Native views in scientific communities, as Nancy Lord notes in *Beluga Days* in discussing "traditional ecological knowledge," or TEK. The inchoate scientific field is particularly taken seriously in ethnological studies, which may then be employed in collaboration with the work of biologists, physicists, etc. (Lord, 172–3).

Interestingly, Lopez' preferred technique in *Arctic Dreams* is to interweave simultaneous ideas in a meandering essay form not unlike the style of Emerson (Buell, *Literary Transcendentalism*, 157). When reading Lopez' work on polar bears, Arctic peoples are highly intermingled in the arguments, with multiple themes. Discussions of Arctic peoples and their viewpoints are indeed dispersed throughout the work, even though they are not offered their own chapter, unlike the polar bear, or the musk ox, or even the narwhal. The effect is a discussion of animals, from the distinct perspective of the Native

Eye, among others. The chapter "Tôrnârssuk" is a great example of this. One section begins on page 89, with a discussion of the dens of mother polar bears. On pages 91–2, Lopez pivots, introducing two scientists who have long examined bear denning habits in Svalbard, and what they say about their observations. This demonstrates Lopez' point about spending vast amounts of time observing, learning the Native Eye, while also continuing his strain of factual information with regard to the actual habits of the bears. By page 93, Lopez is inserting more scientific observations of bear habits by different people in the Arctic, in a shift toward discussing ursine habits of wandering. By page 96, Lopez is discussing the individual identities of bears and not being able to completely understand "all it [the bears] can do." He then shifts to the first person, discussing his own experience wanting to follow a polar bear with a friend in a helicopter, philosophizing on looking out at the expanse of ice and wondering if he could even surpass the epistemological boundaries of what he would see were he to indeed follow. This is when Lopez pivots hard to the idea of the Native Eye in the works of Arseniev and van der Post. He again returns briefly to the bear seen from the helicopter in the reverie brought on by information in the general, factual discussion of bears and their being viewed by scientists and the Inuit. He then once again pivots back to the Native Eye with the quote, "To follow a bear, or simply follow in its tracks, is to 'reeeally learn something,' as the Eskimos say, smiling." From there, he pivots once again to a general, factual conversation on bear tracks. And so on goes the chapter, a constant weaving of discussions of the animal's life and habits, mixed with the Inuit viewpoint and the modern, scientific viewpoint, with first person accounts thrown into the mix. In this way, unlike so many works in nature or environmental writing, the reader is offered the viewpoint of Native peoples, the developed world, the animals themselves, and very often the landscape affected as well. The interplay, potentially confused with a lack of focus, is a successful technique in inclusion and holism.

One could of course view Lopez' overall expression as too emotional, too New Age, as Raban has, which may risk overly emotionally romanticizing traditional lifestyles in a saccharine, aesthetic sense, minimizing community problems and the actual "realist" version of Arctic peoples, which is partially why Wenzel attacks Lopez in *Animal Rights, Human Rights*. When so many other writers are focusing on physical people, however, there does seem to be a need for expressing the Alaska Native viewpoint itself, which is much more in line with the observation and understanding of animals than that of the developed world. The animals themselves rarely receive the spotlight in travel writing in anywhere near the quantity that Lopez offers them, which is a boon for human understanding of them in such a well-publicized and -received work.

To view animals in their direct habitats, as dynamic persons who are part of communities, gives them weight in the eyes of humans, as subjects instead of objects, unlike the Cartesian view of animals as robotic and structurally predictable (Wulf, *The Invention of Nature*, 16–7, 23). With this holistic view, the argument can also be made as to why the landscape is so valuable, and should be protected. The interplay of animals amongst themselves and with their habitats, the growth patterns and resilience of Arctic willows, the interplay of plants in forests, the water, ice, etc. (Lopez, *Arctic Dreams*, 25–41), makes one see the world as a whole, dynamic ecosystem, instead of as a "frontier" to be conquered, and from which to extract metals and fossil fuels. In Lopez' work, therefore, animals come off as noble creatures, and the Western world, other than through the few dynamic scientists who watch with a Native Eye, are mostly insinuated in the modern world as the Arctic oil workers who have no interest in the landscape outside the confines of the oil combine (397), preferring drugs and alcohol, pornography magazines (398), television, etc. (396–7). There is some poignance in demonstrating this viewpoint, as the juxtaposition forces people to see the nobility of creatures they are taking advantage of, and themselves as unwitting marauders of a sort. The symbolism is a bit of a forced view on the reader, and possibly unfair in harping on some of Western Culture's least thoughtful demographics, but the technique may also be effective in creating an intended emotional reaction among its general audience.

As a play of devil's advocate, however, this view of almost complete nobility of animals can also be unrealistic in what it leaves out. Timothy Treadwell viewed bears as noble creatures, anthropomorphized them and was eventually eaten alive by one, along with his girlfriend, becoming the unfortunate protagonist of Werner Herzog's 2005 documentary *Grizzly Man*,[9] which is full of the nuance of nobility and savagery in the Alaska grizzly bear community.[10] An argument can be made that people are dangerous, and bears are dangerous. Likewise, nature itself is dangerous. Humans, however, are overpopulating the Earth and pushing bears out of their natural habitat and forcing landscapes into the barren monotony of invasive species in what has become known as the Anthropocene (Kolbert, *The Sixth Extinction*). Overall, Lopez is able to demonstrate a rare, highly reflective and dynamic work, whose meandering

---

9 Herzog, Werner. *Grizzly Man*. Santa Monica: Lion's Gate Films, 2005.

10 Treadwell also wrote his own work on his experience, Among Grizzlies, before he died, which undoubtedly lacks the nuance and professionalism of Barry Lopez, but which also shows much humanity, serving as an interesting case study for man's potential relationship with bears, attempting to quite literally become part of their community.Treadwell, Timothy and Jewell Palovak. *Among Grizzlies: Living with Wild Bears in Alaska*. 1997. New York: Ballantine Books, 1999.

technique allows him the artistic freedom to connect multiple ideas and viewpoints into an overarching interplay, evoking a poignance in the Arctic while also remaining concretely informative at once. The animals are often the focus, but the result is a strong understanding of their relationship with Native communities as well, as that relationship is truly interspersed throughout.

In contrast, Nancy Lord has followed a much different agenda. Her work itself, to be discussed in the following section, has been openly conservational in recent years. She has employed multiple techniques in at least four overtly environmental works, all very different from one another. However, it is her work as editor of the collection *Made of Salmon* that will be the primary focus of the section, focusing on Lord's decades-long relationship with salmon, as a fisherman.

2 Nancy Lord is Made of Salmon

Whereas Lopez has done the almost unthinkable in creating a long, descriptive, meandering work on the biomes of the greater Arctic, Lord has tended to stick with the specific in her techniques of symbolism. Starting with *Fishcamp*, the actual, physical fish camp itself serves as a symbol for Nancy Lord's Alaskan dream, which she defines in the book-length work, explaining the draw of both fishing and Alaska, while limning her physical space and what it means to her personally. In *Beluga Days*, she explores the idea of the downfall of the Cook Inlet beluga, which, at the time, was on track to becoming endangered (27–8). These two forms of symbolism, however, do not appear to be intentional, in the sense of the employment of an animal purposely representing a broader phenomenon. *Fishcamp* most likely arose from Lord's organic personal experience in her element, and in *Beluga Days* Lord describes the organic investigation that arose from an appreciation for "living in a time and place that sustained fish, whales, and fishermen – all of us together" (XXI). Moving forward into the writer's most recent two works, however, both appear to intentionally employ symbolism as a means of accomplishing a particular environmental outcome. In editing a 2016 literary project *Made of Salmon*, Nancy Lord employs the eponymous phrase multiple times in her introduction to the work (XVII–XXII), which notes that "Alaskans, nourished by salmon in this way, are quite literally made of salmon" (XVII). Being focused on various local communities, and comprised of a combination of non-professional, personal anecdotes mixed with non-fiction short stories by professional writers about the importance of salmon physically, culturally, and socially to Alaskans, this driving home of an intentional trope engrains the importance of the natural food source into the reader's consciousness. Another aspect Lord conveys is that salmon is not only what Alaskans are physically made of but what comprehensively symbolizes their local communities. As she notes, a salmon is swimming around the outer edge of the state seal (XIX). By repeatedly driving forward this idea, reminding communities of that importance, the reader is potentially more likely to support environmental regulation in lieu of short-term economic profit when the time comes to vote or interact politically or ecologically. The long-debated Pebble Mine outside Juneau, near one of the most important salmon sources in the world, is specifically discussed in dissent, as it would have a disastrous effect on local communities who rely so heavily on salmon in so many ways (*Made of Salmon*: Schooler, "Salmon Invictus," 3; Wilson, "Bristol Bay Youth Doing Their Part," 138–41). Therefore, the technique, whose effectiveness remains to be seen in such a recent work, is nonetheless careful and ethical with regard to the consideration of the multitude of unique communities that simultaneously exist with conflicting needs and desires in the state. This feat is rarely accomplished, as most works either broadly generalize, focusing on some theme without exploring the multitude of communities themselves (Lopez), or they stick to a small niche of a certain region (Lowenstein). In fact, this technique recurs throughout Lord's later, activist work. As previously noted, Beluga Days involves travel to multiple parts of the state, involving various demographics and peoples, and the novel *pH* likewise attempts to capture an overarching sense of diversity through positive stereotyping performed in such a nuanced way as to be sensitive to each of its respective characters (while also intentionally subverting those stereotypes at times) (249–52). Interestingly, in the less than six concise pages that Lord employs for her introduction to *Made of Salmon* she not only presents the work and a personal anecdote, but she invokes the multitude of Alaskan communities and its general history. First, she notes how Alaska Native communities have long thrived on salmon, noting that a mere cursory glance at a map will indicate how Alaska Native communities are "lined up on salmon routes." She then continues, saying, "Alaska's not an easy environment in which to live, and it's hard to imagine how there would have been much human success without salmon; salmon constituted much of Alaska's original wealth" (XVII–XVIII). On the following page, Lord describes Alaska's 20th century history with fishing, by stressing the importance of salmon to Alaska's statehood. Under "federal mismanagement," which allowed large-scale fish traps and canneries from Seattle, "'Abolish fish traps' became a statehood rallying cry. When the push for statehood succeeded in 1959, the new state of Alaska immediately banned fish traps and took over salmon management." "Sustained yield" became the new law of the land.

What Lord is doing here is juxtaposing Alaska's Native history against the history of its developed-world statehood fight, pointing out that all demographics are inextricably linked with this one fish and its survival. And, whereas so many works either focus on some aspect of white culture or likewise some aspect of native culture, Lord is careful to employ both perspectives. She points out how salmon is important to all people, whether involving native subsistence or that of urban, white professionals or that of those many in between.

The two citations also demonstrate how Lord has become more and more influenced by the *New Yorker* style of investigative travel writing, in that she chooses a singular topic and builds a dynamic investigation around that topic, traveling and organizing her work based on subject matter rather than a chronological journey. *Beluga Days* roughly follows John McPhee's and Rachel Carson's models in this manner, but *Made of Salmon* is unique. To begin with, Lord's earlier works were more closely related to her personal experiences. *Rock, Water, Wild* was a memoir in essay form. *Survival* and *The Man Who Swam with Beavers* were collections of short stories mostly related to her own personal experiences and local folklore. *Fishcamp* also undoubtedly came out of personal experience. And also, *Green Alaska*, though not personal experience, was travel writing on personal interest in the 1899 Harriman Expedition and particularly John Burroughs' involvement. *Rock, Water, Wild*, then, is the only work of Lord's oeuvre that comes after *Beluga Days* and does not involve an overarching environmental theme encompassed by a multitude of voices from Alaska's varying communities: *Beluga Days*, *Early Warming*, *Made of Salmon*, and *pH*. With that narrowed, *Early Warming* is still a bit of an exception in that it is a much more traditionally environmental work: discussing how climate change is affecting different parts of Alaska before it affects most of the rest of the world. *Beluga Days*, *Made of Salmon*, and *pH*, however, are all similar in that they employ a particular animal as a symbol to move forward an agenda. pH is different in that it is in novel form, which leaves us with the comparison begun above: *Beluga Days* and *Made of Salmon*.

Beluga Days begins with Lord at her fish camp asking why she sees less belugas than before, sparking a multi-year investigation into their disappearance and management. She interviews neighboring fishermen to ask about their relationships with beluga whales decades previous, as well as visiting conferences, meetings, the city of Anchorage, Native villages, and even Quebec, as well as the Lower 48. This is what New Yorker writers like McPhee have long tended to do. *Made of Salmon* indeed employs a similar format, but it does so as a multiple-perspective project instead of a single piece of investigative journalism. By piecing together a mosaic of different voices, different writers, from different communities, and from different demographics, Lord accomplishes much the same goal as if she had done said investigative work herself, or had John McPhee written a similar work and employed his prolific anonymous quote structure all the way through. John McPhee has stated in interview that he leans toward quotes too often in his search for journalistic objectivity, and his wife, Yolanda, has to reel him in because it is then no longer even his own work (Haynes, "The Size and Shape of the Canvas," 73). But, this tendency reveals McPhee's belief that objectivity can be accomplished to a much greater extent through other voices. As *Made of Salmon* is not of one author, with a single presumed agenda, but consists, rather, of multitudes with varying politics and agendas (and many non-professional writers), the book potentially becomes more accessible and convincing. This is achieved through the relative objectivity of the multitude, to readers expecting a voice different from that of Lord. In politically divided climates, evoking this one animal, the salmon, in order to unite the multitudes through the employment of their voices themselves could potentially be very effective for such a diverse community as Alaska. This is also something Lord navigates well in her introduction to *Made of Salmon*:

> This is the way of Alaskans and salmon. We share with one another; the fisherman divides his one or many fish among others, who cook or smoke or can and feed others, the circles widening through our communities. In our cases, the image of a circle expanding across the surface of water, perhaps from a leaping fish, seems … appropriate. (XVII)

She then continues, explaining the circles on water reference, first as the health benefits to individuals, then the development of friendships and communities, which stem from the sharing of salmon, and finally leading to common societal values. "Follow the circles out" she says, and all of Alaska is "made of salmon." This introduction begins with a personal anecdote evoking the culinary wonders of salmon, and the connection it builds with her partner, Ken, and with her community. This very personal anecdote then leads to the much more fact-based passage cited earlier, drawing the reader into multiple communities through fascination with historical context, evoking politics but in a unifying, unemotional way (again, after the appeal to the "felt life" of culinary and communal satisfaction).

There is no doubt that this "made of salmon" passage is meant to create solidarity and community. She says, "This is the way of Alaskans and salmon," in a fashion that surely links all Alaskan communities, and the following sentence is much more potent than meets the eye, because it discusses the idea of sharing. In Alaska, where frontier individualism is a prevailing ideology among the white community, she reminds its citizens that sharing with one another is indispensable. Bob Marshall clung to this idea in his *Arctic Village* (1933) as well, to the point where he considered Alaskan sourdoughs of the Interior to be exemplars of socialism (Cole, "Preface to the 1991 Edition," XX–XXI). Alaskans may be considered, ideologically, rugged individualists in many cases, wary of government intervention, but they simultaneously share with one another, which creates a social-minded atmosphere. This is not government socialism, then, but community plain and simple. This aspect of Alaska's individualism is not discussed nearly as often as the individualist frontier mindset, but it has existed since at least Bob Marshall among the sourdoughs, and recurs periodically.[11] Nancy Lord chooses to give it centerpiece in her work, reminding Alaskans of a not-quite-obvious aspect of themselves, in both white and native communities, that can be important in creating solidarity through this one fish they all love and love to share. Likewise, the tone in *Made of Salmon* is always careful, never condescending, never accusatory, and through this method the animal itself and its relationship with man shines through clearly on each page.

This is where *Beluga Days* both succeeds and falls short. The symbolism of the beluga is clear, and it is important. The beluga, however, represents more than this one animal, particularly as a microcosm of the management of the ecosystem of the entire resource-rich state. Different communities are attempting to be responsible, but *Beluga Days*' most powerful insight is that no one single community lives in a bubble. If all groups act sensibly but do not communicate with other groups, they are not acting sensibly as a wider community. The Cook Inlet beluga population was not declining because one group was misbehaving, but because all groups were not acting in accordance with the needs of the animal: its environment was polluted by industry, it was being taken in small quantities by Native communities, it was taken in small quantities by scientists, etc., and the numbers were simply not sustainable. Not only does Lord succeed in this insight, but she succeeds in approaching all the different mindsets simultaneously: she has developed a technique of speaking to both politically left-leaning, animal-loving communities and politically right-leaning utilitarians at once. This is an aspect described already in

11 Though the Alaskan gold rushes of the turn of the 20th century are often linked to an extension of "The Wild West," as they occurred just after Frederick Jackson Turner's "Frontier Research" (the closing of the West), and authors such as Rex Beach (*The Spoilers*) and Jack London (*The Call of the Wild*) wrote in a masculine manner perpetuating Frontier stereotypes, Robert Service (*Songs of a Sourdough*), among others, noted the community cooperation, particularly needed in getting through the Chilkoot Pass and to the Yukon. James Michener focuses on this idea as well in his work of fiction, *Alaska* (450–570). Michener, James A. *Alaska*. New York: Random House Publishing Group, 1988.

detail, so not worth overanalyzing here, but to recap, by stating her love for the animals, by going so far as to spending multiple years of her life writing about saving their population in the Cook Inlet, but also employing her credibility as a fisherman[12] and mentioning that she is not generally sentimental, this speaks to multiple communities without alienating either and without appearing to be too centrist by actually attempting to accomplish a concrete environmentalist goal of saving a species. Where the book may fall short in comparison to *Made of Salmon*, however, is that *Made of Salmon* includes voices that are obviously much more conservative and utilitarian-minded, partly because, as Lord has noted, all Alaskans are affected by salmon. Beluga are interesting and can attract attention in a "Teddy Bear" fashion in that they can be anthropomorphized and considered "cuddly." But, they are, both physically and symbolically, not nearly as important as salmon. Every community is affected by salmon, including the wealthy whom McPhee notes are generally the only ones who can afford to get away from Anchorage and its ecological footprint (*Coming into the Country*, 10, 25). With such an important, across-the-board symbol, employing across-the-board voices, the same depth of investigation into "The Pebble Mine," for instance, that is employed in understanding the loss of beluga, is not necessarily needed. In fact, such an investigation into a politically contentious topic[13] may even cause a polemic, which would likely drive politics further apart rather than unite them the way The Salmon Project has succeeded in doing, at least in a literary sense.

Most recently, in her 2017 novel *pH*, Lord focused on the pressing environmental topic of ocean acidification, with the disintegration of the shell of the pteropod as its physical symbol. Lord states in an interview at the end of the work that "… I would like, along with my character Ray, for pteropods to be the 'poster animals' of ocean acidification in the same way that polar bears are the 'poster animals' for climate change" (251). In that way, there is no doubt

12 Though the word fisherwoman exists, Lord refers to herself as a fisherman.

13 In 2018, there was a fierce debate, with millions of dollars involved, much of it coming in from Outside ($12 million against and $2 million for), over "Ballot Measure 1" or "Stand for Salmon," "aimed at increasing protections for Alaska's most iconic fish. It would have significantly toughened the environmental permitting process for mines, dams and oil developments in salmon habitat." The initiative failed, but it nonetheless demonstrates a rift between Alaskans over government restrictions in support of salmon. It also demonstrates that Outside interests are still heavily implicated in local Alaskan politics ("Alaska voters strike down 'Stand for Salmon' ballot initiative." By Elizabeth Harball. *KTOO Public Media*. 6/11/2018. "https://www.ktoo.org/2018/11/06/ballot-measure-1-additional-protections-for-salmon-habitats/.").

as to her intention in employing symbolism in the work in order to push an environmental agenda.

In comparison, Matthiessen's *Oomingmak* has no explicit environmental agenda, but it does contain an awareness of the complexity of certain animals in a mostly unknown region to the average person in the Lower 48, particularly in the 1960s, which would have been far enough along after the gold rush and yet before the oil money arrived. Matthiessen does, however, also focus on one particular animal, as Lord does in these works. What is important to remember is the context of the two separate eras. In the early 1960s, merely limning the diversity of animal life and its complexity in the form of the rather exotic musk ox would have been important in the hope that people would find enough interesting aspects about the animal to want to save it, along with its biome. Climate change was not yet at the forefront of concern in the broader scientific community, nor was it at the forefront of academic studies in general, though the late 19th and early 20th centuries had already seen a substantial die-off of various species: the ubiquitous bison was pushed to the edge of extinction, the passenger pigeon was gone, and the Carolina parakeet suffered the same fate, for example (Brinkley, *The Wilderness Warrior*, 9). Therefore, if the musk ox was a symbol for anything, it was the exotic nature of American wildlife that was still out there in the wild to be considered and, hopefully, saved. Meanwhile the 21st century has brought new and pressing challenges that could only have been glimpsed in the mid-20th century. As a result, Nancy Lord has been searching for novel ways to raise awareness for man-made environmental catastrophe. The greatest current risk is that climate change and ocean acidification are in the process of making the planet uninhabitable for a great diversity of species, to the point of mass extinction on a geological scale (Kolbert). With that broad, slow, anticlimactic, climatic issue, the symbolism of a pteropod or a polar bear becomes much more important as a physical representation of an intangible process, whereas the musk ox itself was much more tangible for Matthiessen. Nancy Lord's *pH* lacks that tangibility, as the pteropod lacks the "cuddly" ideal of the "Teddy bear," and she cannot develop a complex relationship with one. Nor can she develop a complex relationship with a single salmon, but the salmon is both tangible and beyond the tangible as an idea. The fish itself is considered delicious and indispensable by all Alaskan communities, many maintaining a physical relationship in subsistence culture with the fish, and yet it also symbolically represents the potential loss of that thing that has always sustained the region's people.

Tom Lowenstein, by contrast, does not employ this symbolism for conservation purposes, but he does indeed examine representations of animals in their duality with the land and other animals. This creates a holistic view of animal/

human/land relations and could potentially lead to a more reflective way of living, with ecological implications.

3 Tom Lowenstein's Whales

The whales of *Ancient Land, Sacred Whale* are almost always viewed through the prism of spirituality and food source with regard to the culture of the Tikigagmiut. The one exception would be the "Introduction" to the book, wherein roughly six pages are devoted to the bowhead whale's biology and migration patterns (xv–xx), though this section follows a discussion on Tikigaq storytelling along with a brief look at the Tikigaq whale hunt. Therefore, the descriptive section on the whale is put in place in order to understand the context of the animal hunt from the anthropocentric perspective. Unlike Barry Lopez, and more poignantly Nancy Lord, Tom Lowenstein does not truly impose symbolism on the whale by linking it, for example, to an environmental agenda. Nor does he directly focus on the whale itself, as the other two writers do, and Peter Matthiessen does in *Oomingmak*. What is interesting from the animal perspective in Lowenstein's work is that the writer, in conveying the nuances of Tikigaq culture through the poetic form, decides to focus on the ultimate respect that the Tikigagmiut themselves have for the "sacred whale" in the process of the hunt. This section will explore Lowenstein's means of respectively linking the Tikigagmiut to the bowhead whale through understanding of tradition before moving along and comparing it to a much more modern work of photography/travel writing by Bill Hess on the same subject. Even though Hess focuses on modern hunting adaptations, he indeed follows Lowenstein in showing the respect the community has for the relationship with its prey.

To begin, like so many other physical objects in *Ancient Land, Sacred Whale*, duality is rife with regard to the bowhead whale. Right from the introduction, the importance of the duality of the whale as both animal and the land itself, in origin stories, is leaned upon heavily. Not only does the bowhead whale form a duality with the land, however, which is extraordinarily heavy in insinuation of its importance to the culture, but another whale, the killer whale or orca, shares a spiritual duality with the wolf. Both mammals hunt in packs (73–4), and they are considered largely a duality of the same person (in terms of personhood). One hunts on land and the other in the sea (the land and sea themselves consequently forming a duality in the work). This is partially because the Tikigagmiut themselves were the rare Alaskan group to hunt inland like wolves for the summers and hunt out on the ice during the winters. For many other groups, this would have been taboo, as they typically considered themselves either land or coastal peoples (72). So, here we see the duality that is "rejoiced in" for a culture that moves seasonally between the two: the land/wolf and the coastal/whale. On the following page, the duality between the two animals itself is approached, as both are hunters, in packs, and both are not unlike humans, who in turn respect them:

> The wolf and the whale in Inuit legend were mystically twinned ... When a hunter ritually slaughtered the wolf, symbolically he killed a whale. The flood of the wolf on TIkigaq nuna presaged next spring's whale hunt, when real whales would be brought to the myth whale's body. (73)

Tikigaq hunters wear caribou skins in hunting for whales, which is taboo among many neighboring cultures for mixing land and sea, according to Lowenstein (73). They also keep wolf skull amulets in their whale-hunting boats. Nonetheless, the point to take away is that the Tikigagmiut and their neighbors have traditionally been keenly aware of the duality of essentially every aspect of their lives with regard to the whale, employing this duality as a form of respect for the creature and for their surrounding world. This idea bears a bit more explanation: if everything is a bit of a mirror, if sea and land animals are considered a duality, as land and sea themselves are, so must everything be. How can we neglect our responsibilities toward others and the environment if all is a mirror of ourselves. Interestingly, this idea falls very closely to Buddhist thought, that each person one encounters is a reflection of the self, all connected to one another (Lowenstein, *Buddhist Inspirations*, 126). In this citation on Buddhism, written by Lowenstein himself, the duality of "male and female, God and the soul" are discussed, as well as that of samsara and nirvana. Therefore, given that Lowenstein has such a spiritual interest, potentially this view is but a prism through which the author views Tikigaq culture. But even if such is the case, it is a perspective worth exploring for its consequent novelty, particularly as it respects the alterity of Iñupiat culture outside the prism of the West, considering the West's insistence on property rights and capitalism (seen in ANCSA), as well as its insistence on Christianity seen by the Russians and Americans alike, to the point of systematization under both Saint Innocent (Garrett) and Sheldon Jackson (Haycox, 184–7).

Aside from this point, however, the two passages presented leave the reader with much upon which to reflect with regard to the idea of respect via duality, even if the impression is merely ephemeral. As we will see shortly with Hess' work, respect for these traditions provides a sense of community as well. And as such, it empowers people to make responsible and novel decisions for the larger community, which leads in a progressive circle when the community is healthy. Alexis de Tocqueville, so often sharp in his observations, said this of individualism: "Selfishness blights the germ of all virtue; individualism, at first, only saps the virtues of public life; but in the long run it attacks and destroys all others and is at length absorbed in downright selfishness" (unabridged version, 104). If this is the case, avoiding government to lean on the abilities and responsibilities of the self being part of Last Frontier ideology, Alaska is in the process of breaking down into selfishness (if nothing changes), which is most certainly not amenable to holistic views of a dynamic environment. Again, this is why the sense of community, and the traditions of the whale hunt (sharing, respect, cooperation), are so important.

A second point to make is that the bowhead whale is revered for its intelligence, as a subject and not an object, offered personhood. Keeping in mind the duality just mentioned, in order for a whale to decide to come to a hunter, the whale must feel respected by the hunter. Sea creatures are thirsty for fresh water, according to both Fienup-Riordan's work (*Boundaries and Passages*, 97, 118) and that of Lowenstein (xxv, 142, 170), and land creatures want sea creatures' oil (Fienup-Riordan, 118). Therefore, when a whale is killed, fresh water is ritually poured into its mouth. Taboo is enforced as well, which means that certain people must follow the advice of their shaman in eating specific ways, in order to appease animal spirits, so that the animal will decide to come to the hunter (Lowenstein, 115–17). Meanwhile, women must stay in the home, performing rituals or acting according to certain taboos and restrictions, so that the whale will come to their husbands, particularly when the women are menstruating (76, 146). All of these actions are portrayed in Lowenstein's work in a way that demonstrates an ultimate respect for the whale, its intelligence, and its importance to their spirituality. Alongside the Tikigaq stories of Raven Man (Lowenstein, 4–9, 21–2, 36–49, 61–70), it is clear that Tikigaq culture views animals and humans all as subjects, as part of the same world and on par with one another in the spiritual realm, unlike Cartesian views of animals in most of the Western world. Lowenstein succeeds in getting this idea across to his Western readers. In this way, the writer is successful in demonstrating respect for animals in his literature.

To follow up on the idea, *Ancient Land, Sacred Whale: The Inuit Hunt and its Rituals* can be compared to another work on the bowhead whale hunt in Arctic Alaska, written six years later, Bill Hess' *Gift of the Whale: The Iñupiat Bowhead Hunt, a Sacred Tradition*. The most noticeable comparison is directly in the title: both authors use the word "sacred" to describe the whale hunt they discuss. Yet, at least ostensibly, the similarities end there. This is to say that the two authors employ almost opposite means of expressing the whale hunt in roughly the same era, but they both ultimately demonstrate the "sacred" nature of the whale hunt and how respect for the whale brings Arctic communities closer together.

As mentioned earlier in this section, Lowenstein discusses the tradition of the Tikigaq whale hunt, employing stories told to him by Asatchaq, a village elder (91–8), as well as using his own experiences in the community, as well as the work of ethnologist Froelich Rainey (xiii, 185). Lowenstein's book is meant to bring failing tradition back to life through connecting the community with its roots and explaining the importance and subtleties of the tradition to the world Outside. By contrast, Bill Hess is not a poet/ethnographer, but a photographer who also writes. Hess' enormous advantage was that, though he was not an Iñupiat, or even an Alaska Native, he spent much of his life among the Lakota and Apache tribes in Utah and Arizona, which gave him an almost automatic respect within the Iñupiat community of Utqiaġvik (formerly Barrow), something not generally afforded to the white community (15–6). This can be compared to the advantage Nancy Lord tends to have going into various Alaskan communities as a longtime fisherman. But, Hess is not interested in enlivening past tradition. Rather, he is attempting to display the living hunt in its current iteration. I have already argued in favor of allowing Arctic and Native cultures to adapt in a self-determined manner, and *Gift of the Whale* shows such adaptation in the modern world. The book challenges Western ideas of the Noble Savage, yet Hess indeed allows the nobility of Inupiat culture to be demonstrated on its own terms. As Lord and McPhee have both employed the 55-gallon drum to describe Alaska's aesthetic (*Lord, Rock, Water, Wild*, 3–4; McPhee, *Coming into the Country*, 380), Hess has employed a similar technique. Though this research is not a study of photography, Hess' photos show a people who are materially poor (84, 103, 137), using whatever is at their disposal to exist in a harsh climate, including exploding harpoons and snow machines (30, 33, 52, 201, 208–9). However, the photos also tend to show people smiling and cooperating with one another, creating community in their relationships to their land and to their hunt (42, 192–5, 204–5, 212–26). Whereas photos of dead whales lined up on the ice or enormous piles of cut-up whale fat may seem anti-environmental (96–100, 106–7, 190), there is always a social relationship evident (174–5, 179, 181–91) that strengthens bonds and understandings of the wider environment (212–26). Such is the prescription given by Susan Kollin in working with Alaska Natives. She also suggests that writers focus more on subsistence than on the beauty being lost in a tragedy such as an oil spill, as subsistence culture is more closely aligned with sustainability of ecosystems than, say, ecotourism, which comes and goes and has a tendency to focus on aesthetics (glaciers, whales, mountains, etc.). Bill Hess indeed served as a journalist for the now-defunct *Tundra Times* (Hess, 2) which Kollin suggests was an exemplar for responsible journalism during the Exxon Valdez oil spill (4–5).

As a strong example of the multiple facets of life and ideology affected by the communal nature of the whale hunt in Hess' work, there is a section where a young man is described graduated from the local high school while his father is in prison and cannot attend. The young man then joins the US Navy, comes home and works a boat crew with his father, and even convinces community elders to stop letting waste from hunts drift into the ice now that waste has become plastic and aluminum. This history leads to a paragraph about the long history of whaling in the family and their frustration with Outside laws not abiding by the respect local Alaska Natives have for the hunt. Finally, to drive home a primary point, the young man is quoted discussing the father's tendency to "black out" during the hunt in what is described as something akin to spiritual union:

> By all appearance, Percy was functioning as normal. Yet he had no physical awareness of himself or his actions. He did not come to until he found himself on the ice, butchering the whale. Percy concluded he had died with the whale and then come back with it. (178)

This excerpt from *Gift of the Whale* discusses communal and environmental responsibility, discrepancies with Western law, the importance of family, Inuit spirituality and its connection to the whale hunt, community, etc. It is typical of Hess to describe a significant communal situation, such as a young man's high school graduation where the young man is described as ceremonially walking beneath baleen arches, while his father is in prison for shooting a whale out of season, a law created by the Western world for a problem the industrial Western world itself created, which is out of line with Iñupiat traditions. Hess so respectfully gets these points across without explicitly saying such in the first person. Instead, he lets moments, photographs, and stories of others' lives fill in the chiaroscuro so often left out (not unlike John McPhee). And further, the passage shows how the responsibility given to the young man allows him the personal responsibility to effect environmental change, which interestingly leads to a literary pivot toward the spiritual aspect of the hunt, the author intertwining the two, the ecological and the spiritual, as so many Native cultures naturally do. Hess never shies away from the gruesome in the hunt, either. However, without sublimating or romanticizing it he allows its details to connect to an utterly other sense of spirituality (from the perspective of the West). Every time the hunter's father took a whale, which means to be chosen by the whale, the man blacked out, died along with the whale in a spiritual duality. This short paragraph respectfully limns a personal, spiritual moment in a way that, for a travel writer, gives that person, that family, and that community control over its own narrative. Furthermore, this is done in a subtle manner that may have never occurred to the writer to even attempt to accomplish had he himself never spent so much of his life among Native American communities.

To pivot slightly away from this specific passage, the reader is offered a unique cultural view through the story of Hess' interaction with the hunt, as the author is shooting photographs while the others are shooting guns. Hess is accepted by the people. He is also expected to work with the group, and, importantly, keep up physically, to the point where in one episode he thinks he is going to die out on the ice. This was a test, and having passed, the author is accepted more closely into the community than before, due to his resilience (23–32). Again, what may seem borderline criminal to Western society (an actual life-threatening test) is an opportunity for the author to prove his mettle and earn his own respect in the eyes of the community, as well as be put into a situation of humility compared to the abilities of Alaska Natives. In this way, said life-threatening situation becomes a blessing, though it could have just as easily turned for the worse. Hess is at first upset, but he eventually does see this event as a blessing, as their laughing at him makes him laugh as well. "I started to laugh. They laughed even louder. This made me laugh harder still. It hurt terribly, to laugh in that condition. Still, it was good to laugh" (32).

He also displays scenarios that are more directly environmental in nature, more directly linked to animals and involving what can seem to be paradoxes. For instance, the most blatant example is a lengthy section in the work describing two whales who become trapped in a water hole, on the ice floe, unable to find their way back to open sea where they may breathe freely in open water (110–37). The entire event becomes national news (126, 133), and the local hunters go out of their way to aid the whales back to safety (113–4, 121, 122, 125, 134–5), even if the whales are likely to die anyway, having serious lacerations from the ice in closing air holes (120, 123, 127). In doing so the hunters offer the whales support as persons in the local community. Certain hunters go out to talk to the whales, to soothe them in their straining moments stuck at a particular hole (111–29). Such an event contrasts against the hunt itself, which discusses and shows photographs of exploding harpoons, of these same men quite literally slicing whales to pieces (179–93, 201–3). The addition of the trapped whales was obviously a choice to show the dynamism of thought toward whales, and that the whale hunt truly is sacred, even if Westerners cannot see the sacred nature in what can be viewed as brutal murder. And, not only that, but the communal nature as described in the lengthy passage above serves to keep people connected to the roots of their worldview. Overall, however, describing the gathering of birds' eggs (50–3), the hunting of seals (9–17), the hunting of beluga (82–109) and bowhead whales (154–209) are all described and photographed with tact. Arctic hunters are offered nobility on their own terms, just as Lowenstein's hunters are, while Hess displays modernity, and Lowenstein reinvigorates near-bygone tradition. The two works, then, support one another, describing two aspects of the same, a temporal duality that seems fitting for the community.

If Lowenstein and Hess both avoid symbolism in a search for the sacred relationship between man and whale, Matthiessen directly discusses the musk ox for the means of research on the animal itself and its understanding by the broader human community Outside. There is no subtlety, no nuance, no community relationship, nor is there overt environmental symbolism with regard to discussion of the animal. The work is, rather, a concise piece of entertainment in corralling a certain number of musk oxen helicoptered from Nunivak to Fairbanks to be studied, while offering a wide range of facts regarding the musk ox itself and its behavior.

4 Matthiessen's Means of Animal Communication

Matthiessen's *Oomingmak* is, as has been mentioned, a short, concise book on the musk ox of Nunivak Island in Alaska. As in *The Snow Leopard*, Matthiessen is asked to go along on a voyage to an exotic land of tribal peoples to perform biological research on an exotic animal about which the average person knows little. The snow leopard was, according to Matthiessen, "that rarest and most beautiful of the great cats," (1–2), living high in the Himalayan mountain range, in places very few Westerners had ever been and, notwithstanding its notorious elusiveness, it apparently nonetheless remained an extraordinary creature ("GS knew of only two Westerners – he was one – who had laid eyes on the Himalayan snow leopard in the past twenty-five years" – 1–2). Except, the snow leopard was not even the ostensible goal of George Schaller's scientific expedition that led to the work *The Snow Leopard*. Rather, the goal was better understanding the also-rare "Himalayan blue sheep" and the biological links that that animal may create with sheep and goats. Likewise, relatively little was known about the musk ox at that time, as it lives only in certain corners of the Arctic (in parts of Greenland, Canada, Norway, and Spitzbergen, and on Nunivak Island, in Alaska – *Oomingmak*, 28–9). This is because the musk ox, once more widely dispersed, was hunted to near extinction by whalers, not to mention Peary's expeditions (28). Writing a book about the musk ox, then, gave Matthiessen the opportunity to discuss various facets of the quite rare creature, whose biological provenance academics could barely agree upon (Matthiessen says that it appears bovine, but is actually related to both sheep and goats, and its scientific name is ovibos moschatus, meaning "'sheep-cow musky,' though it is neither sheep nor cow, nor has it musk glands" – 27–8).

This scientific expedition also gives Matthiessen the chance for a travel writing adventure. This section will first explore *Oomingmak*'s "Western" genre narrative as a technical decision, and its implications on landscape and animals. Next, the decision to employ the name *Oomingmak* will be examined for its offering at least part of the narrative to the Alaska Native community, and to the animal himself. Finally, the work will be put into perspective with regard to its place in the mid-20th century environmental movement, and how its chronological place affects the work, how the work should be viewed with regard to the moment in which it was written. Indeed, environmentalism has had multiple priorities in different times and different places, so that the current lens should not be the one to scrutinize a pre-1960s work of what could be considered "nature" or "environmental" writing.

To begin, *Oomingmak* is based around the musk ox, with in-depth description of the animals and their habits, but particularly when in defensive mode against predators, and that is because the whole reason for being on the island discussing the musk ox is that the scientist, John Teal, wants to figure out how to corral some calves without hurting their parents, taking them back to be raised domestically at the University of Alaska Fairbanks in the center of the state as an experiment, as the musk ox has rare, warm, and potentially profitable wool (10–1). The adventure, then, comes across as mildly of the Western genre, with men chasing around surly musk ox instead of cattle. In fact, they struggle to herd the animals, so they eventually hide in a line and wait for an airplane to herd the animals toward them so that the men could finally trap them:

> … where the lava ridge tumbled down into the water, they were met by a line of yelling men, jumping up like goblins from the rocks. The animals wheeled, but the gate was closed. We ran down the hill behind, swinging the line into a semi-circle and cutting off retreat along the lake edge. (41–4)

Without this chasing, with numerous failures leading to more creative thinking, there would have been no actual narrative, and the book would have been mostly scientific in nature, a much different work. This passage shows an interesting perspective on the Last Frontier, as an extension of The West, with cowboy-esque maneuvers and helicopters added, along with a more exotic animal than the bull and a modern desire not to harm said animal, either psychologically or physically. This extrapolation of cowboy culture into Alaska is interesting in that it continues with the theme of the exploration of the Last Frontier, so that we may examine where the similarities actually end. In Frontier, or cowboy, culture, even in the 20th century, pickup trucks are the generally the most modern machinery. McCarthy's *All the Pretty Horses* (1992),[14] for example, looks to Mexican ranching, which is almost entirely lacking in discussion of modern technology. Indeed, throughout much of the 20th century, the Western film genre, seen through the works of John Ford and Clint Eastwood, for example, looked back toward the 19th century, toward the American Frontier. All of this is to demonstrate the appeal of a traditional view of the American West as a low-tech Frontier. Alaska, however, has had a different relationship with its Last Frontier culture. Though the low-tech American Frontier has indeed been extrapolated onto it in many ways, conquering oil and homesteading, for example, the airplane has long been the symbol of Alaska (Coates, 56). Anchorage has the largest float plane airport in the world,[15] and when the politician in McPhee's *Coming into the Country* discusses moving the capital to the foot of Denali, he suggests the construction of a float plane airport at the door of the governor's mansion (128). The Gelvins, of Coming into the Country, get around by plane, as there quite literally are no roads to many places in Alaska, and planes are a necessity. Therefore, in *Oomingmak*, the plane pilot is the one who interacts negatively with the local Alaska Natives, and helicopters are brought in to corral the musk ox. This distinction is not unique to the helicopter or the plane. Alaska, due to its size and distance from the Lower 48, as well as its lack of suitability for cattle farming, though Vilhjalmur Stefensson famously tried (!) (Coates, 35–6), large machinery is often adapted. Again, the Gelvins bought a D-9 Caterpillar tractor and drove it all the way from Fairbanks deep up into the Interior, in a memorable scene of McPhee's work (205–8). However, there is indeed a sense that Alaska's Last Frontier allows for some appreciation of wildlife not seen in "The American Frontier" of "The West," where horses and steers are corralled, but where most other animals are considered pests attacking their livestock and farms. Alaskans, by contrast, though they have had an antagonistic history with wolves

14 McCarthy, Cormac. *All the Pretty Horses*. New York: Alfred A. Knopf, 1992.

15 "Lake Hood Floatplane Base Walking Tour." *alaska.org*. http://www.alaska.org/guide/lake-hood-walking-tour-a-look-around-the-worlds-busiest-seaplane-base. Viewed 5/6/2019.

at times (Lopez, *Of Wolves and Men*, 195), have always prided themselves on living around animal abundance, including bears, moose, salmon, etc. For instance, moose still wander daily around Anchorage, bears around the city are protected, and salmon are adored by all, as we have seen (Lopez also notes that the bounties did not affect wolves in Alaska the way they did in the West and in Canada – *Of Wolves and Men*, 195). It is no surprise, then, that these men in the passage above are involved in a scene invoking cowboy culture, yet they seem to also display compassion for the animals in the process. That is not to say there is no animal compassion in the West, but rather that it is among the primary considerations in Alaska.

Comparatively, Matthiessen's *Wildlife in America*, of roughly the same epoch of his career, succeeds through its "on the road"-style appeal, as the writer travels throughout different parts of the country describing interesting fauna. Matthiessen indeed had a long and illustrious career, spanning over half a century and crossing many genres, including fiction and non-fiction. Yet, some of his later work lacked the narrative that made much of his earlier non-fiction work successful. *End of the Earth* (2003), for example, almost inherently keeps the reader's attention due to its subject matter, as it takes place at the literal end of the earth: Antarctica. Yet, there is no adventure to be had, as in *Oomingmak*, and there is no spiritual search, as in *The Snow Leopard*. *Tigers in the Snow* (2000) could be considered of the same strain. Like *End of the Earth*, there quite frankly is no narrative arc, descriptions of the Siberian tigers themselves being the major draw. Again, comparatively, John Vaillant also wrote on Siberian tigers, describing them as Matthiessen did but adding the narrative of the Russian authorities in charge of hunting down tigers once they become man-eaters, as well as adding interviews with neighbors of a man who is eaten. All of this is to note that Matthiessen, who would employ various means of narratives for specific reasons, chose to focus the bulk of *Oomingmak* on the chase itself, which gave the work an extra dimension while simultaneously allowing the author the opportunity to add discussions on the highly unique biology of the creature.

However, one aspect apparent in both *Oomingmak* and Vaillant's *The Tiger*, which could potentially be criticized from the 21st century standpoint is their "Western" or "Cowboy" attitudes, in a predominantly patriarchal strain that could be considered overly-industrial or -technocratic, lacking in "felt life," as Nancy Lord has said of Western Culture (*Fishcamp*, 70). For instance, Vaillant's Russian authorities are offered a great amount of respect in the work, for both their bravery and for the respect they generally give to the Siberian tiger, but they nonetheless ride around the taiga in fortified 4X4s that essentially amount to primitive tanks carrying high-powered weaponry (5–18).[16] Likewise, Matthiessen never asks the question of why we should even attempt to domesticate such an animal as the musk ox. The animals are terrified during the expedition, as well as when they are flown in helicopters to Fairbanks (74–5). The conquering of Alaska, as a frontier not unlike that of the West, comes into play in this scenario. The animals are there to be domesticated, conquered, for industrial purposes, in this mindset, which again, Susan Kollin links to the idea of empire (*Nature's State*, 19–20). Viljhalmur Stefansson, through his boosterism of "The Friendly Arctic," also advocated the importation of cows to replace caribou, while downplaying the difficulties in employing standard pipeline-building techniques in Alaska, neither resulting in success (Coates, 35–6, 74).[17] These are potential criticisms, but ultimately, both Matthiessen and Vaillant, broadly speaking, approach their work with humility in the face of such noble creatures, and that fact potentially offsets such lingering criticism, or at least offers more complexity and nuance to the narratives. These decisions, these potentials for criticism, may be part of the reason why Matthiessen decides not to employ narrative arc in some of his late works, such as *Tiger in the Snow* and *End of the Earth*. Interestingly, despite its fascinating narrative, *Oomingmak* is an often-forgotten work of Matthiessen's, possibly due to its novella length. The fact, however, that the author was able to create such an active adventure while also discussing the animals themselves at length in such a relatively short book speaks volumes, and the work potentially merits a re-evaluation now that the author has recently passed away and his prolific writings are no doubt being poured over by scholars.

Yet, how, more particularly, does Matthiessen describe the musk ox in *Oomingmak*? To begin with, the author indeed entitles his work *Oomingmak*, which means "the Bearded One" in the native Cup'ig language (25), immediately and prominently relating the animal to the community of Mekoryuk on Nunivak Island, whose inhabitants both fear and respect the creature. Matthiessen says that Oomingmak himself is little understood, with some people calling him stupid and others intelligent, some dangerous, others timid, some delicious, others inedible, and the duality apparently continues regarding this animal person (26–8). Yet, as examined earlier, Alaska Natives are notoriously close watchers of the natural world, with stories based around the dynamism of

16 Though a counterargument could be Vaillant's restrained-but-nonetheless-always-present strain of mysticism (98–133). In this case, the example would be discussion of the spiritual connection between man and tiger, even evoking Dersu Ousala (30–3).

17 Stefansson, Vilhjalmur. *The Friendly Arctic: The Story of Five Years in Polar Regions*. New York: The Macmillan Company, 1921.

Raven Man in particular, but also with bears, seals, and other animal persons. Oomingmak is indeed revered for his intelligence among the natives of Mekoryuk, who would know from observation, and therefore, despite the fact that Matthiessen offers little space to Alaska Natives in this work in general, the titling of his book in honor of their name for the creature serves as quite a respectful nod to this reverence.

Earlier in this research, I noted Raban's discussion of Wittgenstein. Evoking the philosopher, the writer says, "The world we live in is the words we use," (Raban, 151–2). If we translate the name into English, it becomes just another creature we see in a zoo and no longer truly appreciate. Matthiessen could have prominently employed the term "musk ox," or he could have jokingly titled the work Sheep-Cow Musky. He could have also titled the work in reference to the corralling and chasing of the animals, which make up so much of the action, but he chose to stick with Oomingmak. It could be argued that the choice was purely made from a marketing standpoint, but whether or not this is the case, the ultimate result is that the literal animal remains at the forefront. The chase of the animal becomes secondary, and direct control of the narrative remains in the hands of the Cup'ig of Nunivak Island. This is an important example of the power of language.

However, the name "Oomingmak" is not only a nod to Alaska Native language. It also carries a connotation with much more weight of respect, elegance, and nobility through exoticism to the West than the familiar sounding "musk ox," particularly, as Matthiessen notes, considering that the musk ox neither has musk glands, nor is it an ox. The name "musk ox," then, is deceptive. It is also likely based around the notion of whatever Western explorers and scientists found similar in terms of their own experience. For example, Barry Lopez dedicates an entire chapter to the musk ox in *Arctic Dreams*, and that author is left in awe at the ability of these creatures to go into an almost zen-like meditation, or what appears to be a "standing hibernation," in winter. They will be completely covered in snow, and their protective pelage and lack of motion will leave them perfectly healthy (66–8). Rather than evoke this unique and exotic ability, the word "musk ox" elicits a common animal of subtropical temperatures.

In directly focusing on Oomingmak himself, is Matthiessen attempting to employ symbolism, as Lord and Lopez did? Most likely not, as the story is much more closely centered around the idea of biological experimentation than around conservation or some other overt agenda. There is indeed the idea that these are the last musk oxen in Alaska, but as Matthiessen notes, it is difficult to reach Nunivak Island (8–14), and the Natives leave the animals alone, so that their numbers on the island are not truly threatened (at least as of the writing of the work just over a half century ago). Rather, it is more plausible that the work was written in the vein of adventure in the sense that Matthiessen wants to convey the rich cultural diversity of the America he has come back to after leaving France. This argument is particularly poignant throughout his work of the same era: *Wildlife in America*. One could say that both of Matthiessen's works, not unlike that of Lopez, allow the existence and complexity of the animal kingdom to serve as an argument in favor of conservationism. Therefore, Oomingmak certainly is a symbol, but it is more likely, or at least a heavier tendency, that *Oomingmak* the book serves as an interest story and adventure. In other words, Matthiessen shows Americans how beautiful and exotic their own country truly is, with both *Oomingmak* and *Wildlife in America*.

5 Do Animals Actually Speak in These Works?

If animals have not gained agency in the past almost century, then have they become but symbols employed for other purposes, described in relation to humans, or do the authors find subtle ways to allow them to speak? Essentially, the answer is that none of the primary works of this research allow the voices of animals to come through. In order to explain why, it is important to define what speaking actually means, as no animal communicates in a linguistic tongue directly comprehensible to human beings. Voice in this case can mean interaction and understanding, however, which is why the term "linguistic tongue" is being employed in lieu of "language," as tongue requires a specific physical ability and "language" can be considered beyond words.[18] Take for instance the most obvious cases in Alaska writing: *Stickeen* by John Muir and Jack London's *Call of the Wild* and *White Fang*. All three of these works allow for actual communication between human and animal, as opposed to the descriptions of animals in all of our primary works.

Jack London does not exactly discuss an extended amount of communication between the two species themselves, but he allows the dogs to become his protagonists and for certain human beings and dogs to understand one another through close interaction (*Call of the Wild*, 107–21).

---

18 Merriam-Webster.com indeed defines language first as: "the words, their pronunciation, and the methods of combining them used and understood by a community." Yet, the third definition reaches beyond words: "the suggestion by objects, actions, or conditions of associated ideas or feelings." Even closer to the point at hand, the fourth definition says: "the means by which animals communicate." Viewed on 20/4/19.

In one part, a man named John Thornton saves the protagonist, a dog named Buck, and the two develop a friendship (108–9). London allows man and dog to communicate more fully through action than with words, such as the playful biting of Thornton's hand, and even an anthropomorphized moment of Buck's "mouth laughing, his eyes eloquent, his throat vibrant with unuttered sound." Thornton communicates with compassion, which is universal communication. Even when Thornton, "tak[es] Buck's head roughly between his hands," the compassion is obvious to the dog. This relationship is very respectful between the two, as London notes that so many other dog owners only take care of the animals from "a sense of duty and business." Yet, from the perspective of this research, there are still many doubts from many communities about how a dog should be viewed. London says Thornton was Buck's master; though an ideal one, he was still a master. From the animist perspective the dog should be considered a subject, part of the community. The counterargument to the enlightened Native American perspective given here, particularly with Arctic peoples, is that many traditional Inuit communities kept dogs as a means of transportation, and as a source of meat in case of starvation. The merits of that particular relationship could be argued endlessly, but the point is that the relationships which different peoples have with the natural world are often complex and not easily reconcilable with the expectations of one another. Within the context of this passage, then, Thornton indeed comes across as an enlightened caregiver, searching to communicate with the animal. London shows much humanity in such a discussion, and an admirable attempt at crossing what seem to be ontological boundaries for Western cultures into the viewpoint of the dog. The perspective offered to Buck in his relationship with Thornton gives the dog personhood, and gives man an admirable model upon which to treat animals.

Except, the problem remains that even in the introduction to the combined works currently under the title *The Call of the Wild, White Fang, and Other Stories*, James Dickey admits that London's actual understanding of the behavior of dogs and wolves is somewhat weak (8, 13). Theodore Roosevelt agreed and made his criticism known publicly (Brinkley, *The Wilderness Warrior*, 308, 781–3). Dickey also notes, however, that London's understanding of canines and their relationship with man and each other was absolutely primal, his relationship with the animals being cultivated in his own mind (12–3). So, therefore, what the reader is offered is a human's perspective of a dog's perspective, with that human being still limited not only by ontological barriers but also by the lack of close, prolonged study, the lack of a Native Eye. Yet, the idea is novel in how it presents perspectives, and the stories themselves have been powerful enough to have been read and admired for many generations.

In *Stickeen*, by contrast, Muir does not attempt to understand the canine protagonist's thinking, though he does indeed portray the dog in an anthropomorphic sense. The dog is considered wise and understanding of John Muir, but this line is walked considerably well because Muir explains how the dog understands him, offering the dog characteristics of nobility and dynamism, without the author ever anthropomorphizing the dog too completely (7–13, 36–43). For instance, Muir begins the short story saying that his friend Samuel Hall Young came aboard a boat to study glaciers, "… followed by a little black dog, that immediately made himself at home by curling up in a hollow among the baggage. I like dogs, but this one seemed so small and worthless that I objected to his going, and asked the missionary why he was taking him" (5). On the following page, Young responds to Muir,

> But his master assured me that he would be no trouble at all; that he was a perfect wonder of a dog, could endure cold and hunger like a bear, swim like a seal, and was wondrous wise and cunning, etc. making out a list of virtues to show he might be the most interesting member of the party. (6)

This bit of foreshadowing proved itself to be true throughout the rest of the story. However, if Young discusses the dog as a daring and capable companion, Muir offers Stickeen a sort of proto-personhood:

> On our trip he soon proved himself a queer character—odd, concealed, independent, keeping invincibly quiet, and doing many little puzzling things that piqued my curiosity … he spent most of the dull days in sluggish ease, motionless, and apparently as unobserving as if in deep sleep. But I discovered that somehow he always knew what was going on. (7–9)

The description goes on for several pages, discussing Stickeen's stubbornness, his "equanimity," his ostensible laziness mixed with his courage and capabilities, and his incomprehensible provenance as a mutt, which only add to his enigmatic aura. Proto- is employed in this case as the idea of personhood had not yet penetrated the Western conscious, particularly with the pervading mode of thought being to sublimate "Indians" to Christianity, even in John Muir's writing (*Travels in Alaska*, 297–309), but also because personhood is offered to Stickeen in a sort of halfway, or almost backhanded, manner. Indeed, Stickeen is described as having many characteristics that a human person might have: courage, stubbornness, independence, athleticism, hunting ability, curiosity, etc. Yet, all of these characteristics are described in surprise, as if a dog should not be able to possess these characteristics, should remain a human companion to do a human's bidding, to be trained and adapted to the human "master's" needs. Muir seems to delight in Stickeen's independence and capabilities, however, and the two become inseparable companions to one another. By the climax of the story, Muir and Stickeen adventure as equals, even if Muir has recently attempted to prevent the dog from coming along, worrying about the dog's safety, and even if Muir still discusses Young as the dog's master (18–20). In pages 40–8, Muir crosses a ledge which Stickeen the adventurous animal refuses to cross. Muir describes in detail the dog's intelligent refusal to jump, and finally his gathering of courage, his look of knowing what danger lay before him and his jump to safety. When Stickeen was published in 1897, there had not been anything like it in terms of human relationships with animals in Western culture. There was a unique understanding between the two protagonists in this story, presented as equals at times in a friendship, and Stickeen's comportment, psychology, etc., come through completely, to the reader and to John Muir. There is no need for literal speech between the two. The friendship is affected in other ways. Yet, John Muir allows Stickeen to speak through his close description of their relationship and of Stickeen's personality and physical manner of being, particularly in tough situations.

How can we compare the depictions of animals in John Muir and Jack London, then, with those of this research? For one, unlike the imaginations of Jack London, Stickeen is accepted as having actually happened, being non-fiction, particularly as Muir's close friend and character in the story, Samuel Hall Young, later verifies the friendship between the dog and the man (*Alaska Days with John Muir*, 61–75). In this way, while London is novel in his attempts at reaching into an animal's mind, Muir quite literally allows an animal to speak in a non-fiction adventure story, through direct action, emotion, and comportment. On the other hand, Lopez, Lord, and Matthiessen allow animals to speak in an abstract manner, by describing the intimacies of their lives, their habits, their movements, their food sources, etc. None of this macro biological information is available in Stickeen, which is not about the habits of a particular species but about the personality of an animal. Broader study of a species is, however, a topic in London's works, as the psychologies of multiple canines are limned, but that writer's descriptions are not truly based on fact. The primary works of this research, then, allow fact-based descriptions of species to speak in a secondary manner. What Muir and London have accomplished is, indeed, important for introducing new methods of thinking about the dynamism of animal perspectives, and yet, equally important is the unravelling of inner lives of animals through a process of observation, which several of the authors of this research have also accomplished.

For most of the 20th century, as mentioned above, scientists felt uncomfortable with the anthropomorphizing of animals (Kimmerer, *Braiding Sweetgrass*,[19] 39–47; Lopez, 250). Animal intelligence has often been downplayed, as well as stories about the intelligence of animals by Native Americans and First Nations peoples. Yet, when science, with its strict method of observation and proof, cannot truly validate an idea, story may step in and fill the gaps.[20] Stories are not bound by data collection. What they can do, particularly in the cases of many of these travel writers, is to allow for intimate and deep connection through observation, in the way of the Alaska Natives in their patient hunting, for example, and to tell about those experiences in a way similar to how their authors have learned to see from The Native Eye perspective. Lopez has expressed this idea quite clearly. In this way, the sense of intelligence of the Raven, the musk ox, the polar bear, etc., is conveyed to the reader gutturally, through close observation, relationships and personhood, free from data and proof through the scientific method. One perfect example of this is a story Knud Rasmussen tells in a short story entitled "The Bear in the Ice Hole." Rasmussen was in the process of hunting the bear with his dogs when the bear reared up and purposely slammed his paws down, breaking the ice in order to have the advantage of the water. Rasmussen was ready for this maneuver, noting that he had seen it before but still watched it with respect, but Rasmussen then accidentally fell through the ice into the freezing water alongside the surly polar bear. The author feared for his life in a moment of vulnerability with a creature known for a particular capacity for swimming in Arctic waters.

---

19 Kimmerer, Robin Wall. *Braiding Sweetgrass: Indigenous Wisdom, Scientific Knowledge, and the Teachings of Plants*. Minneapolis, Minnesota: Milkweed Editions, 2013.

20 Robin Wall Kimmerer, a Native American, has discussed this idea thoroughly in her work *Braiding Sweetgrass*. For Kimmerer, a professor of Botany by trade, the rigid science she was forced to learn at university, knocking poetics out from her mind, was compared to "the Indian" being knocked out of her grandfather ("Asters and Goldenrod," 39–47). "… it was happening all over again, an echo of my grandfather's first day at school, when he was ordered to leave everything – language, culture, family – behind. The professor made me doubt where I came from, what I knew, and claimed that his was the right way to think. Only he didn't cut my hair off" (41). In the same essay, Kimmerer discusses how she eventually came back to poetics, to "the indigenous," merging the two. Interestingly, here, Kimmerer equates what Lord calls "felt life" with "the indigenous" (47). This is not exactly The Native Eye, but it does represent a different relationship with the world from that of the technocratic West, one which enters into linguistics as well, as the author discusses in her essay "Learning the Grammar of Animacy" with regard to how her verb-heavy Native language, Potawatomi, allows for the subjectification of all things, animate and inanimate, unlike the noun-heavy English language which tends to objectify, as one of her characters says, "give[ing] us permission to disrespect nature" (48–59).

The bear, meanwhile, was menaced by a pack of dangerous, snapping dogs. Rasmussen says this of the situation:

> I, who had been accustomed only to kill, had never before known that a bear's eyes could be so expressive. At first I saw only fear and anger in them, but as it gradually accustomed itself to me in the same way as I had accustomed myself to it it stopped showing its teeth. I now regarded it even more attentively than before. And it struck me that I no longer looked upon it as a piece of big game to be killed, but as a thinking and intelligent creature that was in the same distress as I was. It was almost as if I could see its thoughts take shape. With its eyes alternately on the dogs and on me, it seemed to be wondering why I too had jumped into the water. It knew now that I meant it no harm. (Rasmussen, "The Bear in the Ice Hole," 14)[21]

Eventually, Rasmussen calls his dogs off, and the bear, who could have easily killed Rasmussen if he had chosen to do so, swam away.

This could be considered an anthropomorphic story and potentially dismissed offhand from the scientific perspective. It could also be read for the depth and compassion in the polar bear, long respected by Arctic peoples. Rasmussen gets across in story form that the polar bear is not just a number, a machine without complexity who is statistically likely to do one thing versus another. On the following page, Rasmussen even goes so far as to suggest that the bear looked at him in the same way he looked at his own dogs as he petted them. There is complexity in this story, personhood, in the bear itself. Bears, then, are represented here with personalities as complex in their personhood as those of humans. The protagonist of Werner Herzog's *Grizzly Man*, Timothy Treadwell, also wrote a book entitled *Among Grizzlies*, in which the author describes the individual personalities and complexities of the lives of grizzly bears in Alaska. Treadwell does not come across as mentally stable in Herzog's film, as he is shown with moods varying wildly, potentially suffering from bipolar disorder, but the complexity of animal life in *Among Grizzlies* is clear and sharp. The animals may not speak the pure way they do in London's works, but Treadwell names them, and discusses them as friends with individual personalities (27–47), which is a rare and daring feat to have accomplished with such dangerous animals. Therefore, the reader is offered a complex view of the grizzly in Treadwell's work, itself a kind of exotic travel writing to bear

21 Rasmussen, Knud. *The Bear in the Ice Hole*. Translated by Reginald Spink, edited by Bent Rying. Copenhagen: Royal Danish Ministry of Foreign Affairs, 1962.

country, which reinforces Rasmussen's experience. These stories are as rich as any scientific method research. Rasmussen, however, knew both worlds and took care to publish data, ethnological observation, and non-fiction (*Across Arctic America*). All have their importance in trying to purvey a dynamic and complex view of individuals, peoples, and animal communities, not to mention landscapes.

The hope is that, having read this chapter, the reader will take away the important nuances of discussing animals respectfully. Respectful discussion can entail multiple literary approaches, including symbolism. Likewise, personhood and subjectification are key to respectful representation, particularly when mixing them with fact-based observation, though fiction has its own merit in expanding perspectives. Respectful discussion, however, is not just about being respectful to different peoples, which can indeed be considered its own end. Potentially more important is a sort of trickling effect that may begin with the way subjects, objects, and ideas are communicated, consequently blending into all aspects of life in unforeseen ways, particularly in viewing nature with more dynamism than it often currently is, which is one of the primary points this chapter tends to lean toward.[22] Also, this chapter quite poignantly wraps up a holistic cluster of four chapters, which all have been inextricably connected, affecting one another. Unlike these four chapters, however, the following chapter has no seamless link to the preceding one, other than the late discussion of Kimmerer's ideas on changing viewpoints affecting politics and economics, which will indeed be explored. This research would be remiss, however, if it were not to discuss the concept of liberty, as it is incredibly important to Alaska, to the Last Frontier mentality. It could, therefore, be considered almost as an addendum to Part 1. Yet, in the sense of connection between Alaska Natives, sourdoughs, women, animals, landscapes, etc., liberty is a concept which could also be considered a bow wrapping together the rest of the ideas. There are questions of liberty in who has the right to hunt which animals and when. There are questions of liberty in whether or not animals have the right to live in a human-dominated space. These questions were never part of the traditional Alaska Native worldview, but they have been imposed by the industrialized world Outside, by the sheer rise of the planet's population.

---

22 Kimmerer's work discussed above is absolutely rife with defense of this point. One of the primary ideas discussed in *Braiding Sweetgrass* is that discussion surrounding nature can affect politics, economics, physical farming, spirituality, academics, and social mores.

FIGURE 8 Steese highway
PHOTO BY CHRIS FERGUSON

CHAPTER 7

Liberty

It would not be a stretch to say that many people come to Alaska for the liberty that it offers. But, what exactly is liberty? Influential billionaire Charles Koch often discusses liberty,[1] conflating modern social and economic ideas and equating liberty with the classical, economic definition now known in the US as classical liberalism or economic liberalism.[2] In the US, liberty has also often been linked with Thomas Jefferson's line in the Declaration of Independence: "Life, Liberty, and the Pursuit of Property." Except, the Pursuit of "Property" must have seemed overly avaricious, so those words (influenced by John Locke's Second Treatise on Government[3] – Ellis, *American Sphinx*, 65–9) were revised to "Life, Liberty, and the Pursuit of Happiness" and sent off to King George (Ellis, 63–70). This declaration of freedom from the British Empire thenceforth linked the words "life" and "happiness" with the term "liberty." And "liberty" has perhaps become the most important of the three in American minds. "Life" seems important enough and self-explanatory. There is not much to be debated in the definition of the right to life (except with regard to abortion). "Happiness" could indeed be debated endlessly, particularly as to whether the word connotes temporary satisfaction or long-term contentment, but it is "liberty" that is so often harped upon in American politics. Yet, hardly is the literal meaning of this term debated in the US.

"Liberty" has become a catch-all term to describe ultimate freedom, manipulated by whoever decides to hold sway over the term, a metaphorical and

---

1 "Big Questions. Short Videos." Learn Liberty. "*learnliberty.org*." Viewed on 4/4/2020.
"Defending the faithful," *Charles Koch Foundation*. "https://www.charleskochfoundation.org/story/stanford-law-religious-liberty/." Viewed on 22/1/2019.

2 In order to differentiate it from social liberalism:
b: a theory in economics emphasizing individual freedom from restraint and usually based on free competition, the self-regulating market, and the gold standard
c: a political philosophy based on belief in progress, the essential goodness of the human race, and the autonomy of the individual and standing for the protection of political and civil liberties specifically: such a philosophy that considers government as a crucial instrument for amelioration of social inequities (such as those involving race, gender, or class)
"Liberalism." *Merriam-Webster*. "https://www.merriam-webster.com/dictionary/liberalism." Viewed on 17/6/2019.

3 Locke, John. *Second Treatise of Government*. 1689. Edited by C. B. Macpherson, Hackett Publishing, 1980.

 | DOI:10.1163/9789004734807_009

pre-emptive high ground for ideological battle. In fact, *Merriam-Webster*, the primary US dictionary, defines liberty in its very first definition as "the qual ity of state of being free," followed by 1a): "the power to do as one pleases." The French Larousse, by contrast, defines liberté firstly as: État de quelqu'un qui n'est pas soumis à un maître. Tellingly, the second Larousse definition is: Condition d'un peuple qui se gouverne en pleine souveraineté.[4] Whereas in Europe, freedom and liberty have been bound by the rights of others, particularly since the end of WWII when hate speech was much more closely monitored, the same can obviously not be said in the US, which defines liberty more as an ability to move and act according to whim versus a state of not having a master. The US definition, then, involves a far more encompassing connotation of freedom, and this fact that has often created tension within the US.

Furthermore, the US definition is important in understanding Alaska and its frontier mentality. Throughout US westward expansion in the 19th century, Native Americans were pushed off their lands and even murdered, as large swaths were taken over by European settlers. Available spaces to conquer and settle have since dwindled, however, rendering Alaska "The Last Frontier" (an ideology whose roots lie with the gold rush, which roughly coincided with "The Closing of the West" and Turner's "Frontier Theory"). For many, Alaska has consequently become that place where liberty, ultimate freedom to do as one pleases, can be had. Yet, as some of the writers of this research, notably John McPhee, have explored, the freedom of the Last Frontier is essentially still connected to everything else and is but a mirage. Salmon streams are managed through hatcheries, which give the appearance of open, free space, but which in fact are carefully controlled by the government, as there otherwise would be no fish left in the streams, or other resources for that matter, such as the forests which so many Alaskans enjoy. Even in the Interior, John Haines found that there were issues with trapping too much and out of season, by too many people for the land to remain self-sufficient (*The Stars, The Snow, The Fire*, 14–5). So, in the sense that one may continue to live a subsistence life, a goal of many Alaskans, the state itself has had to carefully manage the dream. Nancy Lord has equated this scenario to a grown-up Never Never Land of the Peter Pan story she read as a child ("Being Peter," *Rock, Water, Wild*, 3–15). It is indeed a mirage, but it is managed as carefully as possible to allow those who still want to get away from the creeping of civilization a place to go.

John McPhee writes much about this American strain of individualism via his passion for Emersonian "self-reliance." Raban likewise seems to be drawn

4 1) "State of someone who is not subjected to a master." 2) "Condition of a people who governs itself with full sovereignty" (Translation mine).

to this ideology, as he constantly wants to escape from what appear to him to be the false communities of Seattle, sailing his boat to Alaska or driving his old muscle car out to the seedy bars and failed homesteads of Montana, actively searching out the individualists (*Passage*, 364–5, 412–3; *Bad Land*, 6–8). Of course, Raban could be chasing a socialist ideological narrative the way Robert Marshall did, as Raban speaks of his own father's history helping underprivileged communities and his own hatred for Tory governments (*Passage*, 263–4, 330). This narrative could also be viewed through a lens of city elites versus rural folk (*Driving Home*, 29–30), which may still carry some credence, though Raban is undoubtedly a "city" person. What is sure is that Raban's travel writing embodies a strong sense of the discovery of the "real" America, or Heartland America, where that unique sense of American liberty reigns. The author gives space to understanding the individualist mindset, coming from oppressive Europe and Eastern American cities, notably thanks to his background research,[5] while he is particularly unforgiving when discussing the 1990s tech boom in Seattle ("Boomtown USA," *Driving Home*, 404–12).

As for Nancy Lord, she seems to work in between two ideologies in a way. She has always believed in the hard work of self-reliance, coming into The Country herself with almost no material wealth or possessions and building an independent business (*Fishcamp*, 40–2). Yet, she always obviously understands there are limits to individualism, as noted in the "Nancy Lord" section of the "Masculinity v. Femininity" chapter herein. The other three primary writers examined in this research emphatically do not have that strain of American "self-reliance" or individualism. All three, Lopez, Lowenstein, and Matthiessen, are the more spiritual of the six, and come into The Country in search of community, Native culture (though Matthiessen's merits here are debatable), and biology.

The one situation in which we find Matthiessen comment on liberty and individualism is in a short section on Anchorage. Just like in the situation where the author comments on Inuit men, young and old, and their relationship with the white community, in roughly one page of the already short, concise work, Matthiessen breaks down the city of Anchorage. Remembering

5 For example, Raban gives much space to George Vancouver's voyages, particularly with regard to class hatred, which he suggests led to Vancouver's downfall. He says, "I felt a growing kinship with the man. I was annoyed to overhear his midshipmen talking of him as their natural inferior" (*Passage*, 139). Raban also discusses the deceitfulness, or at least recklessness, of those who would send poor Europeans and Easterners to "dry farm" in obviously inhospitable places, based on unproven pseudo-science (*Bad Land*, 16–48, 230–2, 276–7).

this is roughly a decade before John McPhee's work, and oil is already being searched out instead of gold:

> Despite its efforts at Main Street respectability, and town chimes that play Some Enchanted Evening, it still has pawn shops full of guns, hides and snowshoes, and saloons jammed with Indians and prospectors (no longer gold but oil). Violent death is not an unusual event in Anchorage, and its jail is full of wild young men. (*Oomingmak*, 11–2)

Matthiessen goes on to say that the town has a "striking location by the sea," and by the mountains, yet it will soon become "just another provincial town to which zoning came too late." This passage is interesting, like Matthiessen's comments on the Inuit, because it seems to appear out of nowhere, as a short anecdote which stands out on its own and is fleeting, leaving almost as quickly as it has come but resonating in the mind throughout the reading of the work. This concision is necessary, as the work is not about Anchorage. It is about Oomingmak on Nunivak Island. Interestingly, McPhee talks about how the glass towers are already going up in a downtown where cordwood is still being stacked (*Coming into the Country*, 114). That juxtaposition is endearing to McPhee. Anchorage was becoming a moneyed city but also maintained its frontier edge. Ten years before, there was prospecting for oil, but it was a frontier situation, still in the early stages where true money, billions, were not yet involved. ARCO would discover the underground lake of oil in the Arctic in 1957 (Coates, 167, 179). Interestingly as well, this individualism of the frontier town is expressed not just through its novelty shops, but with its jails and the commonality of "violent death." Matthiessen predicts a provincial town lacking in zoning, and in one aspect he is right. McPhee will complain about this same aspect, humorously making fun of the advertising signs everywhere and the fact that the city is pretty much like any other "cookie cutter" American city, such as Daytona or Albuquerque (112–3). Yet, McPhee obviously arrives as the money is arriving, and skyscrapers are going up on a shaky piece of land in the Subarctic. Matthiessen could not have predicted this aspect, but oil is going to run out in the coming decades of the 21st century and the yo-yo of bonanza economics will likely have a heavy impact on the city.

Another comparison could be Matthiessen's short passage on Anchorage measured against the first few chapters of Margaret Murie's *Two in the Far North*, in which that author describes growing up in Fairbanks in the first decades of the 20th century, when respectability is being sought, just as happens in Anchorage, but the violence and prostitution and saloons of the frontier boom town remain one or two streets over from the streets of lawyers and scientists who copy the architecture of the middle class houses of the Lower 48, attempting respectability. That dichotomy still occurred in Anchorage, according to Matthiessen, roughly half a century later. The frontier impulse, then, must have still been strong among so many people coming to Alaska. This search for liberty, which migrated into Alaska during the gold rush at the turn of the 20th century, has never been let go, to the chagrin of Susan Kollin and many others.

Perhaps the most interesting comment by Matthiessen, however, was the conclusion of his short passage on Anchorage that he was happy to see the interesting frontier city but glad to leave. This is a strong statement. The liberty of a frontier town where people carry guns and spend time prospecting and in saloons offers adventure, and an escape from the tedium of the industrialized world with its effective divisions of labor. Anchorage, as a frontier town, was much closer to a Native Eye, even if lacking the long-term, close observation of the Natives, since Natives relied heavily on subsistence practices. Yet, it is also understandable that a person, finding the place extraordinarily fascinating, could quickly become overwhelmed with the extreme liberty: the violence, the alcoholism, the unzoned melee of construction, etc. Civilization unbound is certainly interesting, but living amongst it can be too much for many people. Matthiessen, for his part, moves on to Nunivak Island where liberty still exists, but where civilization has been bound through taboo and tribal mores to keep the liberty in check in its own way (Fienup-Riordan, *Boundaries and Passages*, XIII, 7). Matthiessen seems to feel less oppressed in this atmosphere. He is most at home, however, studying and working with Oomingmak.

An interesting parallel could be made here in more closely examining Bob Marshall's *Arctic Village*. Marshall came into much the same territory as John McPhee, and as a writer McPhee seems to be technically in line with what Marshall accomplished, potentially taking inspiration of his often-employed technique of the anonymous quote structure from Marshall,[6] as well as spending vast amounts of time in local mining communities and interviewing people about every aspect imaginable of their unique existence. Marshall, however, unlike McPhee, was an avowed socialist. Marshall believed that the microcosm of his Arctic village, Wiseman, was an excellent example of how socialism worked (Marshall, 198–203), and how it could work on a larger scale. Part of the purpose of *Arctic Village* was to show that socialism could work, at the height of The Great Depression in the early-to-mid 1930s.

For Marshall, liberty was not so much achievable through self-reliance as through community cooperation, which he witnessed on many levels

6 The difference is that Marshall was not anonymous. He paid out his subjects instead with half of his royalties (Brinkley, *The Quiet World*, 249).

in the leftover miners of Wiseman (198–203). Likewise, Lowenstein goes into Point Hope and studies the local Tikigagmiut community, its history and culture, to unite the culture under its own traditions of whale hunting. Through the togetherness of the community hunt, and the uniqueness of its traditions, the community finds purpose, resilience, and strength. Liberty, therefore, in the Arctic, is apparent via cooperation and not individualism in many cases (the exception in this oeuvre being the few who thrive as trappers in McPhee's *Coming into the Country*). This liberty through community, however, is a liberty of actually living successfully, free from the need of procuring food, security, etc. It is a much different liberty from that of our normalized, industrial-world understanding of the term. Furthermore, how are we to define liberty when resource depletion is a constant threat?

1 Raban's Fishermen

Much of the key to *Passage to Juneau*, so similar to Raban's *Bad Land*, is the concept of misuse. If Lopez approaches environmentalism through a lens of morality, ethics, and mysticism, Raban's humanist side avoids morality in favor of demonstrating the direct effects of humanity's misuse of its resources. *Bad Land* is based on the dreams of homesteaders tricked into "dry farming" a land that should have never been farmed on a large scale. The homesteaders only made the land worse, digging up the thin topsoil, which turned to dust and blew away. *Passage to Juneau* approaches the issues of resource exhaustion due to overfishing by young men with dreams similar to those of the families in *Bad Land*:

> In the United States, wherever young men hang out together, on college campuses as in homeless shelters, this story went the rounds: if you could get to Seattle and talk your way aboard a fishing boat bound for Alaska, you could make $1,000 a day. Or more. Someone always knew someone who'd taken home $100,000, sometimes $200,000, for just two months' work. (*Passage*, 6–7)

And, in a sense, with the fishermen, as with the homesteaders of Montana, Raban associates freely in his writing. He himself feels like an immigrant light on his feet in the New World and moving westward (*Bad Land*, 21–3). He puts on a Seattle Mariners ball cap (*Passage*, 412–13) and tries to move on, fit in, wherever he goes, with the exception of maintaining his heavy accent.

In the process of this exploration of himself and others in the two works, the author indeed notes the foolhardiness of certain actions made by people looking for better lives, and yet he does not blame them for pursuing their dreams. He knows that this is human nature, "the whirling abyss" that is life. In this way, he opens up discussion of issues without placing blame, or releasing finger-pointing discussions of morality. For instance, in a discussion of getting caught up in an argument between British Columbians and Alaskans, claiming neutrality as a Brit, he notes that the British Columbians are upset by the "greed" of Alaskan fishermen, who exploit salmon at the mouths of rivers in Alaskan waters before the fish can make it back home to British Columbian streams. Yet, one fisherman admits that it is also the British Columbians' fault for overfishing in the first place. One man tells Raban, "'We like to blame the Americans, but that's just trying to shift the responsibility. We're all to blame for it. We were too greedy for too long. We wrecked the fishery. And now we're beachcombers'" (*Passage*, 149). Raban then shifts to an anecdote on sailing with his wife and having to try hard to avoid the ubiquitous salmon nets. He says, "I thought then that a salmon would have to be Houdini to evade the miles of suspended nylon mesh that barred its access to the spawning grounds upriver" (150). The writer subsequently notes the consequences, saying,

> I never again saw the inshore fleet in full fig. Most of those boats were now rotting to death in harbors up and down the coast, with FOR SALE signs glued to their forlorn windows. In Puget Sound, Juan de Fuca, and Georgia Strait, the sea had been fished clean out. (150)

This section of *Passage* finally concludes with a tone of emotion, through the voice of a former teacher:

> 'It's the boys at the back of the class that I feel sorry for,' the beachcomber said. 'They were never going to be rocket scientists, and fishing was their one big chance. They could make something of themselves when the fishing was good ... for him – he nodded his head in the direction of his own boy – it's his future, gone, just like that. (150)

The subject matter approached is important, but so is Raban's underlying tactic: here he is explaining the human element, and how the ultimate result of mismanagement affects it, mixed with the personal perspective of an average beachcomber, and the well-timed explanation of what human actions actually occurred to create such a dismal situation. This technique puts the reader into the omniscient position, able to see the negative consequences of the ostensibly laudable actions, while also avoiding judgment against the fishermen, who were in some ways logical in their actions. In both works, the dreamers were told they could change their entire station in life, and in the case of the fishermen, through one summer of hard work, after which they just may have enough money to start a small business and a family. Rather than place blame upon them, Raban demonstrates how they were all acting in self-interest, while also indirectly demonstrating that not one single individual among them created the disastrous issue; all of them did so, together. In fact, not even the fishermen can be blamed as a single group, because entire communities must be implicated in creating the issue and allowing it to continue to happen. This is the same problem which Nancy Lord discusses with regard to the Cook Inlet beluga in *Beluga Days*. It is also a microcosm of global pollution issues heavily affecting climate change.

Another point to raise regarding these young men who value their liberty in order to achieve their dreams of social mobility via exciting outdoor adventure is that this is essentially Raban's entire environmental approach in a nutshell, particularly versus nature writing in general. A point this research has subtly attempted to get across, but will go into in further detail in much of Part 2, is that Raban recognizes that not everyone is affected to the same extent by the same literature. Raban recognizes that there is a chasm in industrialized culture, often symbolized by the dichotomy of Seattle versus its surroundings in his later work. Environmentalists tend to have a plausible narrative, but they are often not getting through to the people who most need to be reached, who are the heaviest polluters and least reflective regarding conservation.[7] But, he is insinuating that these people are not bad; that, instead of declaring war against them or trying to shame them into becoming environmentalists, others should listen to their stories as well. Raban often offers the reader those stories. He explains to the well-read classes to whom his books are targeted, who tend toward ecological responsibility, what the average person, "the [kids] at the back of the class," are faced with in life.

This is Raban "walking down Pound Lane," as has been stated. But, he does not offer a pragmatic solution as much as to highlight the cultural chasm and display the valor of a forgotten class that, to academia, could be considered

7 In her recent work, *Ecology and Literatures in English*, Françoise Besson argues that children can be targeted through literature, convinced to go out and create environmental change (359–60). This is an important point, neglected throughout this research, which concentrates effectively on adults, but the fact that children are easily influenced and capable should definitely be acknowledged.

an alterity. Displaying this type of perspective in literature can create an effective shift that may consequently lead to ecological solidarity through bridging gaps within communities (though such faith in one work of literary nonfiction would be presumptuous, it may serve as part of a broader movement).

Liberty itself, as a concept, has not been discussed very often in this section, and it may not seem directly related, but it is. To tell these young men that they no longer have the liberty to fish, the liberty to make their way into the middle classes in a manner that many white-collared workers would never dare, the middle classes where the educated who tell them they are not environmentally-minded are already well-established, is a difficult discussion to have. Such is the case with so many natural resources, in so many communities. McPhee is much like Raban in this way, demonstrating how laudable communities of people in Alaska have developed around a desire to escape from industrialized society, seeking liberty through self-reliance.

Yet, this research also proposes the idea that "self-reliance" and "individualism" are two different concepts: the self-reliance so often lauded by McPhee connotes a capability of survival outside industrialized society, while individualism as discussed in modern politics tends to connote a separation from government and community. Benjamin Franklin, for example, preached self-reliance, but he also preached community, building libraries, and creating clubs to bring people together and lift up people around him (Isaacson, 2–4),[8] such as John Bartram (Wulf, *The Brother Gardeners*, 98–9).[9] As noted in the previous chapter, de Tocqueville considered individualism to be hazardous to the health of society (unabridged version, 104), because, as we see throughout this chapter, when liberty is taken as an ultimate freedom to do as one pleases resources become depleted and society suffers. Community cooperation is ultimately needed for a society to work, something Robert Marshall knew. Raban never says this outright, but his observations do point out that cultural chasm. McPhee, by contrast, attempts to give a broader, or "multiradial" view, which lets the reader make their own decisions, with a journalistic approach on the peoples in and around Eagle, Alaska, and elsewhere in the state. This journalistic approach is the biggest divide between the two men, who both see the usefulness of the less educated classes who tend to possess the kind of trade (144–74), and often local geographic (130–43), knowledge that Barry Lopez lauds in *About this Life*. The following section will explore McPhee's

8 Isaacson, Walter. *Benjamin Franklin: An American Life*. New York: Simon and Schuster, 2004.

9 Wulf, Andrea. *The Brother Gardeners: Botany, Empire and the Birth of an Obsession*. London: Windmill Books, 2009.

writings on the people living along the Yukon River in Interior Alaska, who are evicted from their self-built cabins as a result of ANCSA.

2 McPhee's River People

In Part 3 of John McPhee's *Coming into the Country*, the reader is faced with an enormous dilemma: should a land that has long been considered open to those who wish to return to a Native Eye perspective, or an individualist, subsistence ideology, be suddenly removed from the land by the federal government in the name of preservationism, but as part of a deal to effect the installation of the pipeline? The deal the federal government essentially made with Alaska Natives and environmentalists was to close off enormous swaths of land for preservation, and to split up Alaska Native communities into corporations (converting them to capitalism), giving the entire community roughly $1,000,000,000 as a whole (McPhee, 132), as well as forty million acres (214).

John McPhee went to investigate personally. What he found was a rather small community of people stretched out along the Yukon River and economically and socially based in the city of Eagle in the Interior, close to the Canadian border. McPhee has never been a rugged individualist, but, as noted previously, he has always been interested in self-sufficiency, or self-reliance. Therefore, subsistence lifestyles in Interior Alaska were a natural fit for him.

McPhee's investigation explores how Interior Alaska life is in many ways one of the most sustainable lifestyles a human can have. The topic was not yet at the forefront of environmentalism in the mid-to-late 1970s, but one's carbon footprint in the Interior can be extraordinarily low. Many of the characters in McPhee's work killed their own food, sewed their own clothes, operated boats without motors, built and heated their own houses, etc. Nonetheless, everything else had to be flown in. Planes were, and still are, a common form of transportation.[10] Furthermore, trees were at times cut down indiscriminately,

10 The Alaska Highway leads from the US Midwest into Canada and eventually to Interior Alaska, at Fairbanks, where one can drive all the way up to the Arctic coast, to Prudhoe Bay where oil is being extracted, or south to Anchorage, and a roughly five hour drive further south to the bottom tip of The Cook Inlet, at Homer (Coates, 64–77). It is well-known among certain circles that one of the longest direct drives one can make in North America is from Key West, Florida to Homer, Alaska, and many do (Homer Spit is almost always full of camping RVs). And, oddly, both towns form a duality: both are scenic fishing villages known for heavy partying and isolation, now known to tourists. One is subtropical, with brightly colored reefs, and the other is sub-Arctic, with visible glaciers. That aside, the point here is that places like Juneau are isolated from automobile traffic, as is much of the

and animals killed in large numbers at any time of the year, without respect for sex or age. The dilemma, then, is who is in the right in their lifestyle? Should the government have the right to come in and tell these people that they can no longer live a life which many consider to be the last place to live a free, subsistence lifestyle, which to so many is an honorable way of living? And in a place that these people have for so long been told they can come to in order to live such a life? In a passage from *Coming into the Country* McPhee explores the idea by discussing how everything changed in the 1970s. Until then, the government not only did not interfere with the dream of the homesteader, but "it even assisted this dream." When oil was discovered at Prudhoe Bay, however, that was the big moment when homesteading ended, due to the need for the "demarcation of Alaskan land." McPhee says of the effects:

> … long-established forms of freedom would disappear – the sort of freedom that drew a family like the Gelvins a generation earlier into the country, the sort of freedom envisioned by young people who set off to live in the wild of the upper Yukon. If the oil had never been discovered, there would not have been an eviction notice prepared for Brad Snow. (213)

McPhee goes on to explain the specifics of the ecological agreement. He obviously feels for these people being pushed off their land (though no parallel is discussed regarding Native Americans being pushed off their lands across the country). But, the above passage is highly interesting for this research because it suggests a sort of harmony existing before oil was discovered, allowing for people to co-exist in a place of escape from civilization, a theme that is not uncommon for McPhee in general. Marshall's *Arctic Village* in a way attests to that as well ("He thought the secret of their happiness was that they lived in harmony with nature, in what Marshall labeled a 'pre-industrial civilization.'" Terrence Cole, "Preface to the 1991 Edition," *Arctic Village*, XVI–XVII). Even if McPhee's villagers did not exactly get along greatly with the local Alaska Natives, the Athabaskan Hungwitchin, they at least got along well enough to live in general proximity to each other. The scale was always small as well, so the devastation (other than during the gold rushes, which abated after roughly a decade) was essentially miniscule. And, there is indeed something to be said for scale, but also technology, which was at a turning point. McPhee had likely not yet realized the effects to come, with regard to the technology that families

rest of the state, such as Bethel. To get to these places, an airplane is absolutely necessary (Coates, 56–9).

like the Gelvins were bringing into the country (205–8). As is frequently the case, McPhee allows individuals to have their say in his anonymous quote structure. Here is how the local villagers of Eagle speak to McPhee on the topic, which is perhaps even more pertinent than McPhee's own words:

> 'It's not as if we're building fifty-thousand dollar houses with asphalt driveways and stinking cesspools.'
>
> 'Our cabins are more like tents than like most people's homes. They are made with native materials – white spruce, earth, moss. They are biodegradable …' (215)

They continue on to blame the Native land deal, people from Outside, and the government for their misfortunes. They believe they are in the right for their low-carbon-footprint lifestyles that are out of the way of society and fitting in with the environment around them.

The relation from the first passage to the second is direct. As noted, the second is a jumble of first-person reactions to the issues described in the first. The Yukon people are angry. At least one of them sees him- or herself, interestingly, as of the landscape, the local biome of the region, almost as part of an indigenous group. This being of the land also carries benefits for society in their minds, particularly as they see themselves as the first line of defense for society in a dangerous environment. This section, which discusses the importance of building cabins on the frontier, leads into McPhee discussing a military man who went down in a plane and almost died. The man was saved by a cabin, which was kept up, stocked with food and wood, as is etiquette in the region, specifically for those in need.[11] McPhee searches for this man, and finds the family. The man had moved on with his life, gotten a degree from MIT, and had had 6 children who had become successful members of their communities. None of them would have existed had not multiple stocked cabins been waiting in Interior Alaska (216–24).

And yet, all of this is wrapped up in a land deal related to oil and Native rights. Notice that McPhee squarely places blame on the shoulders of the oil industry, but the people he interviews place blame on the Native land deal itself, which to them is of two parties: the Alaska Native community they obviously do not

11 I talked with an Interior Alaskan in 2018 (12/2018) who told me he had built a cabin in the wilderness in the 1980s, and had abandoned it, left it stocked, and had thought about going back many times but had not done so. This is quite common in the state. Most expect that someone else will come along and take up the cabin, or that it may save a life someday.

trust (345–6), and people from faraway states telling them what to do, as noted in the second passage above. They are afraid Alaska will become just like everywhere else in the US, which is exactly what they have been escaping. McPhee's work is prescient in this sense, because now this idea is not only important for Alaskans, but it has become part of the rural v. urban political dichotomy seen in much of Raban's work, and more recently in broader politics.

In the first passage above, McPhee says many of these people were searching for "… a life that remembered the past." This is essentially the issue in current political debates, the conservative (literally "conserving" the past) against those who would move forward, urbanizing and physically preserving areas outside cities where many political conservatives live. The ecological debate has come to the forefront, so that the reactionary focus on extreme liberty has become the primary issue among many conservatives, who want to actively utilize the land that urban liberals (often the demographic dynamic, according to environmental historian, Roderick Nash – 343) hope to preserve, while many US conservatives avoid large cities for the profligacy found therein (Thomas Jefferson said "cities were pestilential to the morals, the health and the liberties of man," – Brogan, 257 – and the Republican versus Democrat election map tends to follow a rural versus urban demographic to this day).[12] With the rise of Sarah Palin as a politician-cum-reality television star (*Sarah Palin's Alaska*), as well as "reality" television shows representing a false sense of Alaska as a frontier (*Alaska: The Last Frontier*), the state has come to represent the ultimate symbol of individualism to many Outside who have no understanding of the dynamics of governmental intervention required to keep the frontier illusion alive in the 21st century. Yet, even in the Country, people can be ignorant to these dynamics.

In order to tackle such a difficult, yet interesting, situation, McPhee speaks with multiple people along the Yukon river and around Eagle, as exemplified in the previous excerpt. Some are lone men, some men with women, and some entire families. One family diverts an entire river bed for gold, yet McPhee finds in them the perfect frontier family from the American cultural perspective (379). For McPhee, American family values are something worth fighting for and keeping. Yet, what is it about this family and this community of mostly misfits (179, 184–6, 188–91) that draws in McPhee the most? It is exactly that individualism and self-sufficiency. Yet, McPhee, in other parts of his work, explains how he understands that Alaska is no longer, by any stretch of the imagination, a frontier. He sees that Alaska seems large and remote, but he

12 "Presidential Election Votes: Donald J. Trump Wins." *nytimes.com*. "https://www.nytimes.com/elections/2016/results/president." 9/8/2017.

also sees that civilization is never far off. That is the crux of the issue with the frontier: it is a mirage created for people who think they are individualists, but who depend on the government to keep their dream of individualism alive. Otherwise, they would quickly deplete their own resources and ability to live that dream.

Ultimately, McPhee has no answers. He does understand, however, that the need for a frontier on which to live out a subsistence lifestyle is important for a culture. There will always be people who need this kind of space and relationship with the wild, and there will always be a need for spaces where humanity is able to maintain a close relationship with the wild to be diffused to the broader culture, as a sort of moral compass. But, it is also important to recognize the facts as they are and not as we wish them to be. McPhee hopes to preserve the values of a family that grows and hunts its own food, fixes its own machinery, flies its own planes, etc. There is much in the way of Lopez' argument of finding the Native Eye in this way of life. One is kept in touch with a greater reality beyond divisions of labor, industrialization, and cities of men where an understanding of the natural world that shapes their existence is pushed to the fringes. Being in touch with the land, then, is considered healthy for a nation.[13] This dilemma is obviously unsolvable for McPhee. His technique, then, is not so much to sit on the fence and refuse to "pull the lever" as was discussed earlier in this research. He absolutely does not refuse to pull the lever. Instead, he carefully weighs the importance of a preservation of values and a preservation of land, and the need for both, and refuses to pick one over the other. In this way, he allows the reader to take in the facts and, this is a very important distinction, not decide between the two for themselves but decide on a case-by-case basis, or even have a nuanced understanding of the issue without resulting to militancy against people with opposing views. Obviously, the situation has changed regarding the urgency of the environmental movement since 1978, as well as understandings of minority rights and Environmental Justice, but nuance is still important, particularly if we wish to unite a multitude of voices.

McPhee's technique of respecting nuance would still be welcome today, yet the 21st century is arguably more rife with the twisting of words on a large scale by think tanks and interest groups (MacLean, *Democracy in Chains*,[14]

13 Jefferson certainly believed in the notion, though Roderick Nash notes that nature must be conquered and coiffed first in the Jeffersonian ideal: "... wilderness is just as dead in the garden as it is in the concrete wasteland" (380).

14 MacLean, Nancy. *Democracy in Chains: The Deep History of the Radical Right's Stealth Plan for America*. London: Scribe, 2017.

XIII–XXXII). Descriptive nuance, then, arguably requires more precision, at potential risk of losing a bit of literary aestheticism gained through remaining vague and open as a method of stimulating reflection. For instance, McPhee often moves fluidly between stories, profiles of people, leaving aspects of arguments out at times and coming back to them in order to leave impressions. This technique is seen throughout *Coming into the Country*, particularly in Parts 2 and 3. Liberty is a fickle idea, and yet McPhee has found a way to remain faithful to it in his literature without giving in to the politics of one side or the other. In this way, the author is also able to maintain his own liberty.

Throughout his career, McPhee has played with the positives and negatives of government intervention as an exploration into this nuance. A notable example is *Control of Nature*, which shows that even community cooperation can lead to negative results. In a section on Iceland, entitled "Cooling the Lava," houses are saved from a volcanic eruption, but the flow of lava nonetheless destroys a port, which is the primary economic producer of the town. As we will see with Nancy Lord, however, without cooperation between government and local communities, all working together, a "tragedy of the commons" will be experienced, leading to resource exhaustion, which will be the end of the self-reliant "outdoor" types that McPhee has tended to profile throughout his career. In this way, we may move fluidly from discussion of McPhee's River People into discussion of the numerous parties involved in the endangerment of the Cook Inlet beluga in Nancy Lord's *Beluga Days*.

3 Lord's "Take One" Alaska Native Distrust of the White Man

The idea of "liberty" in Nancy Lord's work *Beluga Days* essentially comprises the entirety of the issue she is discussing. The oil companies feel they should have liberty to explore for oil (22, 37). Fishermen should be at liberty to fish for salmon without schools of beluga whales taking their source of income (4). Native communities should be able to hunt for subsistence whales, as they have done so for thousands of years, though Alaska Natives were never settled in the numbers they currently are in and around Anchorage (23, 29–31). The lists of liberties groups of people in Alaska expect goes on, including the rights of conservationists and everyday people to enjoy seeing both belugas and salmon in proximity to civilization (225–6). Yet, when the liberties of one party infringe upon those of another, arguments and legal entanglements tend to ensue. Not only that, but when an entire sub-species is headed toward extinction because each party expects its own liberties to be respected, something must be done. Lord's book essentially documents that something, to be diffused locally, and she becomes part of that something in the process. She went to meetings where she was flabbergasted at the lack of discussion surrounding the fragility of the beluga whale population of the Cook Inlet. Simple calculations of what fishermen and scientists were telling her meant that the beluga would be gone within a matter of a few years. Yet, no one seemed to want to come out and say it, what she calls "the whale in the room," meaning that there was a very large issue to be discussed which everyone seemed to be avoiding (28).

Essentially, the Native population is justified in not trusting the white population with regard to whale hunting. Industrial white populations are to blame for the devastation of entire populations of bowhead whales which the Inuit of Alaska for so long hunted and relied upon for subsistence (Hess, 1–16), and more recently, the whales that they do catch are likely to be full of mercury that poisons humans to the point where certain Alaska Natives with whom Lord speaks in *Beluga Days* refuse to eat any subsistence foods whatsoever anymore (Lord, 150–1). And, even more recently, since *Beluga Days* was written, ocean acidity has been making oceans unlivable, a topic on which Lord wrote her novel *pH*, and plastics, whose effects are still not fully explored, are ubiquitous throughout the oceans, in all forms of life, even in the depths of the Mariana Trench.[15]

At this point in the section, the global conundrum of industrialized democracy's search for liberty should be very well established, and, in this sense *Beluga Days* has become a microcosm of the greater world. No single group wants to accept responsibility for the majority of environmental problems, and not many groups will change unless all groups change. Lord is ultimately successful in pointing out "the whale in the room" and becoming part of the solution to get groups to work together in her local community. On the global scale, international politics and globalized hyper-consumerism appear to be blocking potential positive change. An important question then becomes whether or not environmental writing can effect change on that broader scale. This question will serve as much of the backbone of Part 2 of this research, but suffice it to say for the moment that Nancy Lord has shown a blueprint for successful raising of community awareness among groups that were ready to cooperate under the right circumstances.

Part of the laying the groundwork for the "right circumstances," however, is respectfully laying out the facts. During the many conferences Lord attended as part of her research for the book, there was a skirting of the idea that the

15 Gibbens, Sarah. "Plastic proliferates at the bottom of world's deepest ocean trench." *nationalgeographic.com*. "https://news.nationalgeographic.com/2018/05/plastic-bag-mariana-trench-pollution-science-spd/.» 13/5/2019.

influx of Alaska Natives to Anchorage was creating an effect of "overhunting," but no one, not even government officials, wanted to approach the subject because relations were already tense between the government and Alaska Native communities. The government, in fact, did not want to restrict access to beluga fishing without proper science, but beluga communities are so elusive in study that precise data was, and still is, tough to come by (26–31).[16] In her chapter, "The Whale in the Room," she explores why people cannot come to terms with the issue. She mentions how there are the Marine Mammals Protection Act and the Endangered Species Acts in place, but there is also the Native hunting exemption to the MPA. She wonders to herself which act and exemption has the most legal sway in this issue before launching into what she sees as the broadest problem:

> In fisheries, I well knew of Garrett Hardin's "tragedy of the commons." When there's a common property – the field where all cattle graze, the ocean and all its fish – from which anyone can take, and not taking only means you get less and someone else gets more, Hardin argued that the result is always and inevitably overexploitation. (32–3)

For Lord, the issue lies in compromise between communities. Alaska Natives simply cannot be blamed, but neither can they go on taking belugas unrestrained. "Community controls" must be implemented, but community controls also almost always lead to social issues between government and Native communities, and rightly so. Yet, she is implying that it must be done. She is also suggesting that the communities themselves be directly involved in the decision-making process of how to do so. This leads back to the broader theme of this research, that all voices must be heard and communities must be brought together, in order to effect change. Many writers promote this ideology through literary devices, of perspective, of quotations, nuances in

16 The Cook Inlet beluga lives in muddy water, tough to be seen, and where it is often shallow. This helps the whales because they can avoid becoming prey to orcas who cannot access the shallows (Lord, *Beluga Days*, XVII). Attempts to pin them with tracking devices often fail because the whale can rub the devices out of their fat, possibly on the shallow floors, unlike many other whales (48–9). Cook Inlet belugas are the only belugas in the world, other than the St. Lawrence river belugas, who do not migrate out into the open ocean, but their migration is tough to follow (20). Studies discussed in *Beluga Days* found them wintering up in the north end of the inlet, which surprised many researchers for the proximity to Anchorage where the water is colder than down near Homer (56–7). Numbers, therefore, are almost impossible to determine, but Lord's point is that the decline is rather obvious, despite the lack of accessibility to precise data.

language and narrative, etc. So does Lord, as we have seen. Yet, here Lord is approaching the subject head on. Interestingly, Lord is almost never, if ever at all, politically divisive, because that would go against her inclusive methods, but here she quite freely insinuates that liberty cannot overstep the bounds of community. Communities themselves must work together to protect their own interests, and they must do so as united groups, who must work together with other united groups. This implies both white and Alaska Native communities, but Lord discusses the implications for Alaska Native communities specifically, which is likely an explanation to the white community, who may remain ignorant to aspects of Native culture they do not understand. She says, "Native people who moved to Anchorage did stop being Native" (34). Even working in the city, they continued to hunt and gather and share their foods among one another. Furthermore, she notes that Native hunters continue to sell wild foods to the Alaska Native Medical Center for Native patients to eat. What she is saying is that hunting is a social, cultural, health, and economic factor for Native peoples. And finally, she discusses a kind of self-actualization of the hunter:

> There was status to being a hunter; there was respect to be earned and to pay; there were responsibilities to family and to elders. It meant something to be a hunter – something more than bringing home food, which was, in itself, not a small accomplishment. (35)

Lord is delving into the modern Native relationship with their food in these discussions, but, as she has done elsewhere in *Beluga Days*, she adds, "I knew all this …," which is likely meant to imply to the reader, "you, reader, likely know this aspect, but …" This offers her the opportunity to expand and discuss the fact that, though the reader may know the importance of Native foods to the Native community, they may not know that hunting is connected to community duty and status, which is not a sense of egotism but a sense of responsibility. This responsibility does not weigh the person down, either, but it offers the person the opportunity to feel strong and connected to the larger community, which is important. Bill Hess discusses this aspect in his work, *Gift of the Whale*. That author notes that an Iñupiat father gives his son a boat, as captain, upon graduation from high school, and the son shortly takes the personal responsibility of convincing the local elders to no longer send their trash out onto the ice flow after whaling season, as the trash is no longer organic. The personal responsibility, then, offers the connection between community and ecology, directly (Hess, 178). The following conclusion can thus be drawn: both authors work toward the idea that liberty must be connected with community.

If liberty falls into individualism, it will inevitably break down into disputes, which will lead to resource depletion and a lack of general livability for all parties.

Most of the primary works of this research, and a few others, have been discussed herein regarding the idea of "liberty" in the extreme US sense of the term. The reader, who was likely at least vaguely aware of the discrepancy between a US sense of liberty and European notions thereof, should take away more precision and nuance in how the US connotation physically affects the place, not just the ideology, that is Alaska. Oil companies claim liberty. Alaska Natives claim liberty. The sourdough community claims liberty. Recent emigrants influenced by pseudo-reality television claim liberty. All claim liberty in this one place, which is physically capacious but is also vulnerable to this panoply of claims. Some environmental writers and critics would increasingly say that the land, the plants, the animals themselves, the entire ecosystem of Alaska should have a claim to liberty as well, which should be respected equally to that of the multiple human parties, guiding the latter to the same extent that human parties guide them/it (Besson, XI–XIII). Another takeaway from this chapter should be that this will not happen unless individualists begin to view Alaska from the broader perspective that they are all damaging its potential to be what they all want it to be: a wild place and a provider of resources. As one of McPhee's river people notes, however, the cabins along the Yukon will have trees growing out of them and will likely collapse back into nature within a few years of abandonment (*Coming*, 215–6). Raban says much the same, impressed by the power of nature to reclaim civilization in *Passage to Juneau* (34, 69–70, 106, 214–5, 316–7). Many individualists, having experienced "wilderness," would make similar arguments in defense of their actions in polluting and destroying environments through gold mining, logging, overhunting, overfishing, and oil extraction, but they would also likely be underestimating the permanent damage that scale can bring. The pure scale of industrialized civilization absolutely requires that communities gather and work together as they have never done before, re-examining their definition of liberty, and this is, subjectively speaking, the single most critical issue facing humanity in the 21st century; enough of an issue to call it a truly existential crisis.

PART 2

Writing beyond the Last Frontier

FIGURE 9 Tlingit mask, 1840-1860, on view at the metropolitan museum in New York https://www.metmuseum.org/art/collection/search/319074

Introduction

The objective of Part 2 is to keep the literary analysis of Part 1 in context as we delve into the following question: what responsibilities does an Alaskan travel writer actually have in relation to environmentalism? To narrow that down, another question will be asked: how has this responsibility changed over time? And furthermore: responsibility toward whom? As briefly mentioned, writers often address different groups, sometimes simultaneously. There is a responsibility to Alaskans, and to different demographic groups within Alaska. There are also those Outside, in the Lower 48, for whom Alaska serves as a symbol of the Last Frontier and a place of natural resources. There are those farther Outside, internationally, who also have a particular view of the United States and Alaska, and their viewpoint may be kept in mind as well. For instance, Tom Lowenstein's work, which was published in England, transmits a strong understanding of traditional Iñupiat culture[1] to UK residents. Furthermore, there is an increasing awareness toward an animist view of responsibility to the plants, animals, and larger ecosystems in the region itself. This last group will likely be considered debatable for some time to come, as there is a historical precedent in many western philosophies and religions placing man above nature (Nash, 9–11).

Another idea to be explored is that environmental writing potentially derives from proximity to these animist societies, which have traditionally viewed both man and nature as subjects, with man, plants, and animals all living as fellow community members. This idea will require further exploration, as it lies at the root of American and European views on environmentalism, as well as debates between "Deep Ecology" preservationists and "Wise-Use" conservationists, and has been accentuated herein via Raban's comments on the NW writers.

To begin, Part 2 of this work unavoidably must be prescriptive. An argument could be made that a line should be drawn between the scholarly and the prescriptive, the activist work, or that activism has no place in scholarly literary criticism. This point has much traditional merit, but the argument here is that literary criticism in an age of mass activism requires examination, metacriticism, of modes and techniques equal to that of the fast-paced scale of the evolution of the current technological revolution. Nicole Seymour, for instance, whose criticism will be highlighted throughout much of Part 2, suggests,

1 "Tom Lowenstein." Wordpress. "https://tomlowenstein.wordpress.com/about/." Viewed on 29/7/2019.

 | DOI:10.1163/9789004734807_010

"Ecocriticism, like environmentalism, has largely avoided self-awareness, self-reflexivity, and metacritique" (*Bad Environmentalism*, 26).[2] That is a heavy statement, which will be examined in depth. But, remaining with the thread of activism for the moment, is the majority of criticism not ideologically motivated, even when it is couched in academic syntax and methodology that pretends objectivity? Is literary criticism itself not inextricable from politics, class, and race? In the "Introduction" to *The Environmental Justice Reader*, Joni Adamson, Mei Mei Evans, and Rachel Stein quote Valerie Kulte who "suggests that, too often, researchers and scholars have been content simply to present a 'postmortem of an already disastrous situation.'" Instead, they, and Kulte, argue for a "'proactive scholarship' that points out patterns of environmentally destructive and socially unjust activity and also identifies resistance movements" (7). For them, this is a way to promote action and avoid "hand-wringing."

This research, then, takes the broader Environmental Justice movement seriously, and will attempt to work both within it and without, by examining and critiquing the technical aspects and stylistic affects of the movement itself while also becoming part of the movement by taking on its attributes and suggesting ways that travel writers to Alaska may integrate the ideals of the movement into their respective works, therefore advocating for the proactive instead of the "postmortem" or the "hand-wringing." In terms of John McPhee, we could, of course, refuse to "pull the lever," but McPhee's work is journalistic and meant for a broader audience, which this work is most definitely not. This work is geared toward an academic audience. It reflects on the merits of proactive writing in multiple ways, which must be clear and not vague. Unlike McPhee's purposely vague nature, which allows him to insinuate but also gives the reader enough leeway to at least ostensibly make up their own mind, this research must lay out arguments as completely as possible.

In searching for a work of environmentalism that most clearly encompasses the Second Wave Ecocriticism and Environmental Justice of the current moment, Naomi Klein's 2014 work entitled *This Changes Everything*[3] is likely the clearest and most influential example. Whereas Rachel Carson is so often looked to as the 20th century's lodestar of literary environmental activism with her work against DDT, Klein's work is quintessentially 21st century in that it has moved beyond informing the public of the facts of an environmental issue. The scale now involves global climate change, and we now live in an age of active

2 Seymour, Nicole. *Bad Environmentalism: Irony and Irreverence in the Ecological Age*. Minneapolis: University of Minnesota Press, 2018.

3 Klein, Naomi. *This Changes Everything: Capitalism vs. the Climate*. 2014. London: Penguin Books, 2015.

disinformation, so Klein's main focus is pointing out the hypocrisy of parts of her own environmental movement and assembling demographics to work together against that hypocrisy and disinformation. In *This Changes Everything*, she claims that the time has come to shift our perspective and priorities on capitalism (16–26), and on "Big Green" as well (191–221), and employ grassroots forms of organization and civil disobedience (293–466) in order to promote the adaptations needed to slow climate change. Again, the idea is no longer to inform people of climate change but to organize them around a set of facts they already know exist as a problem. Quoting examples in Greece, Canada, and South America (298–305), in what she calls "Blockadia,"[4] Klein claims that grassroots organizations in cities, with their academic, technological, and legal training, should work hand-in-hand with native groups, since indigenous peoples have had the most success in fighting against the large corporations that lack incentives to otherwise not destroy man's habitable environment (367–418, 464–6) (what McPhee calls "corporations beholden to thousands of shareholders" – *Coming into the Country*, 379).

Many native communities often still depend upon subsistence use of their land. Therefore, if the water and animals and plants are too polluted to safely consume then these communities are not only threatened physically, but they are cut from spiritual and cultural traditions that have given their lives meaning (Klein, 367–87; Fienup-Riordan, *Boundaries*, 362–70; Hess, *Gift of the Whale*, 1–16). That may be why native groups have won some hard-fought political battles in recent years: with nothing, or perhaps everything, to lose (meaning: no status quo to maintain, and heavy cultural and spiritual traditions, as well as physical health and space, to lose), they keep fighting when others recede into the nest of their own personal lives.[5] Furthermore, though the connections created by social networks have recently been attacked for tearing societies

4 "Blockadia is not a specific location on a map but rather a roving transnational conflict zone that is cropping up with increasing frequency and intensity wherever extractive projects are attempting to dig and drill, whether for open-pit mines, or gas fracking, or tar sands oil pipelines" (294–5).

5 This may also be due to an increasing awareness of indignities done to minority communities in the social media age. In a note in the last chapter of *The Golden Spruce*, John Vaillant mentions that "In 2003, Starbucks, the international, multi-billion-dollar coffee conglomerate, was forced to abandon a copyright infringement suit against Haida Bucks Café, a tiny restaurant in Masset, after facing an unanticipated wave of negative publicity and boycotts from around the world" (229). Vaillant also mentions on the same page that the Haida had won the right to have Weyerhauser (an industrial corporation) forced to consult with the tribal council before logging certain areas.

apart,[6] they are not intrinsically bad. Social networks have the ability to bring unconnected communities together. Alaska Natives can connect and communicate with Canadian First Nations peoples and Northern Greenlanders, and others as well. With the help of the technologically advanced and the human rights lawyers and organizers, more and more networks are coming together to fight "Big Business" and disinterested governments in a way heretofore unseen.

Klein's primary example of why the current path is no longer working is the US environmental group, The Nature Conservancy, which has actually been drilling for oil around the Gulf of Mexico. The organization itself saw to the potential extinction of the Attwater's prairie chicken (195), which it was attempting to save by raising money through drilling for oil in the endangered animal's reserve! Worse, the organization has not stopped drilling since this fact went public and it issued an apology. In fact, as of 2014, The Nature Conservancy has built a new well, the initial well having gone dry in 2003 (191–201). "Green Business," Klein says, and "Big Business" are working in sync, to the point where many environmental groups themselves even invest in fossil fuel companies, with executives of fossil fuel companies on the boards of certain environmental groups such as The Nature Conservancy (195–9). And, she notes, alternate energy is not moving fast enough (230–55). At the same time, there is a reason things are working out the way they are in the global economy. If people, corporations, or governments move away from the global economy, they will be left behind. The entire world is in an economic footrace, and entire countries could be gobbled up in the next century by not keeping up militarily, diplomatically, and financially. So, fighting back against the negative impacts of globalism is a difficult task. Maintaining strong economies in the process of this transition to clean energy is also important. But, Naomi Klein makes a clear and persuasive argument: change will come when "Suddenly Everyone" gets involved, from native peoples to local co-ops to multinationally-organized movements (464–6), and I would add, every generation.

The question with regard to this research then becomes: what about writers? What responsibility does literature, travel writing, and the context of Alaska have to this movement?

First and foremost, Naomi Klein has done her part by creating a spark without being overly empathic, nostalgic, aesthetic, or demagogic. Her arguments are thorough. She could be criticized for being pedantic by those accustomed

6 Sulleyman, Aatif. "Facebook is destroying society and making users feel vacant and empty," *independent.co.uk*."https://www.independent.co.uk/life-style/gadgets-and-tech/news/facebook-society-destroy-social-media-network-users-damage-communicate-connect-chamath-palihapitiya-a8105131.html.'» 12/12/2017.

to the concision of social media and network news, but her thorough work leaves no visible holes in her logic, even if one disagrees with her premises. She is also persuasive and relatively readable, employing numerous narratives, despite her thorough work. As Lawrence Buell notes that Rachel Carson benefited from previous experience in fiction when she wrote *Silent Spring* (VII, 4), Klein also benefits from her experience in documentary making (*The Shock Doctrine*), potentially engendering readability. *Silent Spring*, for example, moves back and forth between data-driven science and descriptive scenarios:

> The town lay in the midst of a checkerboard of prosperous farms, with fields of grain and hillsides of orchards where, in spring, white clouds of bloom drifted above the green fields. In autumn, oak and maple and birch set up a blaze of color that flamed and flickered across a backdrop of pines. (2)
>
> Strontium 90, released through nuclear explosions into the air, comes to earth in rain or drifts down as fallout, lodges in soil, enters into the grass or corn or wheat grown there, and in time takes up its abode in the bones of a human being, there to remain until his death. Similarly, chemicals sprayed on croplands or forests or gardens lie … (6)

Here we see that Carson's work is able to project actual scenery of an apotheosized small town comprised of "a checkerboard of prosperous farms." And amidst such scenery, which one presumably relates to their own experience, particularly in the "Middle America" being most directly affected, the long-term effects of chemical usage are interjected as villainous. Such a mix of understanding the scientific effects of something together with the ability to relate it in story form should not be discounted. Klein, by contrast, potentially suffers from a lack of accessibility in this regard on the larger scale, as her work is much more logically organized, dealing in specifics:

> Recent years have been filled with moments when societies suddenly decide they have had enough, defying all experts and forecasters—from the Arab Spring … to Europe's "squares movement" that saw city centers taken over by demonstrators for months, to Occupy Wall Street, to the student movements of Chile and Quebec. (Klein, 464–5)

Here again, Klein is clear, and perhaps inspiring, while also remaining much more comfortable in numbers and broader organization than of the actual visualization that Carson effects. Klein, however, is of the "Internet Age," a different moment when perhaps this type of work resonates with a new generation. Her work also possibly benefits from this age in that one may read it in-depth if truly interested, but the writer can also utilize concise snippets in her path toward Environmental Justice. Likewise, she employs copious interviews to her advantage, as the modern media, if often a bit too concise and flippant (with meme culture for instance), is also rampant and multiform. Her actual, physical forms of civil disobedience, then, only add to her well-managed image.[7] This allows her to employ a more data-driven, academic structure in her works without her message getting lost.

What about others, though? Are there traps others should avoid? Klein's work is a strong starting point for Alaskan discussions on environmentalism because 1) she advocates working with First Nations peoples (an important aspect of responsible Alaskan literature) and 2) she could essentially be seen as the Carson of the 21st century, her techniques and degrees of advocacy fast becoming the weight against which others are weighed.

What this part of the research will attempt to argue is for a multitude of voices theme, expressed in different ways, evolving at different times in the modern environmental zeitgeist, some broadly-scoped and others targeting specific communities. All of these voices then create a mosaic with which to engage individual communities and the world at large in order to bring them together to address the broader issues extraordinarily pressing to the 21st century. Alaska then becomes a target because 1) Alaska is a microcosm of the wider world, with diverse communities and geography, 2) climate change has been affecting the Arctic first, and 3) Alaska's natural resources are being targeted by oil companies and mining operations.

7 "This Changes Everything." *This Changes Everything*. thischangeseverything.org. Viewed on 30/6/2019.

FIGURE 10 Goldstream valley
PHOTO BY CHRIS FERGUSON

CHAPTER 8

Perspective on Myopic Writing

Perspectives in the Country and from Outside are constantly at odds in Alaska. On the one hand the region has an abundance of rich natural resources that support local biomes and communities, and on the other hand there is a long history of exploitation from Outside, creating a tangled, often fraught, relationship. This chapter will examine how travel writers approach the Country/Outside dynamic, particularly Jonathan Raban's technique in *Passage to Juneau* of drawing himself back out toward wider civilization. A link will also be made to the Environmental Justice Movement, particularly in Janis Johnson's assertion of a genocide of Alaska Native culture. In this case, several questions arise: to which communities are travel writers beholden?; How have travel writers failed in sensitivity toward Alaska Natives?; and what techniques have travel writers employed to take multiple communities into account? Raban's discussions of history and anthropology will remain at the forefront, as I argue that despite his ostensible insensitivity at times, his multiple perspectives are dynamic and approach novel questions.

Due to geographic size and isolation, as well as cultural and natural diversity, Alaskans consider themselves a unique conglomeration of peoples. Nancy Lord, for instance, has written eleven books, and not one takes as its subject the world Outside. Likewise, McPhee employs colloquial vocabulary throughout his work on Alaska. Travel writers have an obvious advantage with regard to a less myopic view by being from elsewhere, but once "in the Country," they tend to get caught up in the Country itself, which tends to gobble one up in its grandeur of pure scale. For instance, nowhere in John McPhee's *Coming into the Country* is there mention of the conversation being had of the effects of Alaska's oil beyond its own borders. Nor is there mention of any other unifying strain with the Lower 48 as Naomi Klein suggests be considered in her 2014 work (367–87). Barry Lopez' *Arctic Dreams* and Peter Matthiessen's *Oomingmak* also do not much approach the relationship with the world Outside.

Out of the six primary works in this research, Raban is the only writer whose work maintains a strong connection with the world Outside. One could argue that *Passage to Juneau* is less about Alaska than about Raban's own, personal, life journey, which takes its physical form in his sailing trip from Seattle to Alaska. They would not be wrong, but nonetheless Raban's personal journey (his failing marriage – 420–4 –, his dying father – 304 –, his abandoning his sailboat to go back home to attend his father's funeral in England – 280–7)

 | DOI:10.1163/9789004734807_011

does not diminish his travel writing. Instead, his personal journey serves as a creative parallel to his physical journey on the waters off the Inside Passage. Both are compared to a "whirling abyss," which lies in contrast to a "tidy and benign" natural design of life (190–1). Furthermore, Raban is able to maintain perspective through the works of previous travel writers, particularly from anthropologists and major explorers (Vancouver – 57–66, etc. –, Cook – 145–7 –, Boas – 203, 216, 225, 231–2 –, Lévi-Strauss, etc. – 22, 30, 37, 217–27). These are constant preoccupations, and he keeps them in mind as he physically moves through space, explaining what happened (sometimes) hundreds of years earlier. Occasionally, he even offers historiographical commentary such as his supposition that totem poles likely grew to the imposing size we now know after the wealth and technology offered by western traders but before the anthropologists arrived (29). This constant frame of reference allows the author to step away from the direct geographical space and analyze the local relationship with the world Outside, over multiple centuries. Lopez arguably does this in *Arctic Dreams*, but that author's discussion of explorers concentrates not on the interplay between the local and the world Outside, but essentially on how the explorers were drawn into the region and how they reacted to such a physical experience (355–406).

Raban's unique perspective involves the physical motion and energy required in the actual travel to Alaska, which commercial airliners have mostly eliminated. In the process, Raban is confronted with the existential perils of the kind of wilderness that constantly encroaches upon civilized life. No other writer in the oeuvre offers this perspective on travel.[1] The reason why Alaska has remained the Last Frontier in popular imagination is partly this geographic isolation. Williwaw winds, debilitating ice and fog, and dangerous, unpredictable tides in the cold oceans on long voyages kept the Russians, the Americans, the Spanish, the English, and the French from truly making headway in the region. In the end, no civilization has ever truly been able to conquer the Last Frontier. In Raban's work, the reader is exposed to these physical difficulties experienced in travel, mixed with the historical perspective not only from the Europeans but from native aspects as well (221–8). Therefore, when Raban seeks out shady bars (*Passage*, 320–1, 349, 364–66, 412), the reader has enough background to understand the context of the negative stereotypes being

1 Joe McGinniss does indeed begin *Going to Extremes* with the actual travel to Alaska itself, in a pickup truck, also getting across the physical difficulties of reaching the region (3–29). In *Coming into the Country*, there is also a short part about how extremely difficult it is to reach the capital, Juneau, because of the fog and rough terrain (138–40), but that is not firsthand travel.

confronted. Except, almost as soon as the reader is sucked into the Country in Raban's work, the reader is pulled back out toward broader civilization by the account of Raban's personal life, back in Seattle and England. The writer constantly maintains a wider perspective through this technique, including the culture shock he feels at re-experiencing his home country (250–1). This tug away from the local perspective offers the reader a guttural shock of travel between whirling, Subarctic waters and industrial civilization.

But, what guidelines should travel writers follow? Should everyone imitate Raban's off-beat, creative model? To whom are Outside travel writers beholden? The answer is multi-fold. Ultimately, a travel writer is writing for an audience outside their chosen region. The writer also typically has an agenda (as Raban, following in the footsteps of the exploratory giants while simultaneously "walking down Pound Lane" – *Passage*, 283). Their goal is often a mix of explaining their exotic, chosen region and furthering this personal "agenda" for an audience back home, but they also must not neglect the peoples and cultures in the places they travel to and through.

In the case of Alaska, the situation can be both tense and tangled because the majority of the population is of white or European descent, and most citizens have not been born in the state.[2] As the state tends to be fairly segregated, or at least Alaska Native culture is often concentrated, one can drive through densely-populated parts of Alaska (from the Anchorage Bowl down along Cook Inlet to Homer or up past the Matanuska-Susitna Valley toward Denali) without even noticing Alaska Native culture (Eggers, *Heroes of the Frontier*). Or, one may exclusively notice Alaska Native cultural degradation in the form of sodden and/or drunken people in Anchorage's parks and streets. *The Environmental Justice Reader* goes so far as to suggest that this is due to a cultural genocide, which is most clearly stated in Janis Johnson's essay, "Saving the Salmon, Saving the People" (271), and in Rachel Stein's "Activism as Affirmation" (204). Genetic susceptibility to alcohol is widely known,[3] as well as susceptibility to illness (Diamond, *Guns, Germs, and Steel*),[4] and Mei Mei Evans also introduces the testimony of Sterling Gologergen of highly-carcinogenic levels of pollution permeating Arctic and Subarctic subsistence communities

2 Silva, Cristina. "People no longer want to move to Alaska for the first time since World War II; Here's why." *Newsweek*. "https://www.newsweek.com/people-no-longer-want-move-alaska-first-time-world-war-ii-heres-why-843609." 13/3/2018.

3 Ehlers and Gizer. "Evidence for a Genetic Component for Substance Dependence in Native Americans." *AM J Psychiatry*. „https://www.ncbi.nlm.nih.gov/pmc/articles/PMC3603686/." 1/2/2013.

4 Diamond, Jared. *Guns, Germs, and Steel: The Fates of Human Societies*. New York: W. W. Norton and Company, 1999.

(*Environmental Justice Reader*, 34–5). Subsistence living, the *raison d'être* and form of self-actualization for many Native peoples, even well into the 20th century, as described in numerous works, including Ann Fienup-Riordan's *Boundaries and Passages* (366–70), Bill Hess' *The Gift of the Whale* (222–4), and Tom Lowenstein's *Ancient Land, Sacred Whale* (XI–XV), is now difficult to come by. Furthermore, depression caused by this loss of purpose, the introduction of alcohol, the mercury poisoning lowering the IQs of children (Hess, 222–4),[5] the women having been abused, have all accumulated as community-wide trauma, and this trauma is subsequently relayed from generation to generation (Stein, "Activism as Affirmation," *Environmental Justice Reader*, 196). Stein notes how a woman's tribe is starved to death after white trappers overhunt and poison meat with cyanide, and the woman is then raped and mistreated. The woman then mistreats her own daughter, who in turn mistreats the granddaughter when she is born, creating a vicious cycle of abuse started by the arrival of European culture (Stein, 196–7). With reason, this generational trauma has made many environmentalists, ethnologists, and, broadly, Native Americans themselves, very defensive. When writers come to Alaska and ignore Natives altogether, focusing on frontier pioneering and homesteading, or only describing Natives as drunk on the streets or in bars without focusing on the issues that led to the problem, the original peoples-cum-minorities are understandably upset. As discussed, McGinniss fell into this trap in *Going to Extremes*, even if he did expose otherwise undisclosed aspects of the state's political and social system (16–18). Later in this part of the research, some of the same issues will be discussed in critiquing aspects of Jon Krakauer's work (who mentions exactly one Alaska Native in *Into the Wild*: "an old toothless Tlingit man," who is the last person in a bar aside from the author himself; Krakauer being there because he cannot sleep, while the Native is not offered subjectivity – 154) and Dave Eggers (who ignores Native Alaskans altogether in lieu of celebrating the Last Frontier). Therefore, despite the fact that the immediate responsibility of the Outside travel writer may be to those Outside, the travel writer is also morally beholden to Alaska Natives themselves. Otherwise, the counter-narrative that native peoples in general were better off under colonial government rule[6] may take root, even though, as demonstrated above, colonial rule was

5 Bell, Lee. "Mercury in women of child-bearing age in 25 countries." *Biodiversity Research Institute*. "https://ipen.org/sites/default/files/documents/updateNov14_mercury-women-report-v1_6.pdf." 9/2017.

6 Sarkar, Ash. "The colonial past is another country. Let's leave Boris Johnson there." *The Guardian*. "https://www.theguardian.com/commentisfree/2018/jul/10/past-boris-johnson-colonial-brexit-colony." 10/7/2018.

the cause of what is increasingly being labeled a cultural genocide, as well as societal levels of generational trauma.

This perspective stems from what philosopher Miranda Fricker has labeled "epistemic injustice."[7] Fricker breaks this idea into two parts: testimonial injustice and hermeneutical injustice. Testimonial injustice involves the testimonies of individuals and groups not being taken seriously due their identity (e.g. sex, gender, or race). Hermeneutical injustice is when injustices are not understood, by the victims or by society, because there is a cultural gap in hermeneutics, or the way that experiences are discussed (e.g. the employment of labels such as sexual harassment that empower women not to put up with unacceptable behavior in the workplace). With epistemic injustice in mind, travel writing carries heavy ethical responsibilities. Travel writers are often the ones who frame a place and culture for outsiders, and a lack of panoptic perspective can lead to misperceptions and narrowed views that are uninformed, such as that of Arctic peoples being backward and needing the support of industrial society to survive, when in many cases Arctic peoples have been thriving for centuries, if not millennia. A strong piece of travel writing, then, can indeed stimulate the imaginations of the world Outside while also educating about local Native communities in a responsible manner. The following chapters will attempt to show multiple ways in which this is possible.

---

7 Fricker, Miranda. *Epistemic Injustice: Power and Ethics of Knowing*. Oxford: Oxford University Press, 2007.

FIGURE 11 University of Alaska Fairbanks
PHOTO BY CHRIS FERGUSON

CHAPTER 9

Psychological Issues and the Middle Way

In the environmental debate, in Alaska or elsewhere, one of the most important things to remember is that a mix of different perspectives, a multitude of voices, is needed.[1] The leadership of Naomi Klein and other modern grassroots activists, such as Joni Adamson and the Environmental Justice movement, is enormously important in this respect. Yet, other authors, speaking directly to targeted audiences who tend to make decisions and tend to be at a political advantage while also holding a deaf ear to environmentalism, could also be considered and not discounted. Concurrently, alongside the attempted mass mobilizations and the targeted writing, we must also keep in mind that most people are just not concerned with important political issues if those issues do not directly and immediately affect the individual. They are more likely to want to live their lives. In fact, it would not be a stretch to suggest that most people in the world are merely attempting to get by, making small decision after small decision to best affect their personal and familial interests, with at times perhaps a slight concern for the environment if it is an easy enough choice to make, such as changing a lightbulb or recycling or riding a bike instead of taking a car. Even that can be too much for the comforts of many, however. As we will see shortly, this group has been labeled "the middle third," and it will not be easily mobilized in a political fight one way or the other.

The problem with the declarations made above is that the evidence is so often circumstantial, but there nonetheless is evidence toward which we may point. In this chapter, Nicole Seymour's work *Bad Environmentalism* will be employed, as her work points out many of the problems with the environmental movement's broader focus. Her ideas can then be extrapolated onto Alaska, pointing out how Alaska's multiple, diverse communities are not taken into account in a global vision of environmentalism.

Seymour advocates for more irony and irreverence in the environmental debate, which stems from her experience in queer theory (where irony is employed playfully to effect change – 23–25). She likewise attempts to create

1 Parts of this chapter were published in Caliban 61.
Ferguson, Ben. "John McPhee and the Technocratic proximity to the Earth." *Caliban 61*. Edited by Françoise Besson. Toulouse : Presses Universitaires du Midi, 2019.

 | DOI:10.1163/9789004734807_012

a "queer ecology," which she explores in her first work, *Strange Natures*[2] (*Bad Environmentalism*, 23), and in chapter 3 of *Bad Environmentalism*: "Climate Change is a Drag, and Camping Can Be Campy." Seymour's views will be explored herein in relation to Jonathan Raban's work. However, it is important to begin by pointing out that her work not only advocates reaching out to different groups of people, but it advocates doing so on their terms. For instance, she evokes such ideas as the economic and social science study of "path dependency":

> "the limitations in choice that, among other things, make car-driving the optimal choice for many of us. Path dependency complicates the idea of free, individual agents who can be swayed by outside knowledge to act environmentally – an idea that so many environmental documentaries seem to take for granted." (*Bad Environmentalism*, 63–64)

This speaks to the idea that most people will not act unless it is within their reach to do so, within the bounds of their normal lives. They are more likely to act environmentally with better options, but Seymour notes that most documentaries avoid these issues, alienating the groups who most need to be convinced, often via unintentional employment of classism. This can take many forms. For instance, people tend to turn away when they are not met with the literary devices employed by their own demographic, such as humor and irony. Seymour even presents the idea that queer culture has been alienated in environmentalism to such an extent that many queer people have begun to lean toward "anti-futuris[m]" ("fuck the future") and align more closely in this regard with "Big Oil moguls" and "the Right" (112–13) than with environmentalists, who traditionally align with the political left. Much of this chapter will approach the idea of reaching out to "the middle third," a broad political group of people, crossing many boundaries not traditionally considered. But, parallel to Seymour's perspective of avoiding hetero-normative and classist perspectives through technique, this part will examine the alienation of technocrats, as well as the rural communities in the urban v. rural divide, and Native American, African American, and other minority communities.

As Seymour notes, irony in a call for change speaks to the queer community's cultural base of understanding, how it has traditionally coped and interacted with the world, a widely adopted affect (111–23). Broadly speaking, this chapter proposes exactly the same idea: effecting change in habits to reflect

2 Seymour, Nicole. *Strange Natures: Futurity, Empathy, and the Queer Ecological Imagination*. Illinois: University of Illinois Press, 2013.

a more environmentally respectful agenda is to be respectful of any particular group's affect, not solely in their literary techniques but where they find their content as well. This is why Seymour focuses heavily on modes of entertainment such as documentary, cinema, music, and theatre, and this idea can include social networking, or even podcasting (*Coffee and Quaq*).[3] For literature, this means that there is an importance in the "bestselling" type works à la John Krakauer and John Vaillant, alongside regional specialty writings such as much of Nancy Lord's work. The 'bestselling' works tend to be syntactically and grammatically simple, readable, and entertaining, while also adding depth, facts, and general conversation about their respective subjects. For instance, Krakauer's *Into the Wild* sparked enormous debate both within and outside Alaska. Werner Herzog provoked a similar debate with his documentary on Timothy Treadwell, *Grizzly Man*. Most readers and viewers seem to have an opinion on the two protagonists because the stories of both Treadwell and McCandless are fascinating in their strangeness, and both are presented by seasoned professionals at the height of their careers, with formulas to their craft which they know will reach intended audiences. *Into the Wild* may not be a literary masterpiece in the same sense that we discuss literary nonfiction, with works such as Raban's *Passage to Juneau*, which is highly informed by historical context of multiple centuries, is multi-layered, includes anthropological academia and personal historiography, but *Into the Wild* is indeed thorough from a journalistic perspective and opens worlds of experience to extraordinary numbers of people. I propose that environmental circles should be supporting these types of works alongside those of the Lopezes of the world (i.e. what Raban calls the US Northwest writers), since what goes on in Alaska immediately becomes adventure, holding the attention of the eyes and ears of the world in search of literary escape.

To sum up: mobilization of the masses through responsible, unifying environmental literature (along with other media), while also targeting politically poisonous groups typically considered a lost cause within the predominant affect of environmentalism, and also attempting to reach the mainstream (or "middle third") in the process, are all equally important. A combination of these three thirds to complete the whole can make a difference together in a way that individual, warring communities and groups of people cannot, even

3 A podcast for Alaska Natives, by an Alaska Native, with Alaska Native guests, in order to offer the broader community an inclusive conversation about topics such as Native subsistence food use and a divide within the larger community over traditional tattoos, as well as LGBTQIA+ perspectives, etc. Glenn, Alice Qannik. *Coffee and Quaq*. "https://www.coffeeand quaq.com. Viewed on 4/4/2020.

if they spur on strong militancy through hatred of the other. By the end of the chapter (and, further, by the end of Part 2 of this work), a clearer picture of this argument should emerge. For the moment, I will focus on analyzing a recent psychological assertion that rationalization itself is a "social construct," and what that means for political dichotomy.

1 Introduction to Rationalization as a Social Construct, Linked with Political Journalism and Contemporary Academic History

Dan Sperber and Hugo Mercier's 2017 work on the decision-making process of humans, entitled *The Enigma of Reason*, posits that 100% of the time human beings make decisions before rationalizing. In other words, rationalization is what they call a "social construct" (127). In a world, and particularly a country like the US, where capitalist-funded propaganda is rampant and large demographics of low-information inhabitants are swayed by campaigns of sensationalism and civil discord, this is a daunting proposition. Alexis de Tocqueville noted during his early 19th century travels through America that democracy crushes individual thought, not by government, as in Europe of the day, but by the masses themselves (*Democracy in America*, abridged, 102–5). Whereas in the metropolises of America climate change and its ramifications are unquestioned truths, are even a social pressure to believe, in majority-rural areas like the US Southeast and Midwest groups are often swayed toward disbelief by direct propaganda (*Dark Money*;[4] *Democracy in Chains*). These disbeliefs are subsequently backed up by the group tendencies of local communities, people finally rationalizing their decisions in ways that from the outside may appear blatantly ignorant. This widens the gulf between the two US political parties, constantly pushing on the "us versus them" narrative, swaying political conservatives into the corner of environmental exploitation with what Nancy Maclean calls "junk pseudoscience" (*Democracy in Chains*, 216).

One of the important underlying theories in research of American environmental conservation in the 21st century is that two polar opposite political camps are simply not listening to each other in American discourse anymore. This is nothing new, but Nancy Maclean and Jane Mayer indeed argue that wealth is being accrued at a heretofore unseen scale, even in comparison to the turn-of-the-20th-century Gilded Age, and that certain super-wealthy families

4 Mayer, Jane. *Dark Money: The Hidden History of the Billionaires Behind the Rise of the Radical Right*. New York: Doubleday, 2016.

are using their influence to purposely tear society apart at its seams in order to stitch it back together in their own image as unfettered capitalists, or what are sometimes called anarcho-capitalists (Maclean, XX–XXXII, 81). Through manipulation of higher education, local and national governments, the media, and the judicial system, wealthy individuals have put in place highly-organized political machines to promote their interests (132–53). The pertinence of this organization to environmentalism is that these political machines have decided that government should be vastly downsized in order to allow a pure, classically liberal economy to exist without regulation. This means industrialists would have free rein to pollute how they saw fit, which the Koch family currently does whenever possible as a means of efficiency in profit (Mayer, *Dark Money*, 120–40). It also means that the belief in climate change itself has been directly manipulated.

MacLean quotes: "'It might be hard to admit,' said the chair of the economics department at George Mason [a Koch funded stronghold], Donald J. Boudreaux, but because public choice showed that a government cure would be worse (from their perspective, of course) than the disease, global warming 'is best left alone'" (*Democracy in Chains*, 216). MacLean goes on to describe how, through the Cato Institute and other "Koch-funded think tanks," "junk pseudoscience" has been spread in the model of the tobacco industry to sew public doubt over climate change. " ... To put all this another way," she says, "if the Koch-network-funded academics and institutions were not in the conversation, the public would have little doubt that the evidence of science is overwhelming and government action to prevent further global warming is urgent" (216–7). She then cites a statistic that the number of Americans who believe in the effect of fossil fuels on climate change dropped from 77 percent to 44 percent between 2007 and 2011. Through propaganda, this collection of "think tanks" is sowing such discord and confusion, that the average person, even the well-educated, has been throwing their hands up and walking away in disgust, thereby rendering this machine more powerful, as disenchantment is promoted by design.

This section serves to introduce the idea of direct ideological interference in the fight against climate change, reinforced by what could be simplified as provoking gut reactions among voters in an organized manner, on the local, state, and federal level. Even if Alaskans tend to be distrustful of Outside US government, cultural, and economic interference, this section nonetheless goes a long way in understanding some key differences between Alaska and other Arctic regions. Now, Jonathan Raban's techniques of pointing out this political dichotomy will be analyzed. Raban does not place blame directly on the US political right, as do Mayer and Maclean, because he examines how local communities have been alienated in the process of gentrification. Classist environmentalism then serves as a further slap in the face. It becomes a divisive issue and takes hold not only due to propaganda but because of the sanctimoniousness (which Nicole Seymour discusses directly in her work *Bad Environmentalism*) of newly arrived classes of wealthy who look down on working class culture as a whole.

2 What in the World Can We Learn about the American Political Dichotomy from an Englishman?

Jonathan Raban, another outsider, discusses the issue of class dichotomy with regard to environmentalism in the eponymous essay of his book *Driving Home* (23–80). Arriving from Britain, Raban fell in love with Seattle's proximity to nature and its nautical and meteorological similarities to his home country[5] (9–11). However, he also made much of his living in the 1990s escaping the burgeoning, palimpsest of a city which was constantly outgrowing itself amidst an unprecedented tech boom (*Passage to Juneau* – to Juneau; *Bad Land* – to Montana; *Driving Home* – all around the outskirts of Seattle). Somewhat like McPhee, Raban tends to view the world multiradially, open to multiple points of view, and he also grew up in a small, conservative town. Possibly because of this, both writers clearly see the redeeming qualities of the conservative working classes. Raban grew up with an Anglican priest of a father (261, 264–5), though he escaped to London to become a bohemian writer, publishing *Soft City* about London of the 1970s as a cultural palimpsest (37–66) before reinventing himself in the 1990s by moving to Seattle. Despite maintaining a somewhat bohemian life in the US, writing on a sailboat and purposely living in an old house with crooked floors (*Driving*, 16–7; *Passage*, 21–5), he spent much of his time attempting to fit into rural, conservative America (*Passage to Juneau*, 3–37; *Driving Home*, 15–80; *Badland*, 1–4). Raban fights back and forth between affiliations in his work, or, seen from another angle, never buys wholeheartedly into a single political ideology with all of its subsequent baggage. To be more precise, Raban seems stable in his politically-leftward leanings from the standpoint of British politics, even discussing them openly at one point when returning to England in *Passage* (263–4), not to mention his works such as *Coasting*[6]

5 Not to mention, "I met someone … the usual story" (9).

6 Raban, Jonathan. *Coasting: A Private Voyage*. 1986. New York: Knopf Doubleday Publishing Group, 2011.

at the moment of the Falklands War and *Old Glory*[7] during the rise of Ronald Reagan, but he also seems to prefer the company of politically-right-leaning working classes outside the cities in the US, at least as a sort of case study.

In technique, McPhee almost always employs white males in his work, and when he does employ female protagonists they tend to be in male-dominant areas of work such as 1980s field geology (*Annals*, 19–20). McPhee's multiradiality, then, comes more from picking multiple perspectives of an argument and discussing the nuance of both sides (such as the case of the Gelvins in *Coming into the Country*), while Raban never exactly picks a particular argument or person to profile, but instead creates discussions around events he comes across in his peregrinations (though his peregrinations admittedly tend to concentrate on areas dominated by white men: the docks of Seattle, for instance). As an example, in *Passage to Juneau*, Raban heads up through the coastal fjords of British Columbia and passes through a small community called Klemtu. Like so many other small communities he passes, he meets interesting people with surprising stories and backgrounds. The guidebooks Raban has brought along tell him to avoid Klemtu "because the natives [are] unfriendly" (*Passage*, 323), yet the author finds just the opposite to be the case. The people are quite friendly, and the inhabitants are actually offended by the guidebooks. The author could stop there, or give a description of the town itself, its geography, infrastructure, and culture, and move on, but what he does instead is bring the individual story into the frame of his work as a microcosm of the local situation, personalizing and yet generalizing at once. From one short conversation, we are given the fact that the community, like so many in and around the Arctic, is a cultural "melting pot," mixed with the economic plight of fishermen turned naturalists, who still attempt to practice subsistence living. We also see how this man lives on Vancouver Island and comes back in the summer. The repeating of the phrase "It's not rocking science" (326) in place of "It's not rocket science," normally employed to point out that a task is not too technically difficult, serves as a linguistic joke, but also as an ironic technique to demonstrate a lack of education in a man whom Raban calls "a marine biologist." The point is likely not to mock the biologist, considering that the author is going out of his way to show the positives of this man and his community, but to demonstrate that his description is not a mere piece of subversive hagiography of the community. Raban, then, is joking about the man, showing a flaw in his education, while also admiring his will, his morals, his sense of community and care for others' well-being. Neither this plight nor

7 Raban, Jonathan. *Old Glory: A Voyage Down the Mississippi*. 1981. New York: Knopf Doubleday Publishing Group, 2011.

the local protagonist are uncommon to small communities, and the multiple aspects involved in the brief story are therefore the subjects so often discussed with regard to stories of small Pacific Northwestern communities, but, importantly, to Native communities as well (McPhee, 331–39; Lord, *Beluga Days*, 38). Raban brings small Western communities and Native communities together, then, discussing them in terms that the other can understand, which is how Raban unites people through his cultural iconoclasm, breaking down bad stereotypes and misinterpretations as he moves through the world, or "walk[s] down Pound Land." Again, we could discuss the fact that the author lacks an eye for the environment, but he also shows opposing spectrums of politics and class reflections of peoples themselves lying in wait in the most unexpected of places, such as Klemtu, British Columbia.

Raban, then, could almost be considered a flaneur of the road, spending vast amounts of time driving aimlessly into rural areas upon arrival in Seattle (if we are to believe his writing – his collection of essays is indeed entitled *Driving Home*), and he noticed quite early in his drives east toward Montana that almost as soon as one left the bubble of the politically left city the politics became very conservative (*Driving Home*, 25). Those on the fringes felt like the city had arrogantly taken over the entire region and stood for everything that they did not (25–30). The city-dwellers were pushing outward as well, wishing to restore habitats of endangered birds they hoped to spot on weekend hikes on the same lands other people already considered for development or other uses, like hunting and forestry (25–30). Raban even bites into the ironic joke that if there is no spotted owl then a forest is not pristine: "old-growth" or "grand cru" (*Passage*, 190; *Driving Home*, 25). The Seattleites, "the Houyhnhnms" (as a rolling, jocular metaphor on Swift's Gulliver's Travels in "Driving Home") dressed differently, ate differently, drove different cars, lived their entire lives differently, and the conservatives outside the city, "the Yahoos," who had already been established in the region, obviously did not appreciate the rapid change (28–32). Raban's discussion of the issues of 1990s Seattle is complicated, which it likely should be. He seems to be pointing out the damaging relationship with the region that the educated liberals are creating, at the expense of collective environmental collaboration.

Concurrent to this argument, as Nicole Seymour notes in *Bad Environmentalism*, the overly didactic, the self-righteous, and the sanctimony (4–5) of the environmental movement leads it toward becoming more of an affect as a movement than an academic methodology (28), with playful irony being left out of environmentalism almost completely. She goes on to propose that there is a difference between irony and sarcasm. Sarcasm, she insinuates, is biting and offensive as an end, while irony hopes to point out an issue to effect change, or at least display nuance (31–2). Irony is what Raban's work employs. For instance, as noted in Part 1 herein, Raban focuses on not being able to genuflect in the right places when reading US Northwest writers such as Lopez and Snyder. He also employs his young daughter to demonstrate how she has been brainwashed to think that a forest is not a forest unless it has never been maintained by humans (*Passage*, 189), preferably with that endangered spotted owl in its midst (190), and that the land he grew up in (England), as green as it was and full of insects, plants, and animals, was not real nature (190). This research would propose that this section of *Passage* is dripping with irony, and not sarcasm, per Seymour's definition of the two terms. Seymour importantly notes that that directness and didacticism that draws many into the environmentalist movement also pushes away others, as they do not want to get caught up in the "doom and gloom," or the consequent anxiety, "solastalgia," of the current climate crisis (3–4). She essentially argues that irony offers a sort of playfulness that would allow environmentalism a sense of humor and an openness to more social groups, including alterities to the movement such as working classes, people of color, and the LGBTQIA+ community.

Raban does indeed criticize the fact that the Seattle environmental community, which he implies is linked with the tech community ("Driving Home," 24–5), is overly zealous about pristine forest versus what is likely a more European ideology of Gifford Pinchot's "Wise Use" policy that the conservationist learned from the École nationale forestière in Nancy, France (Brinkley, *The Wilderness Warrior*, 340–1).[8] The reason this criticism seems somewhat straightforward is that the writer notes the swift retaking of civilization by the forest in the US Northwest and in British Columbia (*Passage*, 185–6, 316–7), though he also chastises the ubiquitousness of clearcutting he sees along the way (175, 237–9). With these multiple and opposing criticisms in mind, it seems fair to say the writer would argue for truly wise-use implementation, which would not involve extreme clearcutting, but would allow more freedom to community development by the working classes outside Seattle, who feel excluded.

Where Raban is particularly ironic, as noted above, is with the insertion of the necessity of a spotted owl in the forest. Here, the author is not at all criticizing the plight of the spotted owl, but rather how the owl has become the single

8 For the record, "Old growth" forests *are* incredibly important, supporting a much more diverse community of insects, fauna, flora than regrown clear-cut forests, which have a tendency to be dominated by invasive brush (Kimmerer, *Braiding Sweetgrass*, 277–92).

representation of the fight against deforestation, as the polar bear has for climate change in the Arctic. To Raban, the militancy toward the spotted owl has become overwrought, and so he employs it as an ironic literary device to point out how its symbolism is broadening a cultural chasm in the region. This same argument is also extended to fights over cultural disparities in recreation such as the riding of ATVs, jet-skis, etc. ("Driving Home," 25–6), which the working classes tend to enjoy more than the tech classes (hence Seymour's classism in the environmentalist affect – 56, 192–6, 203, 206). The irony is extended to the tech classes in the same conversation in that they tend to enjoy golf ("Driving Home," 25), which is a notoriously un-ecological sport, and that they are building entire new suburbs which seem like false cities and have no relation to the natural, or real world: "Thousands of these diagrammatic people were housed along I-90 in … communities in which the villagers met in the evenings at neighboring exercycles in the village health club or at a pre-breakfast tee-off on the village driving range" (23–4).

The physical result of the ever-broadening chasm is that newly-arrived, tech-working Seattleites might go out to the countryside and end up the objects of this pent-up frustration at the hands of locals. The tech workers subsequently become more defensive about their own ideals, re-rationalizing their same pre-digested, "social construct" of a zeitgeist. This urban v. rural issue then escalates in a vicious cycle, potentially without end, and results in rural militias taking over national forests[9] and Ted Kaczynski, a veritable math genius professor at UC Berkeley, who ended up attempting to wrench society back to a pre-Industrial-Revolution epoch through terrorism. Interestingly, Raban focuses on Kaczynski in *Bad Land*, his book on Montana (Raban, *Bad Land*, 292, 299–305), and so do Mercier & Sperber (19–20). The two psychologists focus on the decision-making process itself and how an individual or group is influenced, in this case with large-scale terrorism as the unfortunate result, while the travel writer focuses on the stories and impressions he receives moving through communities affected by such influences of negative feedback, with reflections reinforced by his background in humanities research. Thus, Raban's Seattle work shows us how we may avoid the pitfalls of class hatred and exclusionary tactics due to opposing affects, and, consequently, why working toward inclusion of all groups is ultimately a helpful strategy. This is particularly important in Alaska where so many interests are at play.

9 Bundyville Podcast. *Oregon Public Broadcasting*. Peter Frick-Wright, Robert Carver, and Ryan Haas (Producers). "https://www.opb.org/news/article/bundyville-occupation-podcast/." Viewed on 9/7/2019.

Now, we will turn to this very issue of race and class more specifically, in order to discuss further nuance.

3 Attention to Language and Tone

One aspect that does seem to sway the opinions of individual communities is speaking in local language, in local tone, in local affect. With the psychological research in mind that people make "snap judgements," "echo chamber" judgments, or "tyranny of the mass" judgments in the de Tocquevillian mode of democratic suppression suggested earlier in this chapter (affect), and adjust their logic behind those judgments to fit their theories, the act of speaking in an established affect's tone likely entices at least a certain percentage of adherents to that affect to adjust their judgements in a rationally compassionate direction. This lies in stark contrast to standing steadfast in the urban v. rural mode of political stratification, feeling the necessity (conscious or subconscious) to oppose the other side no matter what. This effect becomes more powerful when the group being targeted for a change in mindset is the group with the most political power.

This strategy is entirely apropos of conservation and Environmental Justice in Alaska because the same process is taking place there. Sometimes, this strategy goes hand-in-hand with that of Naomi Klein, but it differs in one major aspect: though grassroots movements are indeed of the utmost importance in evoking support there is a concurrent need to approach decision-makers, deniers, and the middle third to change their habits. Considering un-mobilized demographics, or demographics with whom we disagree, to be enemies, always employing the combative rhetoric and actions that Jonathan Raban references ("Driving Home," 25), can be unhelpful. This is the pitfall that Nancy Lord avoids so well in her work and into which the rhetoric of the Environmental Justice movement naturally risks falling when not conscientious, considering its activist tendencies. As a pertinent example, in Nancy Lord's "Introduction" to *Made of Salmon* the author talks about the sense of sharing when it comes to Alaskans and salmon. This passage was already analyzed in Part 1, but it continues by talking about how Alaskans are quite literally "made of salmon." She notes the "protein, omega oils, and calcium" in peoples' bodies and the "friendships and communities" created through the sharing. She then discusses Alaska's Native communities:

> The first people to inhabit our part of the world naturally chose the best places in which to live—those places defined as resource rich, where

> they had ample access to food. While some cultures oriented around sea mammal hunting and others around caribou, many of the first people were Salmon People above all … . (XVII)

Nancy Lord has done here what was suggested in the last chapter, namely, avoid victimhood. She has also found the voice, the tone, the general affect, of Alaskans, most often accessed through subject matter. *Made of Salmon* also parallels the Environmental Justice movement by discussing some of the same issues that it typically focuses on, though not employing the movement's typical tactics. Whereas the techniques of a work of Environmental Justice tends to attack the political situation head on, discussing the victimhood of the polluted and overfished salmon waters, instead, Nancy Lord's editing has attempted to pull all communities together through their shared love of salmon. The controversial topic of the Pebble Mine rarely needs to be tackled directly (Schooler, "Salmon Invictus," 3; Wilson III, Vernon, "Bristol Bay Youth Doing Their Part," 139–41). In the passage just above, for instance, Nancy Lord says "Alaskans" in an inclusive way, discussing how all Alaskans have a tradition of sharing through subsistence use. She does not immediately pinpoint any one group. Then, in the next part, she discusses the "indigenous cultures" of Alaska, but not as victims. Instead, she notes their important relationship with, and understanding of, salmon waters. In no way, in this passage, or in her entire introduction to the work, nor in the stories she chooses, do Alaska Natives come off as helpless victims.

As far as her subject matter and tone are concerned, taking on the characteristics, the affect, of all Alaskans, in an inclusive way, we can note the way she discusses the technical: the fishermen's skills, the canning, the smoking, which does not leave out the Natives or the conservative working classes. There is no classism, in other words, in her affect. But, importantly, this also allows her to write about salmon in a holistic way, which is inclusive of the environmental affect as well, be that an Alaska Native environmentalist affect or a white, liberal environmental affect. No other writer of the of six discussed in this research is able to reach across so many spectrums of people in a single geographical region. Yet, it is also important to remember that Nancy Lord is local. Her work has typically enjoyed success locally, whereas the others have had more success Outside. This idea will be examined further in the "Theory of the Local" chapter. For the purposes of this chapter, however, suffice it to say that Nancy Lord's writing shows how adopting local affect can be helpful in reaching local communities.

The next two ideas to be measured in this section will be Paul Bloom's rejection of empathy in lieu of rational compassion, and Mark Lilla's rejection of identity politics in its self-obsession. Both Bloom and Lilla are polarizing academic figures, so, not only will their ideas be discussed, but a metacriticism of their respective messages from the perspective of this chapter's continuing discussion on affect will also be explored.

4 The Bloom and Lilla Controversies

Another psychologist, Paul Bloom, posits in his work *Against Empathy*[10] that, contrary to general tendencies of the modern occidental zeitgeist, empathy is being overemphasized as a necessary tool in decision-making. He argues for rationalization, saying that empathy can lead a person to negative acts as well as positive ones (2–3). Bloom does not call Mercier & Sperber out by name, but he is attacking their strain of thought when he notes, with irony, or perhaps sarcasm, that certain of his colleagues are attempting to prove that rationality is impotent, while they simultaneously employ well-organized, rational studies to prove that point (52). Except, Mercier & Sperber do not argue "that rationality is impotent." In fact, they argue that rationality is important, but that it is affected, 100% of the time, by outside influences, is essentially a "social construct" (an affect), all of our social tendencies coming into play before we rationalize. Bloom uses for reference the idea that no one can agree at a community meeting over small, practical decisions (238), which could use some rational analysis in lieu of empathy. Mercier & Sperber may reply, however, that whether or not one wants to put up a Stop sign, as trivial as this debate may seem, may nonetheless derive from a social and familial construct, which is subsequently rationalized in accordance with the decision one has already made on the subject (again, a cultural affect), so that rational debate would likely have no positive effect.

Bloom also argues that empathy can lead a person to hurtful acts as well as altruism. In other words, understanding the minds and emotions of others can aid a negatively motivated person in exploiting others for one's own purposes, and that the most effective way to make a positive difference in the world, in charity or even everyday acts, is through what he calls "rational compassion" (240–1). This idea translates well into environmental criticism because it can remind us that, through empathy, with a "bleeding-heart" focus on love (and potentially "inclusion" in the current zeitgeist where militants of inclusion are systematically excluding all discussion contrary to their beliefs, to the dismay

10 Bloom, Paul. *Against Empathy: The Case for Rational Compassion*. London: The Bodley Head, 2016.

of previous generations who have long fought for free speech in academia – Lilla, 89–90), certain movements may be turning away as many people as they attract followers (Seymour, 2–3). Jane Mayer notes explicitly that the think tanks of the US political right are highly cognizant of this divide and have been successful in exploiting it. She notes that long-time Koch-network "consigliere" Richard Fink believes that he will get a third of people to support him no matter what and that there will be another third which will oppose him, but that the true war is for this "middle third" (Mayer, 357–8).

With this in mind, David Brower transformed John Muir's Sierra Club into a household name in the mid-20th century and was the protagonist of McPhee's *Encounters with the Archdruid.* He was also highly instrumental in the preservation of numerous large-scale parks in the western U.S., through sheer force of will and charisma, according to McPhee, and, under his leadership, membership in the Sierra Club increased exponentially (*Encounters*, 208–13). But, Brower was also highly controversial and a perfect example of how loose logic, equivocation, and at times outright lies, through means-satisfy-the-end, "bleeding heart" empathy over reasoning, among other issues, can hurt a movement by creating as many enemies as it gathers allies. People were either drawn to or away from Brower, as the sort of anti-McPhee (*Encounters*, 208–13).

I certainly do not argue that the world would have been better off without Brower and needs more McPhees, because Brower was by all accounts an extraordinary force for the environmental movement. McPhee, however, and his type of writing, has something to offer the environmental movement that Brower never could, embodying the reasoning, *Against Empathy*-type that Bloom strongly suggests we consider and may lead to success with "the middle third" alongside the already-mobilized preservationists. Here is a McPhee passage from *Encounters with the Archdruid*, which discusses Brower meeting with Charles Fraser, a real estate developer:

> The size of Fraser's yacht was proportionate to his distaste for wilderness. The yacht was ninety feet long. It contained five staterooms and a floor-through saloon. Its bar was stocked with Tanqueray gin. Fraser's Southern antennae had reached out unobtrusively, suprasocially, and their research had shown that Tanqueray is Brower's gin of gins. (135–6)

McPhee employs frequent irony in discussing both Brower and his foe. Fraser is attempting to present himself as environmentally friendly in his development of biologically diverse islands, and McPhee makes a joke about Fraser's distaste for camping and wilderness, and the size of that millionaire's yacht, effectively "pull[ing] the lever," if ever so slightly. In employing irony, McPhee reaches beyond the environmentalist affect, in Seymour's terms, but he also makes judgments without explicitly voicing them. The writer never accuses Fraser's environmental presentation of being but a facade. In making ironic jokes, which McPhee also does at the expense of Brower, who takes the Tanqueray and falls for Fraser's slick spiel just afterward, McPhee is able to step back journalistically and recuse himself from overt judgement, as the evidence stands for itself. The fact that the writer makes fun of both characters is important here. Yet, interestingly, McPhee's wariness of Fraser may also arise out of Fraser's dislike for camping, something that McPhee is so often physically doing in his works of travel, including *Encounters with the Archdruid* and *Coming into the Country*. So, cultural affect may be at play in McPhee's own judgements as well. Affect may also come into play with regard to Fraser and Brower's relationship. Brower, who is at odds with the two other men with whom McPhee pits him, antagonistically (Charles Park and Floyd Dominy), is swayed by the gin and by the natural scenery (first wave ecocritical perspective) which Fraser presents. McPhee comes across as aghast that the notorious Sierra Club leader is taken in by a plan to develop over such biologically important terrain after merely a gin and a boat ride and some affectionate discussion. Over the next few pages, McPhee describes "the archdruid" as so "invisible" that Fraser has to turn his argumentative attention against other people. Brower is flat out taken by the situation. McPhee wonders whether it is because this is a terrain toward which Brower does not feel quite as attached as that of the West (138). By the end of the excursion, however, Brower and Fraser have joined forces against the local landowner and are making plans on an "environmental" development (140–50). This convincing, by Fraser, of the environmental friendliness of an invasive real estate development, to one of the most important environmental figures of the 20th century, truly shows the importance of affect. Fraser speaks in language that Brower can understand, offering drinks that the latter loves, showing him the perspective of the vision that the environmentalist would want to see. On the other hand, the other two antagonists of Brower never tried to meet the environmentalist on his own terms, and those two trips were full of conflict throughout. This shows how empathy gets the best of Brower in a situation that should likely not do so, according to the man's own moral compunction. Rational compassion would have helped him avoid the pitfall into which he fell, and, having been to Hilton Head Island myself, developed by Fraser, as well as having visited nearby Jekyll Island, which is a protected park, the biological difference is shockingly disparate. This is important to keep in mind from a literary perspective. Fraser employs the affect this chapter has been suggesting, and McPhee mostly attempts to stick with rational compassion. In practice, this means that McPhee makes ironic jokes from time to time, not sarcastic ones, pointing out discrepancies. He also shows a healthy respect for all parties involved, which maintains rationality by pointing out the nuances of both sides of multiple environmentalist debates in the mid-20th century,[11] at a time when Brower is being pushed out of his Sierra Club, which sets up a broader discussion of the issues at hand with the environmentalism of the early 1970s.

To put Brower and McPhee and their respective methods into a political perspective, Nancy MacLean links climate change denial entirely to a coordinated movement of propaganda by the think tanks of a particular wing of the US political right (215–17). Therefore, avoiding the political dynamic is unacceptable here. Rather than taking sides, however, I am presenting an argument within the realm of environmental metacriticism, which must take the current political climate, and political techniques, into consideration.

As such, since environmentalism has been de facto linked to the US political left due to large scale climate denial and political inaction on the subject among the US political right, Mark Lilla's controversial work *The Once and Future Liberal* merits discussion. Lilla essentially proposes that for US political liberals to effect their worldview through politics, they must focus, as MacLean notes that the US political right does, on winning elections at every level instead of waging a culture war (Lilla, 4–5, 13; MacLean, 229–30). Lilla proposes that the Democratic Party move away from Identity Politics and toward inclusion and community. Here are his words:

> Identity liberalism banished the word we to the outer reaches of respectable political discourse. Yet there is no long-term future for liberalism without it. Historically liberals have called on us to ensure equal rights, they want us to feel a sense of solidarity with the unfortunate and help them. We is where everything begins. (117–8)

In this criticism, the author makes a further claim that Identity liberalism is "Reaganism for lefties," as it focuses predominantly on the I instead of the we. Lilla makes some very strong points which are apropos the theme of this research, as this research is metacritically approaching how the ecocritical community should respond to the current political and social environment. Criticisms of Brower, according to the journalism of McPhee, were that Brower became overly focused on the I and not the we. He became overconfident,

11 Brower being a hardline preservationist, and Fraser suggesting that the land would be developed anyway in a few decades, so why not make it an "ecologically friendly" site? (*Encounters*, 95–6, 141–2).

expecting the movement to adapt to him personally, which it partially did, but this is what alienated many people, just as Lilla is suggesting.

Lilla's work has been highly contentious as well, due to its criticism of Identity Politics in academia.[12] However, I would argue that the more important point is not the abdication of Identity Politics, but Lilla's suggestion of leaning toward more inclusive techniques. If we are indeed to focus on inclusion in all possible communities, we could potentially create a multitude of voices, working together and focusing on shared appreciation and common goals instead of focusing on what makes us each different from one another (to clarify: not to force assimilation to a broader culture but to focus on the similarities within our unique, individual cultures). *Made of Salmon* is a great example of this. The path that Lord takes avoids the constant infighting with regard to political issues, reminding all groups in Alaska that they appreciate salmon and should work together toward its prospering. Yet, perhaps the most important aspect of *Made of Salmon* is that Lord does not present some assimilated version of "Alaskans." She does not avoid different groups. To the contrary, she celebrates Alaska's broad diversity, while simultaneously unifying it through this one fish.

The subject of "identity politics," however, remains highly contentious, as it is potentially the most important issue of the 21st century aside from environmentalism. Civil rights and the rights of alterities more broadly are at a tipping point, and the stakes could not be higher. Therefore, the careful use of argument and vocabulary here must be of the utmost importance in its criticism. Being so integral to Second-Wave Ecocriticism, encompassing both Ecofeminism and Environmental Justice, "identity politics" should most certainly be given room for discussion with regard to the development of the multiple narratives that this research has suggested be effected by Alaska Natives. In the podcast *Coffee and Quaq*, for instance, Alice Qannik Glenn discusses LGBTQIA+ issues among Alaska Natives, which carry their own unique cultural baggage, and will likely not be solved through discussion of salmon. Likewise, she discusses a deep cultural divide over whether women should retake control over their own narratives, their Alaska Native heritage, by rekindling a pre-Christian tradition of tattooing their chins, or whether they should assimilate and avoid confronting conservative, White culture.[13] Again, a discussion of salmon will not bridge

12 Conroy, J Oliver. "Mark Lilla: the liberal who counts more enemies on the left than the right." *The Guardian*. "https://www.theguardian.com/media/2017/dec/21/mark-lilla-identity-politics-liberals." 21/12/2017.

13 Glenn, Alice Qannik. "Episode 1: Traditional Inuit Tattoos." *Coffee and Quaq*. "https://www.spreaker.com/user/coffeeandquaq/c-q-episode-1?utm_medium=widget&utm_source=user%3A10663336&utm_term=episode_title." Viewed on 27/7/2019.

this cultural gap, and the lack of "identity politics" discussed in this oeuvre, as it is not often discussed in the comprising works, is perhaps a weakness of this research. The rekindling of face tattoo traditions fall in line with a rise in cultural acceptance of tattoos more broadly, however, including face tattoos,[14] which could bring politically conservative groups and Alaska Native groups together. Yet, when discussing the issue of "identity politics" with regard to Alaska specifically, there could potentially be more reflection given to broader, multi-generational, multi-political communities, in order to combat the normalization of divisive rhetoric. A move toward inclusiveness, then, should be hypersensitive to multiple strains of thought, of identity (particularly important advice for authors attempting to mimic the paths taken by the likes of John McPhee), but also of collaboration between potentially-polarized communities, which would do well to work together, particularly with regard to environmental issues that are constantly at risk in a resource-rich geographic region such as Alaska.

To move on to the final, converging point, Lilla has performed Bloom's call for rational compassion to prescribe the solution to effecting political change, which activist literature implicitly intends to accomplish. His writing itself has been broadly divisive, however, because he is technically divisive: telling people to work together more but not creating a framework to bring people together in his own writing; he describes his work to *The Guardian* (in the article previously cited) as "an intervention," whereas it should rather be unifying according to his own prescription. An interesting comparison could be made with Barry Lopez' *Arctic Dreams*. As noted in the "Native Eye" section of the "Religion and Spirituality" chapter in Part 1 of this research, Lopez rarely discusses Arctic Peoples directly, but rather life in the region from their "Native Eye" perspective. This, then, is taking on the Arctic Peoples' affect. Likewise, Barry Lopez is not directly environmentalist in *Arctic Dreams*. He generally avoids issues of pollution, overhunting, colonization, etc., in lieu of addressing issues such as epistemology and natural beauty. Yet, *Arctic Dreams* is indeed a work of environmentalism, because it carries with it the environmentalist manner of perceiving the world. For instance, Brower tells McPhee that environmentalism is like a religion, and Lopez treats his own relationship with nature as a form of spirituality. In general, as well, even though Lopez is not sanctimonious, and his work is highly reflective, he tends to follow several of the traits that Seymour suggests comprise the environmentalist affect: "In addition to gloom and doom ... guilt, shame, didacticism, prescriptiveness, sentimentality,

14 Kurutz, Steven. "Face Tattoos Go Mainstream." *New York Times*. "https://www.nytimes.com/2018/08/04/style/face-tattoos.html." 4/8/2018.

reverence, seriousness, sincerity, earnestness, sanctimony, self-righteousness, and wonder – as well as the heteronormativity and whiteness of the movement" (4–5). Lopez also constantly writes of appreciation for nature and the different views thereof, particularly with regard to aspects of beauty not often seen above the tree line (*Arctic Dreams*, XXIV). Therefore, Lopez is more politically convincing through his affect than through arguments in *Arctic Dreams*, while Lilla has turned away much of the same demographic he hoped to convince, perhaps due to a lack of proper affect and display of empathy.

Therefore, we could say that Bloom's idea of rational compassion over empathy should be taken further into consideration than it currently is, but that empathy, of which Lilla's writing apparently lacks, should not be discounted either. Healthy doses of each could serve as an effective balance in allowing minority ethnic and identity groups their own narratives in the world while also not alienating them from the broader world. The following section will take in this idea. It will also discuss the no doubt contentious supposition that the technocracy toward whom McPhee's writing tends to reach should not be excluded by Second Wave Ecocriticism in search of inclusion of other groups, as technocrats, though they may view the world through a different lens, are often in a position to effect change and are potentially more closely aligned with avant-garde environmental movements than generally supposed.

5 Community and (Toxic?) Masculinity: What Could John McPhee Do Better?

This chapter has heretofore defended John McPhee's writing against the intolerance of militancy in the environmental movement in the 21st century. The inspiration for this chapter's discussion of inclusion of the technocracy, in fact, arises out of over half a decade working with engineering students, most of whom are both highly cognizant of, and militant toward, environmentalism (in activism and lifestyle), but many also hope to work within the confines of a technocratic system in order to be part of a broader shift to a more sustainable future. Teaching humanities students and theoretical researchers, then, offers a much different perspective from that of students who are actively searching for pragmatic, technological solutions. I would even go so far as to propose that differing academic worlds are prone to falling into respective feedback loops (or echo chambers, both terms working interchangeably here).

With that supposition in mind, scientific and engineering professionals working within a technocratic system could be considered an environmental alterity from the perspective of the Humanities, despite the technocrats being in a position of power, and I would also propose that the technocrats be offered inclusivity from the Humanities instead of divisive rhetoric. A parallel example could be the strategy employed by combatants of the Paris Commune of the mid-19th century, who intentionally, and often successfully, fraternized with the invading military in order to entice mutiny (Bernstein, 117–47).[15] As mentioned previously, the highly-organized think tanks of the US political right are currently sowing discord among "the middle third" through divisiveness. Likewise, during the Paris Commune, Thiers employed what could be called "the middle third" in his military, feeding them lies about the state of the French capital and the behavior of its inhabitants (Bernstein, 137), and inclusivity, fraternization, was much more successful in dissolving Thiers' army than actually fighting it.

Yet, in asking for inclusivity, hagiography must also be avoided. The most glaring criticism against McPhee that this chapter would target would be that, in lauding families like the Gelvins, the author falls into the trap of oversentimentalizing a group of people for their self-reliance: hunting, fishing, building cabins, sewing fur clothes, but even diverting riverbeds with heavy machinery to find gold as well (*Coming*, 379). Furthermore, Part 3 of McPhee's travel-writing triptych, *Coming into the Country*, never focuses on the community cooperation needed to maintain the dream of Alaskan wilderness. McPhee attempts to note that, while families like the Gelvins may technically be living on federal government property (they "have traditionally been able to do almost anything they wished on the land," ever since the Mining Law of 1872–284 –, just after US acquisition from Russia in 1867), and they may pollute the earth by moving riverbeds with bulldozers and by flying single-engine planes on a daily basis, their strong value systems and practice of self-reliance is much more important than supporting multi-national, industrial corporations in a pseudo-environmentally-minded deal allowing the construction of an oil pipeline which spans, north to south, the entirety of the state of Alaska.

In the early pages of the book, McPhee notes, with surprise, that many streams have been fished to exhaustion, are now stocked by hatcheries (9). He does not follow up on this idea, though, preferring to highlight the positive qualities of the individualist families who seem unaware of issues with overhunting and overfishing (see John Haines' *The Stars, The Snow, The Fire*, on resource exhaustion in the Interior, 14–5). In Last Frontier ideology of 21st century, pseudo-reality television shows such as Alaska: The Last Frontier and Sarah Palin's Alaska serve as continuing boosters for the individualist ideology.

15 "The Paris Commune." *Science & Society*. Vol. 5, No. 2 (Spring, 1941), pp. 117–147 (31 pages)

The former even intentionally and entirely neglects to mention the close proximity of a functioning community, Homer, which is a scenic tourist destination and fishing village, including restaurants, bakeries, bars, and supermarkets. If I subjectively agree with McPhee that self-reliance is a laudable trait, and one lacking in the 21st century due to extreme divisions of labor and technology, I absolutely must point out that the frontier mindset of Alaska is an illusion, kept alive by community cooperation, by hatcheries and government intervention preventing exhaustion of resources. It is not maintained by individualist boosters who, when left to their own devices, quite literally created industrial fish traps that exhausted rivers within but a few seasons before the government had to step in (Borneman, 288).

It is important to remember, as well, that the current political climate renders attention to the breakdown of community much more poignant and apropos. Jimmy Carter was president at the moment when McPhee published this work, and Carter had a strong rapport with the idea of community. In fact, that nonagenarian of McPhee's generation was still building homes for the underprivileged as of 2019, mostly in media obscurity.[16] Yet, equally important to the perspective of the 21st century is that of the #metoo movement, gaining steam with reporting done by Ronan Farrow on Harvey Weinstein, published in October of 2017.[17] Since that moment, a broad reflection of relationships between the masculine and the feminine (as well as a broader spectrum of sexual identity) has taken place in Western culture, including reappraisals of historical figures. This is not to say that these reflections were not already taking place, but that Farrow's article served as a catalyst, which provoked the current zeitgeist of gender perspective. With this in mind, this section will attempt to view McPhee's relationship with masculinity from a post-#metoo lens, while maintaining perspective of the hazards of extrapolating modern views onto a 1970s mentality.

John McPhee, then, is essentially a masculine writer, and often surrounded by men in his works. This is not to say that he avoids women, but that the women he encounters tend to be in more masculine-dominated fields, such as geology (*Annals*, 6). It would be odd to say that John McPhee is a "man's man" in popular idiomatic terminology because he does not at all fit the stereotype that comes to US minds when that popular phrase is typically employed. He is not,

16 *Habitat for Humanity*. "https://www.habitatnashville.org/cwp" Viewed on 1/7/2019.

17 Farrow, Ronan. "From Aggressive Overtures to Sexual Assault: Harvey Weinstein's Accusers Tell Their Stories." *The New Yorker*. "https://www.newyorker.com/news/news-desk/from-aggressive-overtures-to-sexual-assault-harvey-weinsteins-accusers-tell-their-stories." 10/10/2017.

for instance, like a Martin Scorsese protagonist in the model of Robert De Niro. Nor is he visibly muscular, or talking constantly of women and booze (Beach, *The Spoilers*). He does not even abscond with cultural misfits and write about LSD trips or motorcycle gangs as his peers Tom Wolfe (*The Electric Kool-Aid Acid Test*) and Hunter S. Thompson (*Hell's Angels, Fear and Loathing in Las Vegas*) did, but he is in his own way an educated "man's man" in that he spends much time outdoors, mostly with other men. William Howarth has described him this way:

> McPhee was a small, trim man with dark hair and no beard – one later appeared and took up permanent residence. He dressed in tweed jackets and thick sweaters, and he spoke rapidly, between intervals of wary silence, in a low emphatic baritone … He called his friends, prep-school fashion, by their last names: McGlynn, Bingham, Crow. (Howarth, "Introducing John McPhee," *Coming into McPhee Country*, 20)

This could be considered the stereotypical description of an effete college professor, but, as stated above, McPhee's subject matter very often tends toward the outdoors, particularly kayak trips, a penchant he picked up in Vermont at a place called "Camp Keewaydin" where he spent most of the summers of his youth (Weltzien and Maher, "Introduction," 11). McPhee almost never writes about city life except when it has to do with civil engineering projects such as how to control the Atchafalaya River or what to do about Los Angeles and its mudslides (*Control of Nature*), or traveling with farmers into the city to work at a farmers' market ("Giving Good Weight"). So, Alaska appeared a perfect fit for McPhee, even for the odd marriage of the urbane *New Yorker* magazine with this primarily rural place and its definitely-not-urbane issues.

Aside from McPhee's obvious aesthetic and thematic proclivities, he much more importantly writes in a particular manner that sets him apart from his peers. His writing is mathematical, apparently spending hours forming an outline suitable to the goal of his current project. For an earthquake or a debris flow, he might purposely write in a fashion of jumbled stories, for instance (Haynes, "The Size and Shape of the Canvas," *Coming into McPhee Country*, 61–4; Philippon, "Academic Air," 290–1). This is an important beginning for the writer and his vision, which the reader may not ever notice at the conscious level. This aspect of the writer's work, along with his unyielding journalistic mores of perceived objectivity, are why this research suggests that McPhee can be considered a "technocratic" writer. Indeed, he takes on an affect that at times crosses over into that of the environmental movement, particularly of First Wave Ecocriticism (often discussing natural scenery, preservation, etc.), but his affect also works with that of a different audience than most "environmental writers," a technocratic one (reference to McPhee's obsession with technocratic vocabulary, for instance – *Annals*, 31–3). Needless to say, this may make some who lie squarely within the environmental affect uncomfortable, and it is likely why the writer is never considered in the same breath as John Muir or Rachel Carson, or considered "Thoreauvian," as so many "nature writers" tend to be.

To add to this debate, this chapter has advocated the consideration of inclusion of the technocracy during a tense moment of minority ethnicity and identity reflection. But, what exactly is a technocrat, and how is this demographic affected by such consideration? Here is how the *Merriam-Webster* dictionary defines the term: "a technical expert; especially: one exercising managerial authority." Notice the terms technical and managerial authority. These words fit well within discussions of "the patriarchy" (not to mention the fact that women are underrepresented in the STEM sciences).[18] And, if we are to believe psychologist Paul Bloom, men, on average, tend to believe in systems and have generally less emotional stimulation and general reaction to suffering than women[19] (Bloom, 121,132–36). This, by the way, serves as a reverse parallel to the basic idea of Ecofeminism, as that philosophy, a broad spectrum of ideas, is based around the encompassing supposition that masculine patriarchy is overly harsh and uninclusive, unreflective toward holistic thinking (Gaard, Murphy, "Introduction," *Ecofeminist Literary Criticism*, 2–5; Buell, 19–20). According to Bloom, empathy can be a blessing and a curse, as it can be paralyzing. He gives an example of doctors in hospitals who need to show compassion but need to avoid experiencing the kind of empathy which may paralyze their judgment and/or make the patient feel worse (Bloom, 142–45). By nature, then, these technocrats, less empathic, less prone to emotion, tend to not rate particularly highly the emotionally-driven arguments of holistic health of the environment around them which Ecofeminism suggests (Buell, 19–20). Instead, technocrats are more likely to lean toward the pragmatism of systems like clear-cutting forests for use and allowing new growth to be properly managed. This is not psychopathy (Bloom, 197–201). These technocrats, like most people, generally want to do good in the world (Bloom, 113–163), but,

---

18 Muller, Carol B. "The Underrepresentation of Women in Engineering and Related Sciences: Pursuing Two Complementary Paths to Parity." *National Center for Biotechnology Information.* "https://www.ncbi.nlm.nih.gov/books/NBK36356/." Viewed on March 5, 2018.

19 Nancy Maclean mentions as well that the Koch political machine in academia views women as adverse to economic thought. Women "tend to anticipate the downside of economic liberty and so support government intervention." She then immediately moves on to efforts at George Mason employing cognitive science in order to warp opinion by taking advantage of "a good deal of irrationality in the electorate" (Maclean, 224). This obviously falls in line with the pitfalls Paul Bloom suggests we resist.

by having followed the educational path they have taken, going through graduate studies and internships and management programs, are naturally going to have the tendency to be tempted by the remunerative payoff of what's best for themselves and their families. If they accept the act of becoming part of a system that makes the small decisions in the daily lives of the general population, they are almost by definition not the David Brower or the John Muir types (though several prominent outliers immediately jump out: Robert Marshall, John Kauffmann, and Peter Wohlleben, for instance). With this in mind, the majority of technocrats are likely not going to be mobilized by either the political right or left. They are sitting in the middle, centrists protecting the status quo, which maintains the way of life this class has earned through years of school and 40 hour-plus workweeks. The problem is that this status quo is proving to be a deterioration of livability for humanity on Earth, as human populations continue to soar, exhausting resources and causing mass extinctions and such quick change that the "anthropocene" will be a clear period in the geological record of Deep Time (Kolbert, *The Sixth Extinction*). The issue has reached points of such urgency that these technocrats must be motivated to make responsible decisions and not, for instance, ignore the phone calls of someone like Nancy Lord who asks for a moratorium on hunting beluga whales or commercial salmon fishing (*Beluga Days*). Or, people may think twice before ending a meeting early on a subject regarding zoning in landslide-prone areas (McPhee, *Control of Nature*). The list of important daily decisions goes on, including what children are eating in schools, the water one may be drinking, and whether to use a certain type of material in a factory over another that may prove to be carcinogenic twenty years from now. The average technocratic decision-maker, if we are to believe Bloom, is likely not reading John Muir at night and planning to rush out to save the redwoods in California on vacation or stop the Keystone Pipeline with civil disobedience, as two exaggerated examples. However, they just may do the right thing with regard to the daily issues that arrive at their respective desks. John McPhee's ostensibly objective style naturally lends itself to a technocratic readership who is intelligent but uninitiated, by the very nature of his employing "rational compassion" over empathy while advancing environmentalism.

This ideology obviously grates on certain militants, but, as we will see, it is not that far off from the Naomi Kleins and Rachel Carsons of the world who employ(ed) rational strategy in order to effect specific environmental goals, which McPhee does in a less direct way, having arguably shifted public opinion on the specific preservation of the New Jersey Pine Barrens, for example (Weltzien, "John McPhee and the Question of Environmental Advocacy," 125–131). With that said, McPhee certainly could have done better throughout his career with regard to minority and identity inclusion. Native peoples, even women, are not inexistent in his work, but they are still perhaps too rare. Yet, McPhee comes from a different time and place. He was not a Hippie in the 1960s, but rather always lived in Princeton, New Jersey (7–8), and he is essentially a product of that place. Being somewhat liberal in its association with the university but also conservative as a wealthy small town, McPhee contains the contrast of both within him. Yet, it is important not to gloss over the avant-garde nature of McPhee in nature writing. He was writing New Journalism on the natural world in an era where there was otherwise mostly natural observation (Snyder, Matthiessen) on the one hand, and a heavy focus on cultural shifts caused by the Baby Boomer generation (Wolfe, Thompson) on the other. The weakness of viewing works through the lens of whatever a current zeitgeist may be (though there are important strengths as well!) is that another zeitgeist will always follow. Cultural and political issues shift, appreciating and dropping art and literature according to generational whim. John McPhee's work will likely remain, on the basis of his careful attention to craft in his writing as well as his poignant cultural and natural observations so often humorously portrayed.

6 Conundrum Conclusion, and a Pre-emptive Defense of the Middle Way

This chapter has heretofore attempted to fairly cover some controversial topics as a form of metacriticism and self-reflection among the environmental affect, particularly with regard to how the argument relates to Alaska. And yet, we are still left with a conundrum because, on the one hand, this chapter is advocating a view of Alaska as the Last Frontier in order to draw attention to the state, to make people care about the plight of its multiple, unique biomes, so that they may be actively affected in a positive manner. On the other hand, this chapter is also advocating against exploitation of empathy, of pure emotion, instead leaning toward compassionate reasoning. That is why environmental writing, as a macro-genre, must walk such a thin line. If we shut off our sensationalist, empathic approach to Alaska and attempt to approach it uniquely from the standpoint of the *Tundra Times*' subsistence, as Susan Kollin suggests (4–5, 55), very few outsiders will be inspired to get involved in important environmental conflicts, as 1) sensationalism sells, and 2) not very many people outside the affected communities knew of the existence of the *Tundra Times* to even read it before it went out of print.[20] Whether we like it or not, whether

20 The reverse argument is that without sources like the *Tundra Times* the people affected may become disenfranchised. This viewpoint is not to be taken lightly. Also, more than pitting the two against each other, the sensationalist outlets of content can learn from the

we agree with the way Alaska is presented or not, helpful people are likely to be enticed, just as unhelpful people are, by the trope of the Last Frontier. With that said, we must nonetheless be responsible in many of the ways that Kollin so carefully and fully suggests in *Nature's State*. Logic and reasoning, mixed with compassion (as Bloom posits), can be used to move forward in the fight for responsible conservation and preservation agendas, and with respect for Alaska Natives.

This concludes the conundrum, but the conclusion has not yet been drawn fully, because there remains a discomfort that many ecocritics have with inclusion of groups already enfranchised and holding much power. This criticism often reflects a general discomfort with a "middle third" being represented by entrenched classes guarded by self-interest. But what is being suggested here is not centrist complacency. Rather, it is "The Middle Way," in homage to the philosophy of Gautama Buddha. Tom Lowenstein describes the idea in his work, Buddhist Inspirations:

The Middle Way

> … the Buddha had discovered that it was best to find spiritual insight by following the "Middle Way" between the luxury of his early palace experience and the asceticism of his six-year search. Having eaten sufficiently … before he sat down to his enlightenment meditation, the Buddha went deeper into an exploration of the Middle Way. (36–7)

This meditation led to the Buddha's eightfold path, a balanced and harmonious way to live amongst the "endless cycle of suffering, desire and rebirth." The eight factors are: Right view or understanding, Right thought or intention, Right speech, Right action, Right livelihood, Right effort, Right awareness, and Right concentration. The idea, then, is to maintain strong focus and compassion without falling into the burning emotions of judgement of "the other," sanctimoniousness toward that which lies outside our individual worldviews. We should be highly diligent, then, but we should also not fall into the asceticism of extremism leading to anxiety and breakdown of society, to "suffering, desire and rebirth." In our diligence as well, we should maintain a sense of humor as a form of balance and self-reflection.

In contrast to "The Middle Way," an argument can be made, and certainly has been, that the time for reconciliation is done, that bellicose literary rhetoric is

---

Tundra Times and become sensitized to Environmental Justice and the plights of Alaska Natives and other potentially disenfranchised and highly-affected communities lacking a voice.

now what is necessary amidst a war where, from the environmentalist perspective, the other side simply does not listen. This is a noble sentiment, and certainly pertinent to an existential crisis on Earth. However, this form of literary rhetoric itself can be dangerous. It has the potential to incite dangerous antipathy and violence, leading not only to hatred but to a strong reactionary effect.

Naomi Klein is indeed walking this tightrope well. She declares that now is the time to fight back. Her rhetoric is fierce, her voice firm, but she is also responsible and extraordinarily prudent and calculating in the same way that Rachel Carson was. For instance, she uses her platform to coordinate the targeting of inclusion of diverse demographics (464–6). One of her recent books, *No Is Not Enough: Resisting Trump's Shock Politics and Winning the World We Need*, is indeed solely based on combatting the politics of Donald Trump. In this process of her career, however, she has emphatically avoided news pundit discussion (i.e. unreflective, conversational chatter); she has been careful at every step, with facts, with strong logic, with clarity when she means to act and explaining why instead of employing demagogic insinuations and the logical fallacies so rampant in modern, polarized politics. All of this is of the utmost importance. The hope of this research is that travel writing, in Alaska particularly, where it can easily serve as a soap box because of the geographic sentimentalization surrounding it, can capitalize and create a unifying voice. The Arctic can indeed lead.

But, one may still ask why to cater to seemingly unresponsive groups ("the middle third") through a dilution of the message? There is much strength to this argument. As mentioned previously, Françoise Besson has argued a case for reaching out to children, who are more open-minded than adults (*Ecology and Literatures in English*, 360). Her most recent work is highly inspirational, evoking broad shifts in perspective toward animals, plants, and the wilder natural world as a prescriptive method. This type of writing is important in catalyzing reflective activism in environmental themes (her subtitle is indeed *Writing to Save the Planet*). Yet, as we have seen, certain types of people are simply not affected by the same forms of literature and other media as their more empathic neighbors, or either they are psychologically closed off by social constructs of what is acceptable in their local communities, so that other approaches should work alongside those laudable forms of reflective literature prescribed by Besson. Seymour notes as well that many people do not respond to the traditional prescriptive methods, viewing them as didactic and overly sincere (1–5). The broader environmental movement, in conclusion, needs more than just Naomi Klein, and even the residual inspiration released by the colossal works of Thoreau and Carson. It needs to be inclusive of a broad range of methods.

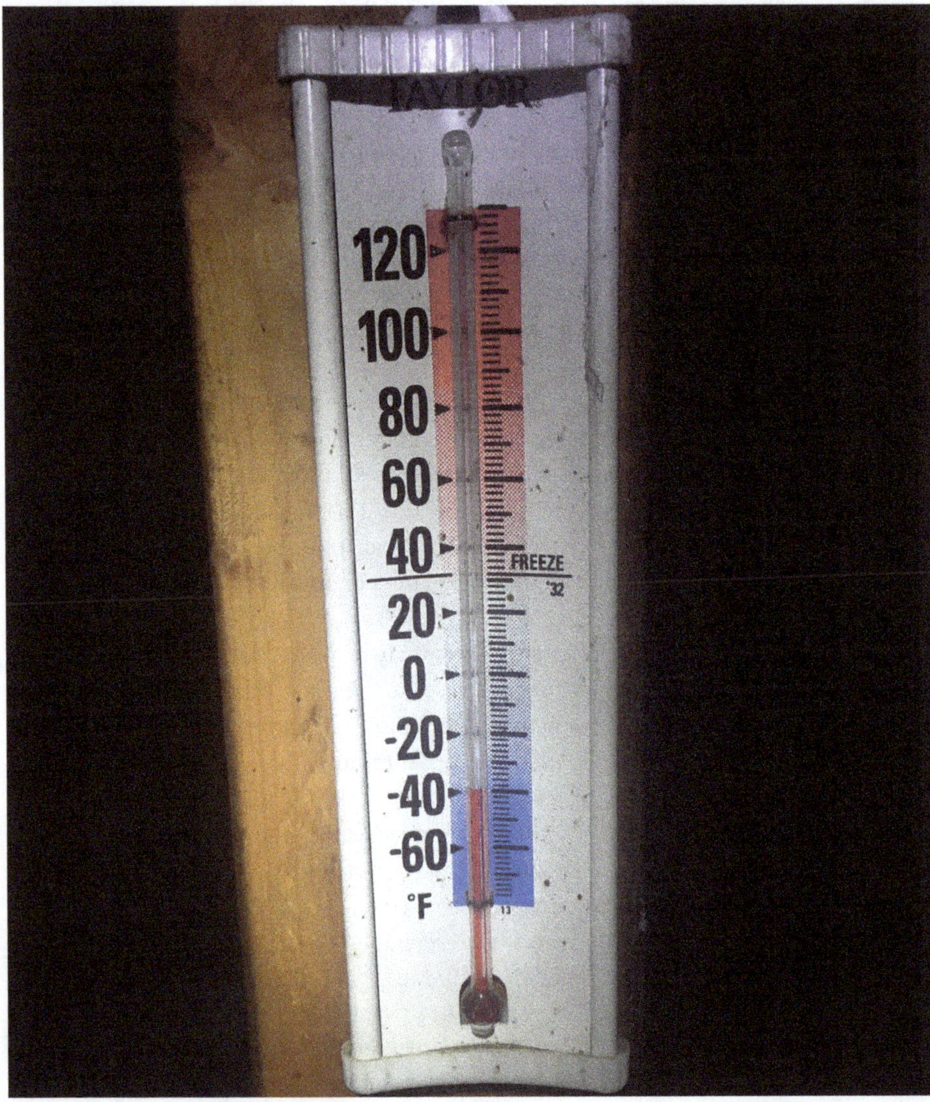

FIGURE 12 Thermometer
PHOTO BY CHRIS FERGUSON

CHAPTER 10

Jon Krakauer, John Vaillant, and Dave Eggers: Different Approaches to Sensational Writing

Alongside targeting demographics and coordinating mobilizations, the art of mass-marketing rounds off that third and final aspect of literary influence which I argue to be so integral to the fight over the environment. This may seem counterintuitive, as not many bestselling modern stories actually directly approach environmental themes. Often, if discussed at all, environmental discussion is a side show, likely not rallying the desperate troops ready to make a difference. But, the subject matter just may spark an interest, even if at some subconscious level. A reader of a conservation- or Alaska-minded work of literature may wonder "Why not me?" or even just change their daily habits in some small way, having caught that contagious bug of empathy. With this in mind, I will examine three primary works in this section: Dave Eggers' *Heroes of the Frontier* (2016), Jon Krakauer's *Into the Wild* (1996), and John Vaillant's *The Golden Spruce* (2005). The first is a work of picaresque adventure fiction, while the other two are non-fiction narratives sharing an investigative style of highly-informative journalism. Vaillant's work is not directly focused on Alaska, but it is about the Queen Charlotte Islands of British Columbia, which are situated proximate to the Alaskan border and directly along the sea route taken by so many to get to Alaska over the past roughly two and a half centuries.[1] (The Queen Charlotte Islands are also home to a warring tribe called the Haida who feature prominently in the book and who also have Alaskan Native heritage.) The goal will be to analyze these three works, showing examples of what is helpful, what is harmful, and what may be effective despite itself in the fight over the environment in modern sensational literature.

To begin with, Dave Eggers is one of the sole literary writers living in the United States today who automatically garners attention no matter the subject matter covered, and, therefore, creates heavy publicity with each new release.

1 From James Cook (1784) (Vaillant, 38) to Jonathan Raban (1999). Vaillant mentions in *The Golden Spruce* that Cook was most likely "The first European to set foot on the Northwest Coast and survive an encounter with the locals …" when he "… stepped ashore at Resolution Cove on the northwestern shore of Vancouver Island on March 29, 1778" (38).

 | DOI:10.1163/9789004734807_013

His dystopian 2013 work *The Circle*[2] was successful enough to subsequently become a film starring Emma Watson and Tom Hanks, and so, at the peak of his fame, after several successful novels and screenwritings, responsibility had begun to build up toward his audience. *Heroes of the Frontier* was the follow-up to *The Circle*, and it is based on the premise of a dentist/single mother from the Lower 48 who has previously worked in the Peace Corps in Central America and cannot handle the plain, boring life of the Ohio suburbs to which she has eventually settled, and so she decides to escape to Alaska for a summer with her two young children. This is an entirely plausible, even promising, premise for a novel, as it involves voices of both children and a woman, who have rarely served important roles in sensational Alaskan literature. The choice of title, however, immediately serves as a red flag from the standpoint of this research. Susan Kollin's assertion that "the Last Frontier" is unhealthy for the state and the way it is perceived and manipulated (22) must be noted. That choice could be overlooked but for the other keyword: Heroes. This word could potentially refer to types like Robert Marshall or Mardy Murie, for their exemplary work on social and environmental issues, or even William Iġġiaġruk Hensley for his work advancing Alaska Native rights in the political sphere, but it does not. The reader is almost immediately informed that the word "heroes" refers to the protagonist's conception of people living in Alaska, in general. She thinks of all citizens of Alaska as heroes who have escaped the thralls of the Lower 48 and have chosen to live a life in the wild, going back to nature. Just after arriving in Alaska the protagonist already asks, "So where were the heroes? All she knew where she had come from were cowards" (5).

Furthermore, Eggers adds to the misinformed mythology by having his protagonist assume that a man lives in a tent simply because he lives in Northern Alaska:

> "Cooper lived in a real house, a red-brick ranch with a black roof, which was surprising, though Josie didn't know why. He'd told her he lived there in town, and he'd been wearing clean clothes when she'd met him, so did she really think he lived in a tent? Something about the hootenanny had her thinking of hoboes." (335)

Keep in mind that this is 335 pages into the book, and the protagonist has been traveling around Alaska for weeks. She thinks of people as stereotypes of fishermen and depression-era drifters at all times. Eggers vaguely attempts

2 Eggers, Dave. *The Circle*. New York: Knopf, 2013.
The Circle. James Ponsoldt. Burbank: STX Films, 2017.

to dispel this myth, but he tends to play on it more than dispel it.[3] Not only that, but blanket statements about people of Alaska being "heroes" for an implied escaping of the Lower 48 means forgetting that, despite the high rate of overturn in the state, many people are still born there, and many of those people are Alaska Natives, who are not mentioned one single time in the book. One may imagine a lack of Alaska Natives in a micro-scenario, but *Heroes of the Frontier* is a novel based on travel throughout large portions of the state. This is a problem from the standpoint of the ethics discussed in this research, and it comes across as disrespectful, particularly from an author known for his cultural sensitivity.[4] There was a real opportunity to write a great novel on Alaska by a well-known and respected author, but instead, Eggers portrays mere fantasy passing for authentic adventure. But, this fantasy is obviously filtered through the stereotypes of Alaska imprinted in his mind by the mainstream cultural zeitgeist against which Kollin warns.

Eggers even succeeds in offending the universally-respected restaurant, Finn's Pizza, on Homer Spit, observing: "it seemed to be some kind of political pizza place, a pizza restaurant embodying its version of utopia" (96). Two pages later, he has the pizza lady "with exhausted eyes" assume that it was the protagonist's son and not her daughter who broke a towel rack in the bathroom, and implies that she got off free because the women were sexist against boys. While Eggers could have focused on Alaska's many social and environmental issues, found ways to be helpful to the state's narrative, or even write a vaguely beautiful homage to the state, he offends its sentiments and spends a large majority of the work complaining about contemporary American society and glorifying wanderers. He does take up the opportunity to discuss forest fires, which are indeed an existential problem in the state and are not often explored in its literature, but that section consists of his protagonist wandering around the Interior, constantly just missing the fires and always luckily coming along an

3 This is in perfect contrast to a Nancy Lord short story in *Survival* (1991) called "Snowblind," which plays on the same stereotype of Alaskans by having a female protagonist in Washington, D.C., fall in love with a married man who tells her all kinds of romantic tales of Alaska (hiking, skiing, dogsledding, etc.), even including snow blindness (hence, the *double-entendre* title), until she talks with the wife at the end of the story and realizes the man has been exaggerating every aspect of his life in order to get her to think of him in a different light, as some sort of "hero" who escapes society and lives out the adventures of which the rest of us dream. The difference between the two is that Lord's story has a point, discreetly advises the reader against perceived notions through embarrassing and slightly humorous repercussions.

4 Eggers made a name for himself with the African story, *What is the What?*, and has attempted to tackle social issues in other works like *The Circle*.

empty cabin or group somewhere who helps her before she ends up wrecking the situation and moving along to another cabin or group. If anything, Eggers has glorified the increasing masses who are taking to "houselessness," living in RVs and finding temporary work as they move from town to town. This may be taken as a positive or a negative phenomenon depending on the person and situation, but there is not really much breadth to their cause and plight in this work. In the end, Eggers' protagonist comes to resemble the angry and frustrated types against which she rails as the problem with modern America.

Despite itself, however, *Heroes of the Frontier* nonetheless brings attention to Alaska, to forest fires, and the subject itself may raise awareness in some readers who could take up an interest in the region and dig deeper, maybe even make a difference someday. The point of bestselling stories of this sort, for our purposes, is to spark the imagination and interest, and in this respect, Eggers may have succeeded.

It could be argued that this book is an adventure story never meant to be what Alaska specialists might demand of it. This is perhaps true. But, there is no question that the novel comes across as somewhat careless. For example, James Michener's *Alaska* (1988) was meticulously researched and carefully crafted. It is sensational and meant for a broad audience as well, and yet the author has succeeded in educating in all sorts of ways: historically, culturally, and scientifically. Michener's work is indeed a historical novel, which attempts to accomplish something different from Eggers' adventures, but well into the 21st century Eggers does not include one single Alaska Native, or even a Pacific Islander. Somewhat similar criticism has been launched against other writers of Alaskan literature, notably Rex Beach and Jack London in the early 20th century. London was lightly criticized in a retrospective introduction to *Call of the Wild, White Fang, and Other Stories* (1986, 7–16) by James Dickey, who notes that dogs and wolves do not at all react the way London has them react. However, Dickey also praises London because the literary value of London's works nonetheless rises above these facts. As for Rex Beach, that writer was famous for several sensational novels about the Alaska gold rush. Standing out among them is *The Spoilers* (1905), based on the libertarian nightmare that took place in Nome during that time. Note that Susan Kollin criticizes both London and Beach for their roles in playing up the Last Frontier myth in the American mind (60–89), but both men were nonetheless influential in the nuance of Alaska, of animal rights issues and problems with lawlessness. By contrast, here is an excerpt from Nancy Lord's review of Eggers' work:

> But his Alaska is imaginatively drawn and should not be confused with the real place. Although the story is set in late August and early

> September, wildflowers are blooming everywhere and it's fire season ... Alaska Natives are also absent; Eggers' Alaska is populated by refugees from the Lower 48 – innkeepers, folk musicians ... [5]

She also points out how fire engines are going after wildfires, which does not happen, snakes are somehow present despite the fact they don't exist in Alaska, and none of Alaska's famous mosquitos show up in the book. Lord is frustrated by a lack of both close observation and study. One may make a few mistakes in an unfamiliar place, but the entire work is full of a lack of understanding of the place through which the author moves, and appears to be based entirely on the "Last Frontier" as a sensationalist idea. This chapter may seem too harsh against Eggers' work, but he essentially commits every ethical error suggested herein.

By contrast, Michener has created a vast and encompassing version of Alaska that is well-researched and examines multiple points of view, including that of Russia. This gives the casual reader a strong geo-historical perspective from a sensational work of fiction. Michener respects Native culture, approaches environmental conservation issues, and also deals with the boosterish side of the state, allowing the uber-capitalists at least a fair say. He displays how over-fishing and generally exhausting resources have affected the region over the centuries, and he does so in a factual manner that sows seeds of environmental understanding without approaching a hardline agenda that may drive away "the middle third." Therefore, readers may not walk away as activists, but at least they will be offered a nuanced understanding of various environmental issues. That is why it works and another reason why Michener's *Alaska* can be qualified as ethically responsible in a number of respects.

1 Sensational Non-fiction

On the non-fiction side, this chapter examines two examples of hardened journalists who have both "cut their teeth" at consumer-based magazines such as *Outside*. Jon Krakauer has been writing for decades, each of his works meticulously researched and with larger points in focus, but he also accomplishes this responsible journalism while maintaining a sensational focus on the

5 Lord, Nancy. "Well-known novelist's latest offers Alaska backdrop, but don't mistake it for the real deal: Heroes of the Frontier, by Dave Eggers." *Alaska Dispatch News*. https://www.adn.com/alaska-life/we-alaskans/2016/09/04/well-known-novelists-latest-offers-alaska-backdrop-but-dont-mistake-it-for-the-real-deal/. 4/9/2016.

narrative. Earlier in this research, Krakauer was softly criticized for falling into the romanticism of the frontier myth, but he does so responsibly, displaying nuance, as Werner Herzog did with Timothy Treadwell in the 2005 documentary film *Grizzly Man*. The two works are somewhat mirror reflections of one another. Herzog took an outcast from society who had fled his apparently-serious drug problem, replacing it with the natural adrenaline of close-up contact with Alaskan grizzly bears (Treadwell, *Among Grizzlies*); a man with not much common sense and hardly any evident redeemable value to society. Herzog, however, finds that something, or *je ne sais quoi*, about Treadwell, the humanity in the man, and focuses on that aspect for the entirety of the film in order to drive home to his viewer an aspect of beauty and value that is often overlooked. Krakauer, on the other hand, explored a highly-educated and sure model of future success, in Christopher McCandless. This young man pulled away from society in the midst of emotional, and potential mental, issues not uncommon during the coming-of-age of a certain, highly-intense personality type. McCandless's only true lack of lucidity in this process was his lack of preparation, which is even debatable, as Krakauer mentions toward the end of the work:

> "It probably misses the point, though, to castigate McCandless for being ill-prepared. He was green, and he over-estimated his resilience, but he was sufficiently skilled to last for sixteen weeks on little more than his wits and ten pounds of rice." (182)
>
> "One suspects that Muir wouldn't have thought McCandless terribly odd or incomprehensible." (183)
>
> "I bring up McCandless's hubris and the dumb mistakes he made–the two or three readily avoidable blunders that ended up costing him his life. 'Sure, he screwed up,' Roman answers, 'but I admire what he was trying to do. Living completely off the land like that, month after month, is incredibly difficult. I've never done it.'" (185)

Chris McCandless is portrayed in a respectful manner throughout but is also judged for why he would be crazy enough to do what he did, which was not all that ostensibly different from what Treadwell did. For Treadwell, Herzog's purpose was to pull him back toward humanity, to his human side. McCandless was already respected for his numerous lifelong accomplishments in society and humanity, but Krakauer needed to justify McCandless's actions as well, particularly with all that the lottery of life had offered him (wealth and opportunity).

Into the Wild seems to accomplish its ultimate goal of explaining what it is that draws certain types of people to Alaska. There is even a chapter from Krakauer's own life, about a solo climb of The Devil's Thumb and the parallels with his own drive toward Alaska, with family problems comparable to those of McCandless. A few chapters before that, Krakauer even writes extensively on several others who have reached mythological status as major blunderers, eccentrics, and audacious risk-takers, and who died "in the wild" as a result.

From the critical standpoint, there is absolutely nothing about environmental conservation in *Into the Wild*, nothing about Alaska Natives (except mention of a single Tlingit in a bar when Krakauer finished his own scaling of Devil's Thumb), and not much really about Alaska at all outside the frontier myth. So, from our macro point of view, this makes the work less interesting, falling short of the ethical considerations discussed herein. But, *Into the Wild* is simply not about Alaska in the same way as other works. Though it has many of the same flaws as Eggers' work, it is responsible and self-reflective about the frontier myth, exploring both that idea's value and its destructive nature. *Into the Wild* ultimately is not about Christopher McCandless, either, but is rather about Krakauer's own life, his way of life as an outdoorsman, about the demons that haunted him and haunt others like him who risk their lives in seemingly foolish outdoor pursuits, and how these demons to be conquered are extrapolated onto the frontier myth in general.

Krakauer undoubtedly wrote a powerful work with *Into the Wild*. In a way, it even carved out a blueprint for a sub-genre of long-form, outdoor, investigative journalism. One thing about investigative journalism is that it contains constant mystery. If Chris McCandless had walked back out of "the wild" and gone back down to the Lower 48 to live a meaningful life, happily ever after, perhaps working in the Dakotas or even becoming an environmental lawyer in Northern Virginia, this story would never have been published. It is the mystery of his death that grabs hold of the reader. Krakauer, then, paved the way for John Vaillant, who follows Krakauer's blueprint very closely. Vaillant has written three books of investigative journalism based around Subarctic wilderness, and he has also written for *Outside* magazine, where Krakauer's short-form article on McCandless originally appeared in January of 1993 (*Into the Wild*, 11). Vaillant's first long-form work is based around a mentally-ill prodigy (not unlike McCandless), who went into the wilderness for long periods of time, living on almost nothing and with insufficient clothing (not unlike McCandless), but this man actually grew up amongst Pacific Northwest's old-growth wilderness (in the lumberjack culture of Vancouver and its surroundings) and knew how to survive it (*The Golden Spruce*, 209). He disappeared (presumably drowned) after cutting down a rare and sacred golden spruce tree that had been standing on the Queen Charlotte Islands on the northern British Columbia coastline near Alaska for over three hundred years. *The Golden Spruce* essentially takes place on the border with Alaska (not far geographically from Krakauer's ascent of The Devil's Thumb), and concerns the Haida people. *The Golden Spruce*, unlike *Into the Wild*, is also directly focused on environmental themes.

To begin with, *The Golden Spruce* succeeds in catering to mass audiences while remaining informative and relatively objective by presenting both the positive and negative aspects of the logging world in the Pacific Northwest. There is a definite environmentalist bent to this work, but Vaillant also explains how and why the destruction happened. He does so by employing a lifelong-logger-suddenly-turned-environmentalist as a protagonist. Vaillant begins his work by explaining the actual layout of the land in a visceral way:

> On the Northwest Coast, there is no graceful interval between the ocean and the trees; the forest simply takes over where the tide wrack ends, erupting full-blown from the shallow, bouldered earth … The air is at once rank and loamy with the competing smells of rotting seaweed and decaying wood. (7)

This sets the tone for the reader, an explanation about why one should care and informing us of a terrain of which most people have almost no knowledge, even as far as mentioning that the British once called it "the backside of America" (37) before that ineluctable train of mass resource extraction took hold. In the process, his intense, descriptive prose sucks the reader in; not through a Muir-like, almost spiritual, aesthetic romanticism but with a harsh and dark beauty. This would appeal to the aesthetic nature-lover but also to the technocratic-minded person discussed earlier who may not see the beauty for beauty's sake but may be excited by the technical experience (specific details of lumberjacking can only further this interest – Vaillant, 81–7).

As soon as the descriptive prose unfolds in the opening pages, Krakauer's mold of explaining the life of the mysterious protagonist begins. Like McCandless, Grant Hadwin grew up relatively wealthy, or "upper middle class," and set out early into the woods around his home in the suburbs of West Vancouver for long periods of time at an early age, despite an obvious precocious youth. He headed naturally toward logging because it allowed constant accessibility to wilderness, to actual old-growth forests; a way of life accessible due to ample opportunity in the then bustling industry, while also having an uncle well-established in the industry. Like Muir and McCandless, Hadwin could survive harsh circumstances with inadequate clothing and foodstuffs for long periods of time as he explored the wilderness. This sets up a battle between conservationist mindsets and the logging industry, as Hadwin, much older than McCandless when he went missing (aged 47), was a long-time "layout man" for the logging industry while also feeling ambivalent about it, and he eventually cut down the sacred tree in a death-defying stunt as an affront to the logging industry who guarded and admired it (Hadwin apparently did not realize that the Haida considered it sacred until after he cut it down – 159).

Another important aspect of this work is that Vaillant digs extensively into the history and culture of the Haida of the region and the background of their particular myth surrounding said golden spruce. His respect for the Native view probably could not have been effected any better within the context. The work, walking that tricky line between being respectful toward Native peoples, lumberjacks, and environmentalists, indeed received the "Governor General's Literary Award for English-language non-fiction" in Canada.

And yet, Krakauer continues to have more of a mass appeal than Vaillant, and this is potentially due to that fact that Krakauer tends to write with concision, leaving out the in-depth details that could help expand his scope but would alienate fickle audiences. Vaillant's work, then, can perhaps be viewed as overly technical for mass consumption on the level of Krakauer. For instance, Grant Hadwin, the protagonist, does not even enter the 239-page book until page 87. The lead-up is a detailed picture of Haida culture, the history of European interaction with Native peoples on the Northwest Coast and how their obsession for fur trade led to logging, and eventually the writer's attention even turns to the history of logging itself. These are the same types of technical subjects that gained John McPhee critical renown and won him a National Book Award, but that also kept him from becoming a household name. *The Golden Spruce* is undoubtedly sensational writing, but this focus on lumberjacking technique does not appeal to the casual reader in the same way that *Into the Wild* does with its swift movement between subjects surrounding the pull of the Last Frontier. In every other way, Vaillant has apparently succeeded, finding a niche that lies between literary non-fiction and mass-marketed journalism in the environmental conservation literature market.

Ideally, Krakauer's mass market appeal and simultaneous unrelenting journalism (not to mention concise prose), mixed with Vaillant's eye for conservation and respect for Native peoples in historical context, will leave a mark on the future of environmental conservation. This would draw readers in toward the subject because, as Vaillant points out, resources are limited and being mismanaged and depleted faster than we are able to replenish them (215).

As a postscript, one could make the comparison of these authors to the mass appeal of Naomi Klein. In fact, her work does have a sort of mass appeal, but to certain groups. She is likely better known for her intelligent manipulation of the media than for the actual, complicated, and well-rounded arguments of her long-form works. The success of someone like Jon Krakauer, by contrast, is predicated on crossing all demographic boundaries. In Alaskan literature, Jack London and Jon Krakauer are the two writers with the most celebrated works. *Into the Wild* is ingrained into the consciousness of the entire Western world because of the broad appeal of its subject matter. The same goes for *Call of the Wild*. The primary point to take away, then, is that the subjects that Naomi Klein discusses could become a major part of worldwide consciousness if the right narrative, approach, and story of individual struggle were properly manipulated. Walt Disney's incomparable eye is one example of this sort of combination on environmental issues, but we will see shortly that that man had his own pitfalls, which have been harmful in their own particular ways.

FIGURE 13 Fritz Creek
PHOTO BY CHRIS FERGUSON

CHAPTER 11

"Local Travel Writing" in the Country, or Broadly Outside?

Having thoroughly discussed issues of environmental conservation in Alaska travel writing, particularly with regard to the pros and cons of promoting the psychological phenomenon of The Last Frontier, another similar but nonetheless unique angle to consider is the scope, and community focus, of a work itself. As noted, writers like Naomi Klein aim for a very broad readership and are consequently able to encompass broad swaths of the environmental conservation movement with hopes of combining them into a global effort, even if connected at the community level. On the other hand, there is a concurrent need for individual communities to be served, and Naomi Klein simply cannot accomplish that particular goal while approaching broader world issues. This chapter, then, will explore how writing for individual communities and their unique, individual needs is important to effective environmental efforts in the 21st century. One issue to be discussed is that individual communities often deal with problems that cannot be extrapolated onto a broader model: a desert community in California, for example, lacking water, faces different issues from a Southwest Alaska rainforest community whose salmon rivers are being polluted by large scale mining operations. Naomi Klein and other broad-minded writers may relate the two in a larger tale of industrial misuse of natural resources, but bringing individual communities together in an inclusive way without alienating broad swaths in a politically polarized climate is almost impossible for her.

Except, pointing out this trend toward the specific, the local, or regional, almost always leaves us with a particular question: who is the target audience of a given piece? Often, as a reader or critic, we cannot know this fact, unless the author comes out and confirms such in an interview or in the work itself. Also, even if the author's intended audience or altruistic focus is not 100% geared toward one community but toward a broader audience, said community or a niche within may adopt the writing as important to its own narrative of itself. Peter Matthiessen's *Oomingmak*, for instance, likely makes no connection with the local community of Mekoryuk. As a subject, it primarily takes on the musk ox and that animal's unique nature, which the few locals would find normal, already know and likely find no point of pride in other than in the

 | DOI:10.1163/9789004734807_014

fact that a famous writer had come to utilize their local community in a non-derogatory manner.

John McPhee focused his eponymous Book 3 of *Coming into the Country* on lifestyles in the Interior Alaskan bush, along the Yukon River. In doing so, he created a micro-mecca of sorts that brought minor celebrity to a community and certain of its protagonists.[1] Other than to vaguely justify a very particular lifestyle, however (and entice literary tourism), McPhee does not do much for that individual community. His influence, then, has been more in making Outside culture aware of certain aspects of Alaska theretofore unknown to a wide audience, as well as to influence wilderness lovers and individualists to potentially take the leap and come into the Country. His work, as well as Jon Krakauer's *Into the Wild* and Jack London's *Call of the Wild* are the frequent citations one hears when discussing with locals why they decided to move to Alaska (from subjective experience – also, Nancy Lord begins her memoir-like collection of essays, *Rock, Water, Wild* with a discussion of McPhee's *Coming into the Country* – 3–4). Lopez' work may be viewed in the same way, as it fills in so many gaps on the local biome for the world Outside. It also likely gives certain, well-educated Alaskans a better metaphysical understanding of their landscape and fauna and flora, but his work is largely conceptual and less about, or for, the actual communities themselves.

Jonathan Raban's *Passage to Juneau* can be viewed differently, as the narrative is extraordinarily personal, on death and divorce, but it does include much ethnographic information and discussion of Eurocentric exploratory history. It remains sensitive to Alaska Natives and Canadian First Nations peoples in its comprehensive understanding of, and appreciation for, their culture and history, but that is not the focus of the work, either. Raban does nothing, essentially, for any single community, including the fishermen, as he portrays their world on the wane. Yet, if we may choose a focus, it would likely be Seattleites and urban Alaskans interested in the connection between the two regions and the waters no longer well-understood outside the fishing community since the introduction of flight.

This brief analysis has carried the aim of clarifying the understanding of the broader objects versus the narrow view, in preparation of looking at the two primary authors of this research who have highly specific community concentrations: Tom Lowenstein and Nancy Lord. Lowenstein's *Ancient Land, Sacred Whale*, for instance, appears to be geared toward helping a broader, western

1 Pemberton, Jennifer. "The lure of John McPhee's 'Coming into the Country,' 40 years later." *Alaska Public Media*. "https://www.alaskapublic.org/2017/01/23/the-lure-of-john-mcphees-coming-into-the-country-40-years-later/." 23/1/2017.

audience to understand the local Tikigagmiut community. It can also perhaps be seen as a gift to the community of Point Hope, Alaska itself, to have a better understanding of itself and its own historical narrative. Nancy Lord's *Beluga Days* attempts to accomplish some of the same goals, though through very different means. Whereas *Ancient Land, Sacred Whale* employs poetic style to get across an ethnographic understanding of a culture and its stories about itself in translation, *Beluga Days* attempts to solve a particular local problem, declining beluga numbers, by approaching various communities within the Cook Inlet community as a whole through investigative journalism. Except, Lord localizes said style to Alaskan needs, which is what excludes her from a broader Lower 48 audience. This focus, however, is important to the local culture, as it approaches a unique problem, with specific interests of particular communities.

Inspiration for this argument came from Barry Lopez' *About This Life*, in which the author discusses the local genius of so many people he has come across living in the Pacific Northwest ("The American Geographies," 130–43). Lopez was defending local communities who have often been derided for a lack of education or broader urbanity, but who, like the Inuit and other Native peoples, understand their local geography and traditions in a manner that perhaps should be better respected by more educated, or wealthier, communities and demographics.

Beluga Days and *Made of Salmon* then stimulated the supposition in my research that a form of "local travel writing" could take hold in specific communities, serving not only to show local genius to the world Outside, but perhaps more importantly, to show local communities the importance of their own forms of genius, bringing them together in their respective talents and uniqueness. In this way, local travel, specifically focused on narrow ideas, such as Nancy Lord's interest in the beluga, could heavily affect local communities. For local travel writing to be effective, this would almost automatically entail writers relinquishing broader readership and acceptance in exchange for local acclaim, meaning that a democratization of literature to smaller, local pieces could allow for more voices than the globalized figures such as Klein, who has served as the symbol for such in the environmental literary movement in recent years. Nancy Lord is the perfect example of this, as she has never gained broader recognition Outside, but she is acclaimed within the state, with an oeuvre of more than a dozen published works.

Before exploring further into this idea, however, stepping back and analyzing the contrary position, its flaws and positives, may prepare the reader to have a better appreciation and grasp of the local. As we have already fairly exhausted Naomi Klein as a literary icon of the broader environmental movement of the moment, the celebrated cartoonist, Walt Disney, will now be explored for his work, particularly in the early-to-mid 20th century.

1 What the Walt?

Walt Disney was a man who had that rare combination of an eye for mass marketing while simultaneously maintaining environmental conservation as a primary objective in his films. One could perhaps call him a capitalist conservationist, and he was so successful that most people never realized they were being manipulated by an environmental viewpoint (Brinkley, *The Quiet World*, 346). From early on in his film production career, in fact, Walt Disney financed wildlife documentaries with political agendas, and his cartoons served as an important conduit for change on American thought. According to environmental historian, Roderick Nash, the US had only recently come to terms with the fact that it was no longer a frontier nation, and was even in the process of shifting from an agrarian economy to an industrial one (*Wilderness and the American Mind*, 345–9). An industrial economy, according to Nash, loses its proximate relationship with nature and begins to appreciate what it has lost. "Nature" can be extended to include non-domesticated animals, and Walt Disney, who came to fame between the two World Wars when the US was in the process of becoming a global industrial powerhouse, was able to capitalize on this shift that led to a whole new nature value system (Brogan, 566).

Disney, a burgeoning cartoonist, opened his first studio in 1923, and developed his Mickey Mouse character in 1928.[2] He had always been interested in nature, as visible in 1932's *Flowers and Trees*,[3] but much of his early work could not be said to be environmentally conservationist, as it was more or less comprised of allegories and fairytales, such as *Snow White*[4] and *The Three Little Pigs*.[5] The release of the film *Bambi*, however, in 1942, marked a pronounced change in both Disney's work, and the way Americans viewed deer hunting. Industrial America, in other words, fully woken by the Second World War, was rapidly moving away from agrarian society (Ellis, *The American Sphinx*, 8–9) at the very moment it was ready to subjectify animals. Deer hunting became taboo to many almost overnight (Brinkley, *The Quiet World*, 347–8).

2 "History." Walt Disney Studios. "https://www.waltdisneystudios.com/history/." Viewed on 6/11/2019.

3 *Flowers and Trees*. Burt Gillet. California: Walt Disney Productions, 1932.

4 *Snow White and the Seven Dwarfs*. David Hand. California: Walt Disney Productions, 1937.

5 *The Three Little Pigs*. Burt Gillet. California: Walt Disney Productions, 1933.

In 1948, Disney moved into live-action films, which are sometimes discussed as wildlife documentaries, but should perhaps be viewed more as corollaries to his early animated films such as *Fantasia*.[6] Susan Kollin notes how Disney had strict rules that entirely restricted any evidence of human impact on the landscapes in his documentaries, making the scenery seem entirely "natural," but in fact making them less natural than they were in reality, as the reach of human impact is quite literally everywhere on Earth, including Interior Alaska, with its oil extraction (TAPS) and military presence (94–102). This impact, Kollin notes, frustrated Lois Crisler while the latter was making documentaries for Disney. From the standpoint of a "realist" mindset in documentary filmmaking, displaying the "true" Alaska could dispel the pristine myths that make people less cautious with regard to heedless exploitation and its effects on regional biomes, as this mindset offers them the false security that their actions are not having a significant effect on such a large and resilient state (Kollin, 96–9). As a related example, the 1958 film *White Wilderness*[7] famously showed a large group of lemmings jumping off a cliff to their deaths in the Arctic Ocean. It turns out that lemmings do not actually commit mass suicide in this manner and were actually coaxed by the filming crew (Brinkley, *The Quiet World*, 445–6). This scene can be viewed, again, not as documentary, but as choreographed nature, a live-action corollary, for instance, to the 1940 Disney film, *Fantasia*, in "The Sorcerer's Apprentice" segment, particularly as the mindless brooms, under the imperfect magic of an adolescent Mickey Mouse, carry water and do not stop, flooding the palace, to the chagrin of the incapable protagonist. The enchanted brooms, then, can be viewed as a similar narrative crescendo to the suicidal lemmings.

In refusing to call Disney's live-action work "natural," much of it in the western Arctic and Subarctic, I am not attempting to chastise the prodigious cartoonist and producer. Instead, I am drawing attention to the fact that his work is flawed from the modern ecocritical perspective, the same way that television shows today, such as *Alaska: The Last Frontier* focus solely on subsistence living and refuse to display local civilization, creating a false utopia for libertarian-minded viewers potentially enticed to the region. In the 1940s and 1950s, however, there was not much precedent for the ethics of documentary filmmaking, which certainly exists today,[8] so that Disney is being judged

6 *Fantasia*. Produced by Walt Disney. California: Walt Disney Productions, 1940.

7 *White Wilderness*. James Algar. California: Walt Disney Productions, 1958.

8 "How To Deal With Ethical Challenges In Documentary Filmmaking." *New York Film Academy*. "https://www.nyfa.edu/student-resources/documentary-filmmaking-how-to-deal-ethical-challenges/." 23/9/2014.

according to our benefit of hindsight.[9] Even so, viewing Disney's work from the modern perspective, it was nonetheless an important stepping stone for both environmentalism and documentary filmmaking, bringing new understandings of nature right into the homes of people who had likely never seen more than pictures of far-off biomes, and did not have the benefit of internet browsing.

Anthropomorphizing animals in a cute and cuddly way and limning nature as pristine can both be very successful tools, evoking sympathy in enticing broad audiences, but it also must be done responsibly. If animals are separated from the "circle of life," from natural cycles of predation, "the food chain," etc., the audience loses its understanding of the natural world in a holistic sense, and only sees animals as pets, friends, companions, rather than as a part of society, as "animal persons." Crisler, for instance, had to constantly fight off the wolves she raised for her documentary because they outgrew her and became too wild to remain domestic but not wild enough to be introduced back into the Alaskan wild itself. She realized she had created an unhealthy, false environment for them and eventually had to euthanize one of them for this very reason, giving the others away (Kollin, 100–14).

All of this can be taken in the context of the responsibility that literature and film both should carry when being involved in environmental conversations. If, for instance, one decides to take advantage of mass-marketing for environmental conservation purposes, particularly at risk of exploiting animals or the frontier myth of Alaska, one should be very careful in the process so as not to create some negative unseen aspect to be regretted in the future. Yet, again, it must be reiterated that Walt Disney's career has been formative for the entirety of industrial society's views on animal subjectification and should not be discounted. Also, as with the theory of the local, which will be explored in the following pages, this broad scale animal advocacy, opening up eyes, may allow for more niche, or nuanced, discussions to occur, just as has happened with people who have been enticed by literature and film to come to the Last Frontier and have stayed to effect their own nuanced discussions, being the ones who effect positive change.

2 Theory of the Local

The reason that this is such a necessary idea to put forth is that discussions in environmental criticism tend to be broad-scoped. Of course, local issues are

9 Susan Kollin indeed notes that there was criticism from the environmental community, and that Lois Crisler wrestled with this criticism in *Arctic Wild* (Kollin, 98–9).

discussed, but mostly when a regional author breaks through to a larger audience, perhaps with ideas that resonate broadly. Environmental Justice in its literary form may seem to focus deeply on local communities, with regional arguments and background, but, again, the movement tends not to involve works such as Nancy Lord's *Beluga Days*. Rather than employing a specific background with universal discussions, her work focuses on one particular region's specific issues that absolutely do not resonate broadly. This distinction can be difficult to convey, but the point being made is that even if a small tribal community's fight is considered, if it is discussed solely with the same talking points as the broader Environmental Justice fight, such as solidarity against impinging industrial forces, one is undoubtedly regarding the issue in Manichean terms and leaving out the nuance of the community itself. For example, Sheila Nickerson addresses the arrival of "Big Oil" in Alaska in her 1996 work, *Disappearance: A Map*.[10] She notes that Alaska Natives were dragged into capitalism through the mechanism of ANCSA, making the majority of Natives shareholders in local corporations, but she also explains how the money has created a dependency on a system which is unsustainable, with Big Oil potentially pulling out of the state at some point in the near future (62–3). It is difficult to overstate to what extent the local infrastructure is so extraordinarily expensive. And now local people do not know how to move forward, maintaining their connections with the 21st century while also recreating links with their own traditional culture and disassociating themselves from oil. Therefore, many support Big Oil for the economic support it offers in maintaining their lifestyles (Nickerson, 63). Otherwise, they will likely need to either move or cut all connection with industrial society, in a now-polluted environment that severely limits the subsistence lifestyles of traditional indigenous peoples. Nickerson even notes that discussion of secession by the northern boroughs has been on the table (63). This example is important in demonstrating the nuance lost when broad-scoped Environmental Justice takes place in local communities, maintaining universal talking points. Nancy Lord even notes in *Beluga Days* that local environmental groups want to keep Outside environmental groups out of the state in many fights because those from Outside tend not to understand the nuances of state politics and can cause more problems than they solve if they come in with good intentions but offend Alaska Native tribes or conservationist hunters attempting to keep an equilibrium in a diverse cultural region (93). However, both local and broad scopes can successfully be approached at once. For example, Christine Shearer's *Kivalina*[11]

10 Nickerson, Sheila. *Disappearance: A Map: A Meditation on Death & Loss in the High Latitudes*. New York: Doubleday, 1996.

11 Shearer, Christine. *Kivalina: A Climate Change Story*. Chicago: Haymarket Books, 2011.

indeed takes advantage of both the local and the universal focus by discussing the nuances of Kivalina, a small Alaskan island, and shifting focus back and forth between the island and the broader fight against organized disinformation campaigns. In this way, Shearer is speaking to Alaskan tribal communities but also the broader Environmental Justice narrative.

Having established the need for a more locally-focused form of environmental travel writing, the ideas dominating the American West in recent years will now be discussed, as that region could likely be considered the dominating region in environmental criticism, the cradle of the movement in a sense. There has indeed been much conversation surrounding the local versus the universal, or global, in environmental criticism, essentially since the field's concretization in academia (arguably with the creation of ASLE in 1992), but the perspective of globalization has long seemed to enrapture the minds of what could be called "Postwestern" theorists. As states of the US West have strong roots in environmentalism dating particularly back to the influence of John Muir and his Sierra Club, and tend to still lead the conversation, this one region's mindset has dominated the US environmental movement and has arguably had an outsized influence over much of the rest of the world's view on the subject in the 20th century. The problem with this domination is that Postwesternists have a very particular lens through which they view the world, and they arguably project that view outward onto regions and situations to which their view may not be apropos. For instance, the US West has been enraptured with its proximity to Silicon Valley for decades, and perhaps the shifting views on Silicon Valley and its extension throughout the US West are changing quickly, outside the region more than within its bubble.

Geographer Edward Soja sits at the center of this debate, and the Postwesternists discuss the world in terms of its physical space, its virtual space, and a third space, the latter being the result of the former two influencing one another. This is particularly clarified in Neil Campbell's essay, "John Brinckerhoff Jackson's Landscapes",[12] (59–60). Campbell suggests that Soja "rejects the 'either/or choice' perspective and recognizes instead a 'both/and also logic' encouraging new combinations alongside the mixing and hybridizing of thought and idea" (60). The neologism "glocalization" is often employed to discuss this idea, in terms of how global markets and cyberspace are affecting the physical space of the American West (Tatum, "Spectrality and the Postregional Interface," *Postwestern Cultures*, 7–8). Stephen Tatum employs the particular example of the "abstract economic phenomenon of 'the world

12 Kollin, Susan. Editor. *Postwestern Cultures: Literature, Theory, Space*. Lincoln and London: University of Nebraska Press, 2007.

steer,'" which finds its physical form in the American cattle of the West, and how "the actual steer live[s] on the ground and gorg[es] on the hormone-laced feed at this place called Wilderado," but its "spectral or virtual double ... over-determines the actual steers and feeder cows ... of local places" (5). The virtual space or animal, then, dominates prices and the amount of the actual space or animal in the West, what is farmed, and what the animals are fed. The work *Postwestern Cultures* was edited by Susan Kollin and published in 2007, as globalism was taking hold, not only through the technological advances of Silicon Valley and the Seattle of Jonathan Raban discussed throughout this research, but through liberal free trade agreements.[13] A little over a decade later, and trust in the inevitability of globalism seems to be fading, as nationalism and new forms of isolationism are quickly gaining ground in political elections (Donald Trump – "America First" –, Boris Johnson – "Brexit" –, Xi Jinping – "Chinese Dream" –, etc.). The point being made here is that, even if we never truly trusted globalism, it has dominated our theory for the first two decades of the 21st century, and we have created a tendency to avoid discussion of grass-roots work in literature for fear it would be co-opted by the global to negative ends, while perhaps also being enraptured by the possibility of the global *reveille* we could employ via the rise of the internet, which is proving to possess as many flaws as it does positive attributes. Shortly before the Postwesternists were telling us that the regional will only be gobbled up by the global, which will take it and form it as it so wills, so that regional cannot truly exist and should not be pursued in its perceived, literal form, Nancy Lord employed the local to attempt to save beluga whale populations. More recently, she has worked to save salmon streams and fight against ocean acidification. As has been noted, she is well respected in Alaska, and yet extraordinarily rarely is the occasion when her work is recognized Outside or studied by the broader environmental community. Her work, again, is the case in point of this chapter that focusing on grassroots issues at the local level is important and can be successful.

This is certainly not an attack on the text *Postwestern Cultures*, however, which raises important points and puts many ideas surrounding the West at the advent of the 21st century into a cohesive perspective. Susan Kollin herself, for instance, discusses an idea similar to that approached herein, which is that affect, or what she calls "lifestyle," "has emerged as a commodity in late-capitalist America, certain relationships to nature ultimately becom[ing] signs

13 "President Bush Expanded And Enforced Trade Agreements To Open New Markets For American Products." *The Bush Record*. "https://georgewbush-whitehouse.archives.gov/infocus/bushrecord/factsheets/tradeagreements.html." Viewed on 27/10/2019.

of class distinction" ("Survival, Alaska Style," 145). Kollin points out that sensationalist television shows, books, and culture advocate a first world lifestyle commodifying outdoor equipment, which seems to be an escape from the industrial West but ends up becoming another form of attached capitalism (145). Capper Nichols explores this theme thoroughly in the essay, "Backpacking and the Ultralight Solution" (127–42), where outdoor gear choice is turned into a veritable political philosophy against industrial capitalism, but the latter eventually adopts the material tenets of the philosophy for its own remunerative purposes. Essentially, the conclusion that Kollin makes in "Survival, Alaska Style" is that the notion of adventure in travel is a class privilege entirely wrapped up in global capitalism (148).

Again, to be clear, the point being made in this chapter is that the threat of the Postwestern discussion differs from the threat discussed in this research. The two are not at all ideologically opposed, but focusing on the manipulation of the West by global capitalism has long overshadowed potential discussion of the positives of local, grassroots work, which, in the mindful travel form advocated in this chapter, on a large scale, could bring a fresh perspective to the local. There is always a risk that regionalism, particularly through sensationalism, will fall to commodification. However, done mindfully the way that Nancy Lord has, and Barry Lopez did with *About this Life*, it could also be a boon, particularly in a world of increasing urbanization and abandoning of spaces deemed backward or unworthy of attention.

FIGURE 14 Soldotna
PHOTO BY CHRIS FERGUSON

Conclusion

Most of the works presented in this research were written in the second half of the 20th century when the pressing concerns of the current moment were not exactly clear. It has become manifest over the past roughly decade and a half, however, that this century's challenges lie in ecology and community. Ecology is obviously important for an overpopulated planet grappling with the gravity of its existential debate like never before, and community is a result of both the former and the continual leaps in technology (Florida, *The New Urban Crisis,* XI–XX).[1] I set out with those concerns in mind, hoping to examine the ethical responsibilities that travel writers carry when entering specifically into Alaska, with its Last Frontier mentality, Alaska Native communities, and history of preservationism and conservationism. Meanwhile, the intention was simultaneously to look backward and measure the successes and, sometimes, missteps of travel writing since statehood.

There has been no one question answered per se, but there has hopefully been clarification of what those ethical responsibilities are. At a moment when polarized and political frustrations are boiling over globally, this research has more or less suggested cooperation with all demographics, allowing for a "multitude of voices" to speak their experiences the way they do, for instance, in the Nancy Lord-edited work *Made of Salmon,* with historical backgrounds taken into consideration as well. From the close readings and reflections required, I would submit that the less misconceptions are offered, the less likely debates are to continue to polarize. Nancy Lord was particularly strong in *Beluga Days* at breaking down the myths created through political and cultural polarization, by attacking them one-by-one at the micro level, with the case of the beluga.

One of the struggles of this research was casting its net wide enough, but also maintaining focus on the task at hand. For me, politics and environmentalism are inextricably interwoven, so that debates over the two tend to flare up simultaneously, their arguments not being so respective, as we have seen with works like Nancy MacLean's *Democracy in Chains.* When we explore anglophone travel writing more broadly, we see that its motives have often derived from economic gains, mixed with a desire for adventure, so that the well-being of individuals and communities has time and again been prioritized

1 Florida, Richard. *The New Urban Crisis: How our Cities are Increasing Inequality, Deepening Segregation, and Failing the Middle Class–And What We Can Do About It.* New York: Basic Books, 2018.

 | DOI:10.1163/9789004734807_015

as subsidiary to the wealth of large cities. Such large cities as Seattle, Vancouver, and Copenhagen have often dictated the cultures, even the architecture, of spaces farther north, from which they extract resources. Therefore, analysis of broader political themes remains pertinent, if not a slippery slope of which to be wary.

1 A Quick Take on Alaska Native Identity

Environmental Justice and discussion of Alaska Native communities, meanwhile, ostensibly come off as the most straightforward of topics, and in a way they are. The problems in those communities run deep, and not many people would debate the fact that their issues have been catalyzed by the shock of being thrown into the metaphorical deep end of industrial civilization without an adaptation period. And yet, this discussion can be difficult, not only for its sensitivity, but also for the fact that Alaska Native voices are so rare throughout this shock. When an Alaska Native puts ink to paper, as Mary Tallmountain and William Iġġiaġruk Hensley have, for instance, they tend to have spent much time Outside, and their literary perspective is affected (no less authentic but perhaps less representative of their communities than many voices who have not written – a highly sensitive discussion in itself). As Jonathan Raban has insinuated, however, essentially all Native groups have been affected, even before the anthropologists came to tell their stories (*Passage*, 29). There is indeed promise in the 21st century, as Alaska Natives are taking the initiative of taking back their own narrative, with support from groups such as The Anchorage Museum, who recently hosted a roundtable discussion of Alaska Natives on how museums can be more inclusive. This discussion was diffused via the podcast briefly discussed herein: *Coffee and Quaq*.[2] Likewise, the Institute of American Indian Arts in Santa Fe, New Mexico, has been training Native American writers to express themselves in novel ways, to find their own personal techniques of bridging the gaps between Native and Western civilization.

The problem still remains, however, that there is little written record for scholars to piece together Alaska Native histories. The choices, essentially, are, 1) to pore through anthropological records, or, 2) to go and listen to oral

2 "Episode 9: Museums & Media." *Spreaker*. "https://www.spreaker.com/user/coffeeandquaq/museums-media?utm_medium=widget&utm_source=user%3A10663336&utm_term=episode_title." Viewed on 12/12/2019.

histories passed down through generations. Of course, historical narratives tend to be controlled by the victors, but this lack of record is perhaps even tougher to overcome than a potentially skewed narrative. Though an argument could be made that academia is running a fool's errand in searching for fact-based answers in Alaska Native history, as Native cultures traditionally do not place the same emphasis on historical record. In the Alaska Native worldview flexible oral histories may make more sense.

2 Conclusions Taken from Close Reading

Part 1 dug deeply into pertinent sets of themes, particularly referencing the six primary works and their writers' respective canons. Being the longest part, the meat of the work, I set about reading each of the primary works multiple times, examining them from the point of view of the chosen themes along the way. The results were enlightening and went even deeper than I could have imagined. The process was no doubt frustrating, adding layer upon layer to my understanding of the themes, regularly shifting the premises of my arguments, but the work nonetheless always remained fascinating.

For instance, I have sometimes quoted Jonathan Raban's iconoclastic comments at conferences, where they tend to cause the intended stir, and yet Raban saw so many of the problems of the 21st century years before his peers did. When we stop and examine his words, we see a long-held fight against classism that invades anglophone environmentalism, and perhaps even academia. Barry Lopez likewise offers a stunning array of depth. Quite literally, every time I look through his works, I find something I missed the last time, and often more insightful commentary than I had considered before. McPhee's work, by contrast, may not offer such gems from the standpoint of philosophical evocations, but his own techniques and simplicity sometimes hide the fact that his writing can be purely beautiful. This is to say that we tend to focus on the writer's unique, journalistic techniques, as well as the information he has to offer us, so that we miss the fact that his often-understated poetics can be powerful. McPhee, like a mature musician, knows when to minimize, to focus on the information, the journalism, and when to bunker down and densely express the pure beauty of a situation, allowing for the text to breathe and intensify alternately. These are the depths of appreciation we may reach with multiple, close readings.

And yet, travel writing can also have the effect of evoking broad, psychological shifts as well. In understanding the world and exotic cultures more deeply, not only does one become more "worldly" in a superficial sense, or in a sense of understanding broader themes of aesthetics and politics, for instance, but one also more clearly digests the bases for philosophical differences between cultures, their *umwelten*. A truly talented, reflective, travel writer possesses this gift of conveyance to the reader. Again, this is not merely a superficial conveyance, simply noticing the aspects of a culture, as a traveler and not a voyager, that others may miss. Such reflection can indeed be enlightening, but, often in a place such as Alaska the conveyance of entire worldviews so different from the industrial world that one's imagination would never arrive there on its own can truly affect the way that one moves through the world, opening up new horizons and shifting the way that information is received.

This is a difficult sentiment to express, perhaps because, with such barriers, there is no adequate vocabulary. An example could be a presentation one of my students gave recently, conflating ideas of veganism with a discussion of speciesism and anti-speciesism. As Nancy Lord notes in *Beluga Days*, this entire discussion from the Lower 48 perspective is dismissive of hunter/gatherer societies. In fact, as noted earlier, what can come across as a novel solution for eco-tourism, whale watching, for instance, can be profoundly offensive for Alaska Natives who may view such actions as disturbing a community of "people" (Lord, *Beluga Days*, 38). There obviously is no correct answer as to whether an industrial vegan or a Native hunter/gatherer is more ethical toward animals, because both are tenable, ethical positions, but opening up the conversation to the Alaska Native community, which includes "animal people," at least offers a more nuanced debate than the one currently being had. One of the problems is that the constraints of Western philosophy in such discussions may not even allow for the worldview of Native peoples, in which case they are merely considered outliers and told to adapt, as industrial civilization is now dominant. I would dare say that this is an extension of the colonial mindset from which the de-colonization movement is broadly attempting to break free. The travel writing discussed in this research approaches themes such as this that lie so far outside discussions being had in industrial society that they simply do not register. Barry Lopez' experience with Native American culture, then, consequently differentiates that writer from the lion's share of other environmental writers over the past half century or so. Likewise, John McPhee's in-depth discussions within the broader Alaskan community may have been instrumental in forming those views of his that have been criticized Outside. And, this is not even native culture, but Alaskan culture more broadly. In an *Anchorage Daily News* video from 2011, for example, discussing dipnet salmon fishing, an interviewee tells the reporter that the idea of dipnetting just does not even register with people Outside. The man has attempted to explain the subsistence practice, "but it just doesn't ring a bell with them."[3] Meanwhile, the practice is widely considered to be one of the most important aspects of Anchorage life and identity.

Part 2 of this work employs these perspectives analyzed in Part 1, essentially attempting to garner Alaska Native, and broader Alaskan perspectives, and allowing them inclusivity in the environmental movement, approaching the multiple viewpoints with an open-mind, offering solidarity to all, including conservative-minded Alaskans, both Native and non-Native. Large-scale, grassroots participation in environmental conversations, then, may be enticed by such approaches.

3 What's Left to be Done?

There is still plenty of work to be done beyond the domain of this research to move forward. Jonathan Raban's assertion that we may truly have the wrong impression of Alaska Native and British Columbian tribes, as the anthropologists arrived after the whalers and the promyshlenniki, leaves much room for examining relations between cultures in past travel writings under novel perspectives. Particularly, new understandings of Alaska Native culture and relationships pre-European contact could give us stronger comprehension of a more primal state and the kind of TEK that that could eventually evoke (an example being writer/psychologist Christopher Ryan's current attempts at composing a picture of life under hunter/gatherer societies pre-civilization, in order to demonstrate the positive qualities that they had, and which we may have lost – *Civilized to Death: What Was Lost on the Way to Modernity*).[4] We often discuss potlatches, slavery, and wealth gathered by the Tlingits and Haida, for instance, but what would their state of existence have been before the Russians came into contact with them? How would their art have expressed itself?

4 Modernity, and Its Possibilities

Modern research, on the other hand, could lean heavily toward de-colonization narratives, as they will likely increase over the next few decades and are among

3 "Alaskans descend on the Kenai River to dipnet for salmon." *Anchorage Daily News*. "https://www.youtube.com/watch?v=AL_fzdpmvo4." 19/7/2011.

4 Ryan, Christopher. *Civilized to Death: What Was Lost on the Way to Modernity*. New York: Avid Reader Press, 2019.

the most novel and interesting genres, not yet having been codified by publishers and dogmatic MFA programs with specific formulas. With the support of the academic and publishing communities, these narratives could create a substantial literary shift before they inevitably become codified. Writers such as Zadie Smith have succeeded in conveying minority experiences in the West, with works like *White Teeth*,[5] and Chimamanda Ngozi Odichie has done something similar for the experience of an African in America, with *Americanah*,[6] but the added dimensions that novel Native techniques have to offer could take identity studies much further: the vocabulary, the subversive narrative structures, the surprisingly refreshing syntax of the multilingual, the undergirding techniques of this movement are so different from the current mainstream that it will likely take decades for criticism to catch up.

Of course, de-colonization indeed runs parallel to both Social Justice and Environmental Justice, with much academic interest through identity, alterity, and intersectionality, but indigenous de-colonization nonetheless remains its own phenomenon because, unlike cultural affects, Native culture is deeply rooted through thousands of years of its own language, lifestyles, worldviews, *umwelten*. What that means for de-colonization is that, not being bound by the framework of existing rules, its freedom has allowed it to become one of the most exciting domains of literature at the moment. Terese Marie Mailhot, Tommy Orange, Tanya Tagaq, Alice Qannik Glenn, and Robin Wall Kimmerer are all doing exciting work with regard to de-colonization more broadly, which appears to be spreading rapidly throughout Alaska. Such work would likely be of interest to the wider population due to an increasing questioning of the political status quo, through what Mark Lilla calls the current "abdication" of "political vision" (*The Once and Future Liberal*, 7–13, 97–8). Yet, there does not seem to be much exploration in the domain of Native Americans or Alaska Natives reflectively approaching de-colonization through travel.

The ultimate chapter of this research suggests "local travel writing," or local people reflectively looking more closely at their own cultures and communities at a time when outward exploration is *de rigueur* and small communities are experiencing extraordinary population flight, as people are headed toward metropolitan and technological hubs. This could particularly be an interesting technique for Alaska Natives in exploring their own de-colonization. Alice Qannik Glenn explores various local communities and ideas with regard to Millennial and Gen Z experience of Alaska Natives, but place, geographic

5 Smith, Zadie. *White Teeth*. London: Hamish Hamilton, 2000.

6 Odichie, Chimamanda Ngozi. *Americanah*. New York: Knopf, 2013.

disparity, and travel are not necessarily an important part of the podcast. Her work can indeed serve as a model, however. Interestingly, third-generation, white Alaskan, Julia O'Malley published a book in December of 2019 discussing Alaskan food culture throughout the state (it is also the first book I have seen that approaches the influence of Pacific Islanders in the state), entitled *The Whale and the Cupcake: Stories of Subsistance, Longing, and Community in Alaska*.[7] This type of work is important, and similar texts published from the perspective of the Alaska Native, particularly with the spirit of the work coming out of the Institute of American Indian Arts, could potentially lead to some powerful narratives in the coming years. And yet, they may arrive via Youtube or podcasts rather than in book form, given that oral history is more important to Alaska Natives than the written record.

5 Conclusion to the Conclusion

With this research, I have attempted to merge the fields of Arctic and Environmental Studies through a study of travel narratives dedicated to Alaska, two domains which seem to connect in so many ways but continue to appear uncomfortable with one another. The Arctic has been heavily affected by industrialization and climate change, and so environmentalists have long employed the polar bear and melting glaciers and icecaps for their own purposes. But, do environmental studies truly take into account what's best for the peoples and places that lie within the Arctic? My students often say when discussing what precious metal mining will likely do to the Arctic in the coming decades in the search for "Green Energy" that the degradation of the lives of a few people who live outside industrial civilization are just the price to pay for the betterment of the cities, where the majority live. And yet, when I suggest that this mentality is more or less similar to that of the colonialists the 21st century loves to criticize they tend to shift in their seats uncomfortably. There are no easy answers, and the ethical dilemmas will only increase in the coming decades with new challenges, so that frank conversations between Alaskan society (and Native societies more broadly) and the environmentalist community are extraordinarily important to be had. This research could very well have carried an almost hagiographic spirit toward the environmentalism of European society, or it could have done so toward the Alaska Native hunter or the self-reliant

7 O'Malley, Julia. *The Whale and the Cupcake: Stories of Subsistence, Longing & Community in Alaska*. Anchorage: Anchorage Museum; Seattle: University of Washington Press, 2019.

Alaska sourdough. But, it does none of the above. Instead, it is a discussion of nuance, hoping to bring communities together through commonality rather than polarizing them even further than they already are. I believe Nancy Lord has shown this path particularly well with her work on the Cook Inlet beluga whale, and I believe that awakening nuanced conversations, globally, but on grassroots scales, can have a great effect, and that literature is a strong tool in accomplishing that feat.

Glossary

The Alaska Purchase (1867) The United States of America purchased Alaska (586,412 square miles 1,518,800 – square km) from Russia for $7.2 million.[1]

ANCSA Alaska Native Claims Settlement Act (1971) – In order to resolve land issues between the US federal government and Alaska Natives, Congress paid out $962.5 million for all but 44 million acres of the state. 12 corporations were created, by region, and the money was split evenly between these 12 corporations over 20 years. Each Alaska Native was assigned a corporation, which were further broken up into villages (Haycox, 283–4).

ANILCA Alaska National Interest Lands Conservation Act (1980) – After decades of fighting over ANWR, ANILCA served as a final agreement. 104 million acres were set aside as national parks and wildlife refuges, and land freezes on government selections of lands to be chosen under ANWR were lifted (Haycox, 297).

ANWR Arctic National Wildlife Refuge (1960) – As part of the statehood act, 103 million acres of land (28% of Alaska) were allotted to the new government as a site of refuge for wildlife, to remain unsettled, with the understanding that oil and minerals could nonetheless be extracted. This closed off potential for frontier but led to fights between environmentalists and economic boosters over the use of the lands (Haycox, 275).[2]

Boosterism The belief that environmental concerns are subservient to economic interests in a given region (Coates, 28).[3]

Conservationism An environmental movement designed to protect parks and forests while employing "Wise Use" of their natural resources (Coates, 28).

The Country A term employed by Alaskans for the state of Alaska itself

Deep Ecology The belief that man must change his relationship with nature. Man must learn to respect the inherent value in nature, subjectifying it and creating a balance with it rather than subduing it.[4]

Ecopoetics Using poetry to raise awareness of ecological concerns. Science and philosophy are often intertwined.[5]

---

1 "Alaska Purchase." *Britannica*. "Https://www.britannica.com/event/Alaska-Purchase." Viewed on 27/11/2019.

2 Haycox, Stephen W. *Alaska: An American Colony*. London: C. Hurst & Co. Publishers, 2002.

3 Coates, Peter A. *The Trans-Alaska Pipeline Controversy: Technology, Conservation, and the Frontier*. Bethlehem, PA: Lehigh University Press, 1991.

4 Madsen, Peter. "Deep Ecology." *Britannica*. "Https://www.britannica.com/topic/deep-ecology." Viewed on 2/4/2020.

5 "Glossary of Poetic Terms: Ecopoetics." *Poetry Foundation*. Https://www.poetryfoundation.org/learn/glossary-terms/ecopoetics." Viewed on 2/4/2020.

 | DOI:10.1163/9789004734807_016

Lower 48 The continental, or contiguous, United States. In other words, all states except for Alaska and Hawaii.

Outside In Alaskan terminology, any place outside Alaska, anywhere.

Preservationism An environmental movement based on leaving nature alone completely, as man cannot be trusted to improve upon nature (Coates, 36).

Sourdough Early miners and settlers to Alaska carried sourdough starters with them for the purposes of leavening bread, so the name is now associated with any longstanding resident of the state.

Statehood Alaska officially became the 49th state of The United States of America via the Alaska Statehood Act of 1959.[6]

TAPS Trans-Alaska Pipeline System (1977) – an oil pipeline system connecting Prudhoe Bay, on the Arctic Ocean, with Valdez in the southern part of Alaska. The pipeline system is 800 miles (1,300km) in length.[7]

---

6 "Alaska Statehood Act." *The Alaska State Legislature*. "https://www.akleg.gov/basis/get_documents.asp?session=29&docid=29890." Viewed on 27/11/2019.

7 "Trans-Alaska Pipeline." *Britannica*. "Https://www.britannica.com/topic/Trans-Alaska-Pipeline." Viewed on 27/11/2019.

Bibliography

Principal Works Studied

Lord, Nancy. *Beluga Days: Tracking a White Whale's Truths*. New York: Counterpoint, 2004.

Lopez, Barry. *Arctic Dreams: Imagination and Desire in a Northern Landscape*. 1986. New York: Vintage, 2001.

Lowenstein, Tom. *Ancient Land, Sacred Whale: The Inuit Hunt and its Rituals*. 1993. 2nd ed. New York: North Point Press, 1996.

Matthiessen, Peter. *Oomingmak: The Expedition to Musk Ox Island in the Bering Sea*. New York: Hastings House, 1967.

McPhee, John. *Coming into the Country: Travels in Alaska*. 1977. London: Daunt Books, 2015.

Raban, Jonathan. *Passage to Juneau: A Sea and its Meanings*. London: Picador, 1999.

Other Works by Principal Authors

Lopez, Barry

Desert Notes: Reflections in the Eye of a Raven. Kansas City: Andrews and McMeel Publishing Company, 1976.

Giving Birth to Thunder, Sleeping with His Daughter. Kansas City: Andrews and McMeel Publishing Company, 1978.

Of Wolves and Men. 1978. New York: Touchstone, 1995.

River Notes: The Dance of Herons. Kansas City: Andrews and McMeel Publishing Company, 1979.

Winter Count. New York: Charles Scribner's Sons Publishing Company, 1981.

Crossing Open Ground. New York: Charles Scribner's Sons Publishing Company, 1988.

Crow and Weasel. New York: North Point Press, 1990.

The Rediscovery of North America. Lexington: University Press of Kentucky 1990.

Field Notes: The Grace Note of the Canyon Wren. New York: Knopf Doubleday Publishing Group,1994.

Lessons from the Wolverine. Athens: University of Georgia Press, 1997.

About This Life: Journeys on the Threshold of Memory. New York: Knopf Doubleday Publishing Group, 1998.

Apologia. Athens: University of Georgia Press, 1998.

Light Action in the Caribbean. New York: Knopf Doubleday Publishing Group, 2000.

Resistance. New York: Knopf Doubleday Publishing Group, 2004.

Vintage Lopez. New York: Vintage Books, 2004.
The Future of Nature: Writing on a Human Ecology from Orion. Selected and introduced by Barry Lopez. Minneapolis, Minnesota: Milkweed Editions, 2007.
Outside. San Antonio: Trinity University Press, 2014.
Home Ground: Language for an American Landscape. Edited by Barry Lopez and Debra Gwartney. San Antonio: Trinity University Press, 2010.
Horizon. London: The Bodley Head, 2019.
Embrace Fearlessly The Burning World: Essays. New York: Random House, 2022.

Lord, Nancy

The Compass Inside Ourselves. Madison: Fireweed Press, 1984.
Survival. Minneapolis: Coffee House Press. 1991.
Darkened Waters: A Review of the History, Science, and Technology Associated with the Exxon Valdez Oil Spill and Cleanup. Homer, AK: Homer Society of Natural History/ Pratt Museum, 1992.
Fishcamp: Life on an Alaskan Shore Island. Press/Shearwater Books.1997 (hard cover) Berkeley, CA: Counterpoint Press, 1999 (paperback).
Green Alaska: Dreams from the Far Coast. Berkeley, CA: Counterpoint Press, 1999.
The Man Who Swam with Beavers. Minneapolis: Coffee House Press, 2001.
Rock, Water, Wild: An Alaskan Life. Nebraska: University of Nebraska Press, 2009.
Early Warming: Crisis and Response in the Climate-Changed North. Berkeley, Ca: Counterpoint Press, 2011.
Made of Salmon: Alaska Stories from the Salmon Project. Anchorage: University of Alaska Press, 2016.
pH: A Novel. Berkeley, Ca: West Margin Press, 2017.

Lowenstein, Tom

Our After-fate. UK: Softy Loudly Books, 1971.
Eskimo Poems from Canada and Greenland (*translation*). Allison & Busby, Pittsburgh: University of Pittsburgh Press, 1973.
Stories from Point Hope. Juneau: Alaska State Museum, 1973.
Booster – A Game of Divination. London: Many Press, 1975.
The Death of Mrs Owl. London: Anvil Press, 1975.
Sea Ice Subsistence at Point Hope, Alaska. Barrow: North Slope Borough, 1980.
The Shaman Aningatchaq, translation & commentary. London: Many Press, 1982.
Filibustering in Samsara. London: Many Press, 1987.
La Tempesta's X-ray. London: Many Press, 1988.
The Vision of the Buddha. London: Duncan Baird Publishers / Macmillan, 1996.
Ancestors and Species. Bristol: Shearsman Books, 2005.
Treasures of the Buddha. London: Duncan Baird Publishers, 2006.

Classic Haiku. London: Duncan Baird Publishers, 2007.

Conversation with Murasaki. Bristol: Shearsman Books, 2009.

Ultimate Americans: Point Hope, Alaska 1826–1909. Fairbanks: University of Alaska Press, 2009.

Lowenstein, Tom and Piers Vitebsky. *Native American Myths and Beliefs*. New York: The Rosen Publishing Group, 2011.

The Civilization of Ancient India and Southeast Asia. New York: The Rosen Publishing Group, 2012.

Buddhist Inspirations. New Jersey: Chartwell Books, 2013.

From Culbone Wood – in Xanadu: Notebooks and Fanasias. Bristol: Shearsman Books, 2013.

The Structure of Days Out: With storytellers, hunters and their descendants in a Native Alaskan Community, 1973–1981. Bristol, UK: Shearsman Books, 2021.

Matthiessen, Peter

Race Rock. 1954. New York: Vintage, 1988.

Partisans. 1955. New York: Random House, 1991.

Wildlife in America. 1959. New York: Penguin Books, 1977.

Raditzer. 1961. New York: Vintage Books, 1987.

The Cloud Forest: A Chronicle of the South American Wilderness. 1961. New York: Random House, 2012.

Under The Mountain Wall. 1962. New York: Random House, 2012.

At Play in the Fields of the Lord. 1965. New York: Knopf Doubleday Publishing Group, 2012.

Blue Meridian: The Search for the Great White Shark. 1971. New York: Random House, 2012.

The Tree Where Man Was Born. 1972. New York: Penguin, 2010.

The Snow Leopard. 1978. New York: Random House, 2014.

Matthiessen, Peter and Hugo van Lawick. *Sand Rivers*. London: Aurum Press, 1981.

In The Spirit Of Crazy Horse. 1983. New York: Random House, 2012.

Far Tortuga. New York: Vintage Books, 1984.

Indian Country. New York: Penguin Books, 1984.

Men's Lives: The Surfmen and Baymen of the South Fork. 1986. New York: Knopf Doubleday Publishing Group, 2012.

Nine-headed Dragon River: Zen Journals 1969–1982. 1986. Boulder, CO: Shambhala Publications, 1998.

On the River Styx: And Other Stories. 1989. New York: Knopf Doubleday Publishing Group, 2012.

Killing Mister Watson. 1990. New York: Knopf Doubleday Publishing Group, 2012.

African Silences. 1991. New York: Random House, 2012.

The Wind Birds: Shorebirds of North America. Shelburne, Vermont: Chapters Publishing, 1994.

Baikal: Sacred Sea of Siberia. 1992. San Francisco: Sierra Club Books, 1995.

Lost Man's River. 1997. New York: Random House, 2012.

The Birds of Heaven: Travels with Cranes. New York: Farrar, Straus and Giroux, 2001.

East of Lo Monthang: In the Land of Mustang. 1995. Boulder, CO: Shambhala/Timeless Books, 1995.

Bone by Bone. 1999. New York: Random House, 2013.

Tigers in the Snow. London: Harvill Press, 2000.

The Peter Matthiessen Reader: Nonfiction, 1959–1961. New York: Vintage Books, 2000.

End of the Earth: Voyage to Antarctica. Washington, D.C.: National Geographic, 2003.

Shadow Country. London: MacLehose Press, 2010.

In Paradise. London: Oneworld Publications, 2014.

Sal Si Puedes (Escape If You Can): Cesar Chavez and the New American Revolution. Berkeley, CA: University of California Press, 2014.

McPhee, John

A Sense of Where You Are: a profile of William Warren Bradley. New York: Farrar, Straus and Giroux, 1965.

The Headmaster: Frank L. Boyden, of Deerfield. New York: Farrar, Straus and Giroux, 1966.

Oranges. New York: Farrar, Straus and Giroux, 1967.

The Pine Barrens. New York: Farrar, Straus and Giroux, 1968.

A Roomful of Hovings and Other Profiles. New York: Farrar, Straus and Giroux, 1968.

Levels of the Game. New York: Farrar, Straus and Giroux, 1969.

The Crofter and the Laird. New York: Farrar, Straus and Giroux, 1970.

Encounters with the Archdruid. New York: Farrar, Straus and Giroux, 1971.

The Deltoid Pumpkin Seed. New York: Farrar, Straus and Giroux, 1973.

The Curve of Binding Energy. New York: Farrar, Straus and Giroux, 1974.

The Survival of the Bark Canoe. New York: Farrar, Straus and Giroux, 1975.

Pieces of the Frame. New York: Farrar, Straus and Giroux, 1975.

Giving Good Weight. New York: Farrar, Straus and Giroux, 1979.

Basin and Range. New York: Farrar, Straus and Giroux, 1981.

In Suspect Terrain. New York: Farrar, Straus and Giroux, 1983.

La Place de la Concorde Suisse. New York: Farrar, Straus and Giroux, 1984.

Table of Contents. New York: Farrar, Straus and Giroux, 1985.

Rising from the Plains. New York: Farrar, Straus and Giroux, 1986.

Heirs of General Practice. New York: Farrar, Straus and Giroux, 1986.

The Control of Nature. New York: Farrar, Straus and Giroux, 1989.

Looking for a Ship. New York: Farrar, Straus and Giroux, 1990.

Assembling California. New York: Farrar, Straus and Giroux, 1993.
The Ransom of Russian Art. New York: Farrar, Straus and Giroux, 1994.
Irons in the Fire. New York: Farrar, Straus and Giroux, 1997.
Annals of the Former World. New York: Farrar, Straus and Giroux, 1998.
The Founding Fish. New York: Farrar, Straus and Giroux, 2002.
Uncommon Carriers. New York: Farrar, Straus and Giroux, 2006.
Silk Parachute. New York: Farrar, Straus and Giroux, 2010.

Raban, Jonathan

Mark Twain: Huckleberry Finn. London: Edward Arnold, 1968.
The Technique of Modern Fiction: Essays in Practical Criticism. London: Edward Arnold, 1968.
The Society of the Poem. London: Harrap, 1971.
Soft City. London: The Harvill Press, 1974.
Arabia: Through the Looking Glass. 1979. London: Picador, 1987.
Old Glory: A Voyage Down the Mississippi.1981. New York: Knopf Doubleday Publishing Group, 2011.
Foreign Land: A Novel. 1985. New York: Knopf Doubleday Publishing Group, 2010.
Coasting: A Private Voyage. 1986. New York: Knopf Doubleday Publishing Group, 2011.
For Love & Money: Writing, Reading, Travelling, 1968–1987. London: Picador, 1988.
God, Man, & Mrs Thatcher. London: Chatto & Windus, 1989.
Hunting Mister Heartbreak: A Discovery of America. 1991. New York: Knopf Doubleday Publishing Group, 2011.
The Oxford Book of the Sea. Oxford: Oxford University Press, Incorporated, 1992.
Bad Land: An American Romance. 1996. New York: Knopf Doubleday Publishing Group, 2011.
Waxwings. London: Picador, 2004.
My Holy War: Dispatches from the Home Front. New York: New York Review of Books, 2006.
Surveillance: A Novel. London: Picador, 2007.
Driving Home: An American Journey. 2011. Seattle: Sasquatch Books, 2013.

Other Primary Sources

Abbey, Edward. *The Monkey Wrench Gang*. 1975. Salt Lake City: Dream Garden Press, 1985.
Aristotle. *The Art of Rhetoric*. Translated by H. C. Lawson-Trancred. London. Penguin Books, 1991.
Arseniev, V.K. *Dersu the Trapper*. 1923. Translated by Malcom Burr (1939). St. Louis: McPherson, 1996.

Bartram, William. *Bartram: Travels and Others Writings*. Edited by Thomas P. Slaughter. New York: Library of America, 1996.

Beach, Rex. *Alaskan Adventures*. New York: A.L. Burt Company, 1909.

Beach, Rex. *The Spoilers*. New York: Harper & Brothers Books, 1906.

Burroughs, John. *Alaska: The Harriman Expedition, 1899*. Mineola, NY: Dover Books, 1986.

Carson, Rachel. *Silent Spring*. 1962. Boston and New York: Houghton Mifflin Harcourt, 2002.

Coetzee, J.M. *Waiting for the Barbarians*. London: Secker & Warburg, 1980.

Collis, Septima. *A Woman's Trip to Alaska – Being an Account of a Voyage through the Inland Seas of the Sitkan Archipelago, in 1890*. 1890. London: Forgotten Books, 2018.

Cook, James. *The Journals of Captain Cook*. 1955–67. 2nd ed. London: Penguin Classics, 2003.

Cournut, Bérengère. *De pierre et d'os*. Paris: Le Tripode, 2019.

Crisler, Lois. *Arctic Wild: The Remarkable True Story of One Couple's Adventures Living Among Wolves*. 1958. Connecticut: Lyons Press, 1999.

de Bougainville, Louis-Antoine. *M. Bideau et S. Faessel, Editors. Voyage autour du monde*. 1771. Paris : Presses de L'Université de Paris-Sorbonne, 2001.

de Chateaubriand, François-René. *Atala, ou Les Amours de deux sauvages dans le désert*. 1801. Paris : Le Livre de Poche, 2012.

de Chateaubriand, François-René. *Mémoires d'outre-tombe: Livres I à XII*. 1848. Paris: Garnier, 1989.

de Chateaubriand, François-René. *Intinéraire de Paris à Jérusalem: et de Jérusalem à Paris*. 1811. Paris: Éditions Gallimard, 2005.

Defoe, Daniel. *Robinson Crusoe*. 1719. Edited by Thomas Keymer. 2007. Oxford: Oxford University Press, reissued, 2008.

de Tocqueville, Alexis. Revised by Francis Bowen. *Democracy in America*. 1835. Translated by Henry Reeve. Hertfordshire: Wordsworth Classics of World Literature, 1998.

Delillo, Don. *White Noise*. New York: Viking Adult, 1985.

Deming, Alison Hawthorne. *The Edges of the Civilized World: a journey in nature and culture*. New York: Picador, 1998.

Diebitsch-Peary, Josephine. *My Arctic Journal: A Year Among Ice-Fields and Eskimos (1891–1892)*. 1893. Cooper Square Press, 2002.

Earle, Sylvie A. *Sea Change: A Message of the Oceans*. 1995. New York: Fawcett Columbine, 1996.

Eggers, Dave. *The Circle*. New York: Knopf, 2013.

Eggers, Dave. *A Heartbreaking Work of Staggering Genius*. New York: Simon & Schuster, 2000.

Eggers, Dave. *Heroes of the Frontier*. New York: Alfred A. Knopf, 2016.

Eggers, Dave. *What is the What?* San Francisco: McSweeney's, 2006.

Emerson, Ralph Waldo. *The Essential Writings of Ralph Waldo Emerson*. New York: Random House Publishing Group, 2009.

Emerson, Ralph Waldo. *Representative Men: Seven Lectures*. Boston: Phillips, Sampson and Company, 1850.

Étienne, Jean-Louis. *Le Marcheur du Pôle*. Paris: Éditions Robert Laffont, S.A., 1986.

Faulkner, WIlliam. *Go Down, Moses*. New York: Random House, 1942.

Ferber, Edna. *Ice Palace*. 1958. New York: Knopf Doubleday Publishing Group, 2014.

Franklin, Benjamin. *The Autobiography of Benjamin Franklin*. 1793. Mineola, New York: Dover Publications, Inc., 1996.

Griselin, Madeleine. *Huit femmes pour un pôle : La première expédition polaire féminine*. 1988. Paris : Éditions Albin Michel, 2016.

Haines, John. *The Stars, The Snow, The Fire: Twenty-five Years in the Alaska Wilderness*. 1977. New York: Washington Square Press, 1989.

Hakluyt, Richard. editor. *Divers Voyages Touching the Discovery of America, and the Islands Adjacent*. 1582. London: The Hakluyt Society, 1850.

Hakluyt, Richard. editor. *The Principal Navigations, Voyages, Traffiques & Discoveries of the English Nation: Made by Sea Or Over-land to the Remote and Farthest Distant Quarters of the Earth at Any Time Within the Compasse of These 1600*. 1589. Glasgow: Yeeres. J. MacLehose and Sons, 1905.

Hanh, Thich Nhat. *Living Buddha, Living Christ*. 1995. London: Penguin, 2007.

Hawkes, John. *Adventures in the Alaskan Skin Trade*. New York: Simon and Schuster, 1985.

Hayes, Ernestine. *Blonde Indian: An Alaska Native Memoir*. Tuscon: University of Arizona Press, 2015.

Hayes, Ernestine. *The Tao of Raven: An Alaska Native Memoir*. Seattle: The University of Washington Press, 2019.

Hensley, William L. Iggiagruk. *Fifty Miles from Tomorrow: A Memoir of Alaska and the Real People*. New York: Picador, 2009.

Hess, Bill. *Gift of the Whale: the Iñupiat Bowhead Hunt, a Sacred Tradition*. Seattle: Sasquatch Books, 1999.

Jameson, Anna. *Winter Studies and Summer Rambles in Canada*. 1838. Cambridge: Cambridge University Press, 2011.

John, Betty. *Libby: The Alaskan Diaries & Letters of Libby Beaman, 1879–1880, as presented by her granddaughter Betty John*. Boston: Houghton Mifflin Company, 1987.

Johnson, Nathanael. *Unseen City: The Majesty of Pigeons, the Discreet Charm of Snails & Other Wonders of the Urban Wilderness*. Pennsylvania: Rodale, 2016.

Kane, Elisha Kent. *Arctic Explorations: The Second Grinnell Expedition in Search of Sir John Franklin*. 1857. Cambridge: Cambridge University Press, 2012.

Kane, Elisha Kent. *The United States Grinnell Expedition in Search of Sir John Franklin*. Philadelphia: Childs & Peterson, 1856.

Kantner, Seth. *Shopping for Porcupine: A Life in Arctic Alaska*. Minneapolis: Milkweed Editions, 2008.

Kent, Rockwell. *Wilderness, a Journal of Quiet Adventure in Alaska*. New York: Knickerbocker Press, 1920. (This print: Leopold Classic Library – no date, original Knickerbocker Press logo prevalent).

Kerouac, Jack. *Big Sur*. 1962. London: Harper Perennial, 2006.

Kerouac, Jack. *The Dharma Bums*. 1958. London: Penguin Books, 2006.

Kerouac, Jack. *On the Road*. 1957. London: Penguin Books, 2000.

Kimmerer, Robin Wall. *Braiding Sweetgrass: Indigenous Wisdom, Scientific Knowledge, and the Teachings of Plants*. Minneapolis, Minnesota: Milkweed Editions, 2013.

Kimmerer, Robin Wall. *Gathering Moss: A Natural and Cultural History of Mosses*. Corvalis: Oregon State University Press, 2003.

Klein, Naomi. *This Changes Everything: Capitalism vs. the Climate*. 2014. London: Penguin Books, 2015.

Kolbert, Elizabeth. *The Sixth Extinction: An Unnatural History*. New York: Picador, 2015.

Kpomassie, Tété-Michel. *An African in Greenland*. Translated by James Kirkup. 1981. San Diego: Harcourt Brace Jovanovich, 1983.

Krakauer, Jon. *Eiger Dreams: Ventures Among Men and Mountains*. 1990. Guilford, Connecticut: The Lyons Press, 2009.

Krakauer, Jon. *Into the Wild*. 1996. Hampshire: Pan Macmillan, 2011.

Krakauer, Jon. *Into Thin Air: A Personal Account of the Mount Everest Disaster*. 1997. New York: Anchor Books, 1998.

Krakauer, Jon. *Missoula: Rape and the Justice System in a College Town*. New York: Doubleday, 2015.

Krakauer, Jon. *Under the Banner of Heaven: A Story of Violent Faith*. New York: Pan MacMillan, 2003.

Lamazou, Zoé et Victor Gurrey. *Une saison de chasse en Alaska*. Paris : Éditions Paulsen, 2014.

Leopold, Aldo. *A Sand County Almanac*. 1949. Oxford: Oxford University Press, 2001.

Lévi-Strauss, Claude. *Mythologiques : Le Cru et le Cuit*. 1964. Paris: Éditions Plon, 2009.

Lévi-Strauss, Claude. *Tristes tropiques*. Paris: Éditions Plon, 1955.

Linnaeus, Carl. *Systema Naturae*. 1735. *Systema naturae per regna tria naturae :secundum classes, ordines, genera, species, cum characterizes, differential, synonymis, locis*. (in Latin) (10th ed.). Stockholm: Laurentius Salvius, 1758.

London, Jack. *The Call of the Wild, White Fang, and Other Stories*. 1903, 1906, 1907. London: Penguin Books, 1981.

MacLean, Norman. *A River Runs Through It, and other stories*. 1976. Twenty-fifth Anniversary Edition. Chicago: The University of Chicago Press, 2001.

Mailer, Norman. *Armies of the Night: History as a Novel/The Novel as History*. New York: New American Library, 1968.

Mailhot, Terese Marie. *Heart Berries: A Memoir*. 2018. London: Bloomsbury Publishing, 2019.

Malaurie, Jean. *Hummocks 1 : De la pierre à l'homme Nord-Groenland: Tome 1—Livre I*. 2nd ed. Paris: Éditions Plon, 1999.

Malaurie, Jean. *Hummocks 1 : Arctique Central Canadien: Tome 1—Livre II*. 2nd ed. Paris: Éditions Plon, 1999.

Malaurie, Jean. *Hummocks 2 : Alaska, avec les chasseurs de baleine, mer de Bering: Tome 2 – Livre I*. 2nd ed. Paris: Éditions Plon, 1999.

Malaurie, Jean. *Hummocks 2 : Tchoukotka, Sibérie : Aux origines mythiques des Inuit*. Paris: Éditions Plon, 1999.

Malaurie, Jean. *Les derniers rois de Thulé, avec les Esquimaux polaires, face à leur destin*. 1955. Paris: Éditions Plon, "Terre Humaine"; 5e édition définitive, 1989.

Malaurie, Jean. *Ultima Thulé : Les Inuit nord-groenlandais face aux conquérants du Pôle (1818–1993)*. Paris: Bordas, 1990.

Marshall, Robert and George Marshall. *Alaska Wilderness: Exploring the Central Brooks Range*. 1956. Berkeley: University of California Press, 2005.

Marshall, Robert. *Arctic Village: A 1930s Portrait of Wiseman, Alaska*. 1933. Alaska: University of Alaska Press, 1991.

McCarthy, Cormac. *All the Pretty Horses*. New York: Alfred A. Knopf, 1992.

McCarthy, Cormac. *Blood Meridian or the Evening Redness in the West*. New York: Random House, 1985.

McCarthy, Rory. *Nobody Told Us We Are Defeated: Stories from the New Iraq*. New York: Random House, 2011.

McGinniss, Joe. *Going to Extremes*. New York: Alfred A. Knopf, 1980.

McPhee, John. "Direct Eye Contact." *The New Yorker*. "https://www.newyorker.com/magazine/2018/03/05/in-search-of-new-jerseys-wild-bears#." 2/2018.

McKibben, Bill. *American Earth: Environmental Writing Since Thoreau*. New York: Library of America, 2008.

Melville, Herman. *Moby Dick; or, the Whale*. 1851. Barnes & Noble Books, 2003.

Melville, Herman. *Omoo: A Narrative of Adventures in the South Seas*. New York: Harper & Brothers, 1847.

Melville, Herman. *Typee: A Peep at Polynesian Life*. New York: Wiley & Putnam, 1846.

Michener, James A. *Alaska*. New York: Random House Publishing Group, 1988.

Morrison, Toni. *Song of Soloman*. New York: Alfred A. Knopf, 1977.

Mowat, Farley. *Never Cry Wolf*. Toronto: McClelland and Stewart, 1963.

Mowat, Farley. *The Polar Passion: The Quest for the North Pole*. 1967. Toronto: McLelland and Stewart Limited, 1973.

Mowat, Farley. *Sea of Slaughter: A Chronicle of the Destruction of Animal Life in the North Atlantic.* 1984. Shelburne, Vermont: Chapters Publishing Ltd., 1996.

Muir, John. *Letters to a Friend.* Boston: Houghton Mifflin, 1915.

Muir, John. *The Mountains of California.* New York: Century, 1894.

Muir, John. *My First Summer in the Sierra.* Boston: Houghton Mifflin, 1911.

Muir, John. *Our National Parks.* Boston: Houghton, Mifflin, 1901.

Muir, John. *Nature Writings: The Story of My Boyhood and Youth; My First Summer in the Sierra; The Mountains of California; Stickeen; Selected Essays.* New York: Library of America, 1997.

Muir, John. *Picturesque California.* San Francisco: J. Dewing Publishing Company, 1888.

Muir, John. *Steep Trails.* Boston: Houghton, 1918.

Muir, John. *Stickeen: An Adventure with a Dog and a Glacier.* 1897/1909. CA: Dawn Publications, 1998.

Muir, John. *The Story of My Boyhood and Youth.* Boston: Houghton Mifflin, 1913.

Muir, John. *Studies in the Sierra.* 1950 (reprint of serials from 1874). Gloucestershire: Dodo Press, 2009.

Muir, John. *A Thousand-mile Walk to the Gulf.* Boston: Houghton Mifflin, 1916.

Muir, John. *Travels in Alaska.* 1915. Apple iBook.

Muir, John. *The Yosemite.* New York: Century, 1912.

Murie, Margaret. *Two in the Far North.* 1962. 2nd ed. Anchorage: Alaska Northwest Publishing Company 1979.

Nabokov, Vladimir. *Pnin.* 1953. New York: Vintage Books, 1989.

Nansen, Fridtjof. *Farther North.* 1897. Apple iBook.

Nichols, John. *The Milagro Beanfiled War.* New York: Holt, Rienhart and Winston, 1974.

Nickerson, Sheila. *Disappearance: A Map: A Meditation on Death & Loss in the High Latitudes.* New York: Doubleday, 1996.

Odichie, Chimamanda Ngozi. *Americanah.* New York: Knopf, 2013.

Okri, Ben. *A Time for New Dreams.* 2009. London: Rider, 2011.

Olmsted, Frederick Law. *Frederick Law Olmsted: Writings on Landscape, Culture, and Society.* Edited by Charles Beveridge. New York: Library of America, 2015.

O'Malley, Julia. *The Whale and the Cupcake: Stories of Subsistence, Longing & Community in Alaska.* Anchorage: Anchorage Museum; Seattle: University of Washington Press, 2019.

Orange, Tommy. *There There.* London: Harvill Secker, 2018.

Packer, George. *The Unwinding: Thirty Years of American Decline.* 2013. New York and London: Faber and Faber, 2014.

Pollan, Michael. *How to Change Your Mind: What the New Science of Psychedelics Teaches Us About Consciousness, Dying, Addiction, Depression, and Transcendence.* New York: Penguin Group, 2018.

Polo, Marco. *The Travels of Marco Polo.* 1302. Edited by Milton Rugoff. 1961 (Translator unlisted). 1st Signet Printing (Mittelmark afterward), 2004.

Pound, Ezra. *Cathay*. London: Elkin Mathews, 1915.

Pynchon, Thomas. *Mason & Dixon*. 1997. New York: Penguin, 2012.

Rasmussen, Knud. *Across Arctic America: Narrative of the Fifth Thule Expedition*. 1927. University of Alaska Press; Reprint edition, 1999.

Rasmussen, Knud. *The Bear in the Ice Hole*. Translated by Reginald Spink, edited by Bent Rying. Copenhagen: Royal Danish Ministry of Foreign Affairs, 1962.

Remarque, Erich Maria. *All Quiet on the Western Front*. Translated by E.W. Wheen. New York: Little, Brown and Company, 1929.

Riel, Jorn. *Un gros bobard et autres racontars*. Traduit par Suzanne Juul et Bernard Saint Bonnet, 1986. Paris : Gaia Éditions, 1999.

Serres, Michel. *C'était mieux avant*. Paris : Éditions Le Pommier, 2017.

Serres, Michel. *Petite poucette*. Paris : Éditions Le Pommier, 2012.

Service, Robert. *Songs of a Sourdough*. 1907. Washginton, D.C.: Westphalia Press, 2015.

Shakespeare, William. *Much Ado About Nothing*. 1623. Edited by R. B. Kennedy. London: Collins Classics, 2011.

Shelley, Mary. *Frankenstein: or, The Modern Prometheus*. London: Lackington, Hughes, Harding, Mavor, & Jones, 1818.

Smith, Adam. *The Wealth of Nations*. 1776. New York: Random House, 2000.

Smith, Zadie. *White Teeth*. London: Hamish Hamilton, 2000.

Snyder, Gary. *The Practice of the Wild: Essays*. Berkeley: Counterpoint Press, 1990.

Solnit, Rebecca. *The Faraway Nearby*. 2013. London: Granta, 2014.

Solzhenitsyn, Aleksandr. *The Gulag Archipelago: 1918–56: an experiment in literary investigation*. 1973. English translation 1974. London: The Harvill Press, 2007.

Stefansson, Vilhjalmur. *The Friendly Arctic: The Story of Five Years in Polar Regions*. New York: The MacMilland Company, 1921.

Steinbeck, John. *The Grapes of Wrath*. New York: The Viking Press, 1939.

Steller, Georg Wilhelm. *Journal of a Voyage with Bering, 1741–1742*. Translated by Margritt Engel and O.W. Frost. 1743. California: Stanford University Press, 1993.

Streever, Bill. Cold: *Adventures in the World's Frozen Places*. London: Hachette UK, 2009.

Swift, Jonathan. *Gulliver's Travels*. 1726. London: Jones & Company, 1826.

Tagaq, Tanya. *Split Tooth*. New York: Viking Press, 2018.

Thesiger, Wilfred. *The Marsh Arabs*. 1964. London: Penguin Books, 2007.

Thompson, Hunter S. *Hell's Angels: The Strange and Terrible Saga of the Outlaw Motorcycle Gangs*. New York: Random House, 1967.

Thoreau, Henry David. *Walden and «Civil Disobedience»*. 1854 and 1849. New York: New American Library, 1st Signet Classics Printing, 2012.

Thubron, Colin. *In Siberia*. London: Chatto and Windus, 1999.

Treadwell, Timothy and Jewell Palovak. *Among Grizzlies: Living with Wild Bears in Alaska*. 1997. New York: Ballantine Books, 1999.

Vaillant, John. *The Golden Spruce: A True Story of Myth, Madness, and Greed*. New York: W.W. Norton & Company, 2005.

Vaillant, John. *The Tiger: A True Story of Vengeance and Survival.* London: Sceptre, 2010.

Van der Post, Laurens. *The Lost World of the Kalahari.* 1958. London: Penguin Books, 1962.

Wallace, David Foster. *Broom of the System.* New York: Viking Press, 1987.

Wallace, David Foster. *Consider the Lobster: And Other Essays.* London: Abacus, 2005.

Wallace, David Foster. *Infinite Jest.* New York: Little, Brown, and Company, 1996.

Wallace, David Foster. *A Supposedly Fun Thing I'll Never Do Again: Essays and Arguments.* 1997. New York: Little, Brown and Company, 1998.

Wheeler, Sara. *The Magnetic North: Travels in the Arctic.* 2009. London: Vintage Books, 2010.

Wheeler, Sara. *Mud and Stars: Travels in Russia with Pushkin and Other Geniuses of the Golden Age.* London: Jonathan Cape, 2019.

White, E.B. *Charlotte's Web.* New York: Harper & Brothers, 1952.

Whitman, Walt. *Leaves of Grass.* 1855. New York: Signet Classics, 1954.

Wolfe, Tom. *The Electric Kool-Aid Acid Test.* New York: Farrar, Straus and Giroux, 1968.

Wollstonecraft, Mary. *Letters Written During a Short Residence in Sweden, Norway, and Denmark.* London: J. Johnson, 1796.

Wordsworth, William. *Poems, in Two Volumes: Volume I.* 1802. London: Longman, Hurst, Rees, and Orme, 1807.

Young, Samuel Hall. *Alaska Days With John Muir.* Peregrine Smith Books, 1915.

Secondary Sources

Adamson, Joni, Mei Mei Evans, & Rachel Stein. Editors. *The Environmental Justice Reader: Politics, Poetics & Pedagogy.* Tucson: The University of Arizona Press, 2002.

Beaglehole, John. *The Life of Captain James Cook.* 1974. Stanford, California: Stanford University Press, 2001.

Beck, Mary Giraudo. *Potlatch: Native Ceremony and Myth on the Northwest Coast.* Anchorage, Seattle, and Portland: Alaska Northwest Books, 1993.

Bendixon, Alfred and Judith Hamera. Editors. *The Cambridge Companion to American Travel Writing.* Cambridge: Cambridge University Press, 2009.

Bernstein, Samuel. "The Paris Commune." *Science & Society.* Vol. 5, No. 2 (Spring, 1941), pp. 117–147 (31 pages).

Besson, Françoise. *Ecology and Literatures in English: Writing to Save the Planet.* Newcastle upon Tyne: Cambridge Scholars Publishing, 2019.

Black, Lydia, Nora Marks Dauenhauer, and Richard Dauenhauer, editors. *Anóoshi Lingít Aaní Ká / Russians in Tlingit America: The Battles of Sitka, 1802 and 1804.* Seattle and London: University of Washington Press / Juneau: Sealaska Heritage Institute, 2008.

Boas, Franz. *Anthropology and Modern Life.* 1928. New York: Dover Publications, 1986.

Boas, Franz. *Tsimshian Texts: A Collection.* Volume III, Part 2. New York: Publications of the American Ethnological Society, 1912.

Bloom, Paul. *Against Empathy: The Case for Rational Compassion.* London: The Bodley Head, 2016.

Bomann-Larsen, Tor. *Roald Amundsen.* 1995. Translated by Ingrid Christophersen. First Published in English, 2006. Gloucestershire: The History Press, 2011.

Borm, Jan. *Jean Malaurie : un homme singulier.* Paris : Éditions du Chêne – Hachette livre, 2005.

Borneman, Walter R. *Alaska: Saga of a Bold Land.* New York: Harper Collins, 2003.

Brinkley, Douglas. *The Quiet World: Saving Alaska's Wilderness Kingdom 1879–1960.* New York: Harper Collins, 2011.

Brinkley, Douglas. *Rightful Heritage: Franklin D. Roosevelt and the Land of America.* New York: Harper Collins, 2016.

Brinkley, Douglas. *The Wilderness Warrior: Theodore Roosevelt and the Crusade for America.* 2009. New York: Harper Perennial, 2010.

Brogan, Hugh. *The Penguin History of the United States of America.* London: Penguin UK, 2001.

Buell, Lawrence. *Emerson.* Cambridge, Massachusetts: The Belknap Press of Harvard University Press, 2003.

Buell, Lawrence. *The Future of Environmental Criticism: Environmental Crisis and Literary Imagination.* 2005. Malden, MA, Oxford, and Victoria, Australia: Blackwell Publishing, Second Printing, 2006.

Buell, Lawrence. *Literary Transcendentalism: Style and Vision in the American Renaissance.* 1973. Ithaca and London: Cornell University Press, Second Printing, 1974.

Byers, Michael. *Who Owns the Arctic?: Understanding Sovereignty Disputes in the North.* Vancouver: Douglas and McIntyre, 2010.

Chernow, Ron. *Alexander Hamilton.* New York: Penguin, 2004.

Coates, Peter A. *The Trans-Alaska Pipeline Controversy: Technology, Conservation, and the Frontier.* Bethlehem, PA: Lehigh University Press, 1991.

Collomb, Jean-Daniel. « Christianity to Ecology: John Muir's Walk through America », *Transtext(e)s Transcultures* 跨文本跨文化 [Online], 4 | 2008, Online since 15 October 2009, connection on 17 April 2020. URL : http://journals.openedition.org/transtexts/254 ; DOI : https://doi.org/10.4000/transtexts.254.

Collomb, Jean-Daniel. « Drill, Baby, Drill ! », *Dynamiques environnementales* [En ligne], 35 | 2015, mis en ligne le 01 janvier 2016, consulté le 17 avril 2020. URL : http://journals.openedition.org/dynenviron/1112 ; DOI : https://doi.org/10.4000/dynenviron.1112.

Collomb, Jean-Daniel. « The Ideology of Climate Change Denial in the United States », *European journal of American studies* [En ligne], 9–1 | 2014, document 5, mis en ligne le 01 avril 2014, consulté le 15 juin 2017. URL : http://ejas.revues.org/10305 ; DOI : 10.4000/ejas.10305.

Collomb, Jean-Daniel. *John Muir, écologie et parcs nationaux*. Bordeaux : Presses universitaires de Bordeaux, 2013.

Colwell, Mary. John Muir: *The Scotsman who save America's wild places.* Oxford: Lion Books, 2014.

Davis, Wade. *The Wayfinders: Why Ancient Wisdom Matters in the Modern World.* Toronto: Anansi, 2009.

De Laguna, Frederica and George Thornton Emmons. *The Tlingit Indians*. Seattle: University of Washington Press, 1991.

De Revere, Paul. "Jason Isbell, Kacey Musgraves, and the 'Duality of the Southern Thing'." *Pitchfork*. "https://pitchfork.com/thepitch/814-jason-isbell-kacey-musgraves-and-the-duality-of-the-southern-thing/." June 25, 2015.

Diamond, Jared. *Collapse: How Societies Choose to Fail or Survive*. London: Penguin Books, 2013.

Diamond, Jared. *Guns, Germs, and Steel: The Fates of Human Societies*. New York: W. W. Norton and Company, 1999.

Diamond, Jared. *The World Until Yesterday: What Can We Learn from Traditional Societies?* New York: Viking, 2012.

Douglas, William O. *Muir of the Mountains*. Boston: Houghton Mifflin Co., 1961.

Dubos, René. *Courtisons la terre*. Paris: Éditions Stock, 1980.

Dubos, René. *So Human an Animal*. New York: Charles Scribner's Sons, 1968.

Dugast, Stéphane et Daphné Victor. *Paul-Émile Victor : J'ai toujours vécu demain*. Paris : Éditions Robert Laffont, S.A., 2015.

Egan, Timothy. *Short Nights of the Shadow Catcher: The Epic Life and Immortal Photographs of Edward Curtis*. New York: Houghton, Mifflin, Harcourt, 2012.

Ehlers and Gizer. "Evidence for a Genetic Component for Substance Dependence in Native Americans." *AM J Psychiatry*. „https://www.ncbi.nlm.nih.gov/pmc/articles/PMC3603686/." 1/2/2013.

Ellis, Joseph J. *American Sphinx: The Character of Thomas Jefferson*. 1996. New York: Vintage Books, 1998.

Emmett, Robert S. *Cultivating Environmental Justice: A Literary History of U.S. Garden Writing*. Amherst and Boston: University of Massachusetts Press, 2016.

Fall, James A. and James Kari. *Shem Pete's Alaska: The Territory of the Upper Cook Inlet Dena'ina*. Fairbanks: University of Alaska Press, 2003.

Fienup-Riordan, Ann. *Boundaries and Passages: Rule and Ritual in Yup'ik Eskimo Oral Tradition*. Norman: University of Oklahoma Press, Norman, 1994.

Florida, Richard. *The New Urban Crisis: How our Cities are Increasing Inequality, Deepening Segregation, and Failing the Middle Class–And What We Can Do About It*. New York: Basic Books, 2018.

Florida, Richard. *Rise of the Creative Class ... and how it's transforming work, leisure, community, and everyday life*. Topeka: Tandem Library, 2003.

Ford, Corey. *Where the Sea Breaks Its Back: The Epic Story – Georg Steller & the Russian Exploration of Alaska*. 1966. Portland, Oregon: Graphic Arts Books, 2003.

Fricker, Miranda. *Epistemic Injustice: Power and Ethics of Knowing*. Oxford: Oxford University Press, 2007.

Gaard, Greta and Patrick D. Murphy. Editors. *Ecofeminist Literary Criticism: Theory, Interpretation, Pedagogy*. Urbana and Chicago: University of Illinois Press, 1998.

Garrett, Paul D. *Saint Innocent, Apostle to America*. Crestwood New York: Saint Vladimir's Seminary Press, 1979.

Gilbert, Sandra M. And Susan Gubar. Editors. *The Madwoman in the Attic: The Woman Writer and the Nineteenth-Century Literary Imagination*. New Haven: Yale University Press, 1979.

Glave, Dianne D. *Rooted in the Earth: Reclaiming the African American Environmental Heritage*. Chicago: Lawrence Hill Books, 2010.

Hamilton, Alexander, James Madison, and John Jay. *The Federalist Papers*. 1788. New York: Signet Classics, 2003.

Harper, Kenn. *Give Me My Father's Body: The Life of Minik, the New York Eskimo*. New York: Simon and Schuster, 2001.

Hartman, Ian C. and David Reamer. *Black Lives in Alaska: A History of African Americans in the Far Northwest*. Seattle: University of Washington Press, 2022.

Haycox, Stephen W. *Alaska: An American Colony*. London: C. Hurst & Co. Publishers, 2002.

Hochschild, Adam. *King Leopold's Ghost: A Story of Greed, Terror and Heroism in Colonial Africa*. 1998. London: Pan Books, 2006.

Hulme, Peter and Tim Youngs. Editors. *The Cambridge Companion to Travel Writing*. 2002. Cambridge, UK: Cambridge University Press, Fourth printing, 2008.

Isaacson, Walter. *Benjamin Franklin: An American Life*. New York: Simon and Schuster, 2004.

Isaacson, Walter. *Einstein: His Life and Universe*. New York: Simon and Schuster, 2007.

Kauffmann, John. *Alaska's Brooks Range: The Ultimate Mountains*. Seattle: The Mountaineers Books, 1992.

Kawagley, Angayuqaq Oscar. *A Yupiaq Worldview: A Pathway to Ecology and Spirit*. Second Edition. Illinois: Waveland Press, Inc. 2006.

Kollin, Susan. *Nature's State: Imagining Alaska as the Last Frontier*. Chapel Hill: University of North Carolina Press, 2001.

Kollin, Susan. Editor. *Postwestern Cultures: Literature, Theory, Space*. Lincoln and London: University of Nebraska Press, 2007.

Latour, Bruno. *Politiques de la nature : Comment faire entrer les sciences en démocratie*. 1999. Paris: Éditions La Découverte, 2004.

Latour, Bruno. *Sur le culte moderne des dieux faitiches : suivi de Iconoclash*. Paris : Éditions La Découverte, 2009.

Le Guin, Ursula K. *Steering the Craft: A 21st-Century Guide to Sailing the Sea of Story.* 1998. New York: Mariner Books, 2015.

Lilla, Mark. *The Once and Future Liberal: After Identity Politics.* 2017. New York: Harper Collins, 2018.

Louv, Richard. *The Nature Principle: Human Restoration and the End of Nature-Deficit Disorder.* Chapel Hill: Algonquin Books of Chapel Hill, 2011.

MacLean, Nancy. *Democracy in Chains: The Deep History of the Radical Right's Stealth Plan for America.* London: Scribe, 2017.

Massiani, Jean-Stéphane. *Les journaux de voyage de James Cook dans le Pacifique : Du parcours au discours.* Aix-en-Provence : Presses Universitaires de Provence, 2015.

Max, D.T. *Every Love Story is a Ghost Story.* London: Granta Books, 2013.

Mayer, Jane. *Dark Money: The Hidden History of the Billionaires Behind the Rise of the Radical Right.* New York: Doubleday, 2016.

Meade, Marie and Ann Fienup-Riordan. *Agayuliyararput: Kegginaqut, Kangiit-llu=Our Way of Making Prayer: Yup'ik Masks and the Stories They Tell.* Seattle: University of Washington Press, 1996.

Meaux, Jean Morgan. *In Pursuit of Alaska: An Anthology of Travelers' Tales, 1879–1909.* Seattle: University of Washington Press, 2013.

Mercier, Hugo and Dan Sperber. *The Enigma of Reason: A new theory of human understanding.* London: Allen Lane, 2017.

Nash, Roderick. *Wilderness and the American Mind.* Fifth Edition. 1967. New Haven: Yale University Press, 2014.

Newell, Mike. *No Bottom: In Conversation with Barry Lopez.* Gambier, OH: XOXOX Press, 2008.

Obeyesekere, Gananath. *The Apotheosis of Captain Cook: European Mythmaking in the Pacific.* Princeton, New Jersey and Honolulu, Hawai'i: Princeton University Press and Bishop Museum Press, 1992.

Pagden, Anthony. *World's at War: the 2,500 Year Struggle Between East and West.* New York: Random House Trade Paperbacks, 2009.

Paine, Thomas. *Common Sense, Rights of Man, and Other Essential Writings of Thomas Paine.* 1776. New York: Signet Classics, 2003.

Rainey, Froelich G. *The Whale Hunters of Tigara. Volume 41: Part 2.* New York, Anthropological Papers of the American Museum of Natural History, 1947.

Ross, John F. *The Promise of the Grand Canyon: John Wesley Powell's Perilous Journey and his Vision for the American West.* New York: Viking, 2018.

Ryan, Christopher. *Civilized to Death: What Was Lost on the Way to Modernity.* New York: Avid Reader Press, 2019.

Rybczynski, Witold. *A Clearing in the Distance: Frederick Law Olmsted and America in the Nineteenth Century.* New York: Scribner, 1999.

Sayre, Robert. *La modernité et son autre : Récits de la rencontre avec l'indien en Amérique du Nord au XVIIIe siècle*. Bécherel : Éditions les Perséides, 2008.

Serres, Michel. *Le contrat naturel*. Paris : Éditions François Bourin, 1990.

Seymour, Nicole. *Bad Environmentalism: Irony and Irreverence in the Ecological Age*. Minneapolis: University of Minnesota Press, 2018.

Seymour, Nicole. *Strange Natures: Futurity, Empathy, and the Queer Ecological Imagination*. Illinois: University of Illinois Press, 2013.

Shearer, Christine. *Kivalina: A Climate Change Story*. Chicago: Haymarket Books, 2011.

Slovic, Scott. *Going Away to Think: Engagement, Retreat, and Ecocritical Responsibility*. Reno & Las Vegas: University of Nevada Press, 2008.

Slovic, Scott. *Seeking Awareness in American Nature Writing: Henry Thoreau, Annie Dillard, Edward Abbey, Wendell Berry, Barry Lopez*. Salt Lake City: University of Utah Press, 1992.

Soja, Edward W. *Postmodern Geographies: The Reassertion of Space in Critical Social Theory*. London and Brooklyn, New York: Verso Books, 1989.

Suberchicot, Alain. *Littérature et Environnement : Pour une écocritique comparée*. Paris: Éditions Champion, 2012.

Sundborg, George. *Opportunity in Alaska*. Birminghan, N.Y.: Vail-Ballou Press, 1945.

Tredinnick, Mark. *The Land's Wild Music: Encounters with Barry Lopez, Peter Matthiessen, Terry Tempest William, and James Galvin*. San Antonio: Trinity University Press, 2005.

Tuan, Yi-fu. *Coming Home to China*. Minneapolis: University of Minnesota Press, 2007.

Tuan, Yi-fu. *Topophilia: A Study of Environmental Perception, Attitudes, and Values*. 1974. New York: Columbia University Press, 1990.

Tydeman, William E. *Conversations with Barry Lopez: Walking the Path of Imagination*. Norman: University of Oklahoma Press, 2013.

Viviès, Jean. *English Travel Narratives in the Eigtheenth Century: Exploring Genres*. 1999. Translated by Claire Davison. Aldershot, England and Burlington, Vermont: Ashgate Publishing, 2002.

Weltzien, Alan O., editor, *Coming into McPhee Country: John McPhee and the Art of Literary Nonfiction*. Salt Lake City: University of Utah Press, 2003.

Wenzel, George. *Animal Rights Human Rights: Ecology, Economy and Ideology in the Canadian Arctic*. 1991. Toronto: University of Toronto Press, 1997.

Wittgenstein, Ludwig. *Tractatus Logico-Philosophicus*. Introduction by Bertrand Russel. 1922. Sweden: Chiron Academic Press, 2016.

Wolfe, Tom. *The New Journalism*. 1975. London: Picador, 1990.

Wohlleben, Peter. *La vie secrète des arbres : ce qu'ils ressentent, comment ils communiquent*. 2015. Paris : Éditions des Arènes 2017.

Wulf, Andrea. *The Brother Gardeners: Botany, Empire and the Birth of an Obsession*. London: Windmill Books, 2009.

Wulf, Andrea. *Chasing Venus: The Race to Measure the Heavens*. London: Windmill Books, 2013.

Wulf, Andrea. *The Founding Gardeners: How the Revolutionary Generation Created an American Eden*. London: Windmill Books, 2012.

Wulf, Andrea. *The Invention of Nature: The Adventures of Alexander Von Humboldt, The Lost Hero of Science*. London: John Murray, 2016.

Film and Television

Ailo's Journey: The Amazing Odyssey of a Newborn Reindeer. Guillaume Maidatchevsky. France: Vendetta Films, 2019.

Alaskan Bush People. Narrated by Asa Siegel. Silver Spring, Maryland: Discovery, 2005–2020.

Alaska: The Last Frontier. Narrated by Bray Poor. Silver Spring, Maryland: Discovery, 2011–2020.

Away We Go. Sam Mendes. California: Focus Features, 2009.

Bambi. David Hand. Burbank: Walt Disney Productions, 1942.

The Circle. James Ponsoldt. Burbank: STX Films, 2017.

Curse of the Golden Flower. Zhang Yimou. Hong Kong: Edko Films, 2007.

Deadliest Catch. Narrated by Mike Rowe. Silver Spring, Maryland: Discovery, 2005–2020.

Dersu Uzala. Akira Kurosawa. Japan, Dalei Film, 1975.

Django Unchained. Quentin Tarantino. New York: The Weinstein Company, 2013.

Fantasia. Produced by Walt Disney. California: Walt Disney Productions, 1940.

Flowers and Trees. Burt Gillet. California. Walt Disney Productions, 1932.

Grizzly Man. Werner Herzog. Santa Monica: Lion's Gate Films, 2005.

House of Flying Daggers. Zhang Yimou. Hong Kong: Edko Films, 2004.

Into the Wild. Sean Penn. Hollywood: Paramount Vantage, 2007.

Molly of Denali. Dorothea Gillm and Kathy Waugh. Arlington, Va and Toronto: PBS Kids/CBC Television, 2019.

The Olympic Elk. James Algar. California: Walt Disney Pictures, 1952.

Ratatouille. Brad Bird. California: Pixar Animation Studios, 2007.

Sarah Palin's Alaska. Silver Spring, Maryland: TLC, 2010–2011.

Seal Island. James Algar. California: Walt Disney Productions, 1948.

Snow White and the Seven Dwarfs. David Hand. California: Walt Disney Productions, 1937.

The Spoilers. Ray Enright. California: Frank Lloyd Productions, 1942.

Super Size Me. Morgan Spurlock. New York: Samuel Goldwyn Films, 2004.

The Three Little Pigs. Burt Gillet. California: Walt Disney Productions, 1933.

True Detective: Night Country. Lopez, Issa and Nic Pizzolatto. HBO. Los Angeles, 2024.

True Grit. Ethan and Joel Coen. Hollywood, Paramount, 2010.
Where the Wild Things Are. Spike Jonze. Burbank: Legendary Pictures, 2009.
White Wilderness. James Algar. California: Walt Disney Productions, 1958.

Digital Sources

Adichie, Chimmamanda Ngozi. "Chimamanda Ngozi Adichie on Liberal Cannibalism | *The New Yorker Festival*." Youtube. "https://www.youtube.com/watch?v=f03OZWPOa3g." 9/10/2017. Alpert, Megan. "'They come, they photograph, but don't help': how ecotourism in the Amazon shortchanges the locals." *The Guardian*. "https://www.theguardian.com/environment/2015/mar/11/ecotourism-amazon-shortchanges-locals-ecuador." 11/3/2015.

Azerrad, Michael. "Grunge City: The Seattle Scene: For Real Rockers, Seattle is the Ultimate Wet Dream." *Rolling Stone*. "https://www.rollingstone.com/music/music-news/grunge-city-the-seattle-scene-250071/." 16/4/1992.

Baker, Nick. "How eco is eco-tourism?" *The Guardian*. "https://www.theguardian.com/travel/2005/nov/12/ecotourism.guardiansaturdaytravelsection." 21/11/2005.

Baynes, Chris. "Anne Frank Centre warns of 'alarming parallels' between Trump's America and Hitler's Germany." *Independent*. "https://www.independent.co.uk/news/world/americas/us-politics/anne-frank-centre-donald-trump-america-president-hitler-nazi-germany-alarming-parallels-warning-a7884731.html." 9/8/2017.

Bell, Lee. "Mercury in women of child-bearing age in 25 countries." *Biodiversity Research Institute*. "http://www.briloon.org/uploads/BRI_Documents/Mercury_Center/Publications/Mercury%20in%20Women%20of%20Child-Bearing%20Age%20in%2025%20Countries.pdf." 9/2017.

Belson, Ken and Benjamin Hoffman. "Trump Blasts Players as Protests Resume During Anthem." *New York Times*. "https://www.nytimes.com/2018/08/09/sports/nfl-national-anthem-kneeling-protest.html." 9/8/2018.

Blakemore, Erin. "The Statue of Liberty Was Originally a Muslim Woman." *smithsonian.com*. "https://www.smithsonianmag.com/smart-news/statue-liberty-was-originally-muslim-woman-180957377/." 24/11/2015.

Brown, Willa."Lumbersexuality and Its Discontents." *The Atlantic*. "https://www.theatlantic.com/national/archive/2014/12/lumbersexuality-and-its-discontents/383563/." 10/12/2014.

Bump, Phillip. "The Oregon occupiers' land dispute, explained in 9 maps," *Washington Post*. "https://www.washingtonpost.com/news/the-fix/wp/2016/01/04/the-oregon-occupiers-complaint-explained-in-9-maps/?noredirect=on&utm_term=.f4dd96f56ebd." 4/1/2016.

Carlson, Michael. “Peter Matthiessen obituary.” *The Guardian*. “https://www.theguardian.com/books/2014/apr/06/peter-matthiessen.” 6/4/2014.

Chayka, Kyle. “My Own Private Iceland.” *Vox*. “https://www.vox.com/the-goods/2019/10/21/20905485/iceland-overtourism-reykjavik-blue-lagoon-northern-lights.” 21/10/2019.

Chapman, Matthew. “Paul Krugman calls out “radical centrist” Howard Schultz.” *Salon*. «https://www.salon.com/2019/02/01/paul-krugman-radical-centrists-like-howard-schultz-retreat-into-a-fantasy-world_partner/.» 2/2/2019.

Child, Ben. “Finding Nemo sequel Finding Dory ‘could damage exotic fish populations.’” *The Guardian*. “https://www.theguardian.com/film/2016/may/19/finding-nemo-dory-pixar-could-damage-tropical-fish-populations.” 19/5/2016.

Collinson, Patrick. “The climate crisis must mean the end of British summer holidays in Spain.” *The Guardian*. “https://www.theguardian.com/money/2019/aug/03/climate-crisis-british-summer-holidays-spain-prices.” 3/8/2019.

Conroy, J Oliver. “Mark Lilla: the liberal who counts more enemies on the left than the right.” *The Guardian*. “https://www.theguardian.com/media/2017/dec/21/mark-lilla-identity-politics-liberals.” 21/12/2017.

Crépon, Marc. “Face à la violence, Martin Luther King, Gandhi et Mandela opposent une politique de l’amour.” *France Culture*. “https://www.franceculture.fr/conferences/bibliotheque-nationale-de-france/comment-gandhi-martin-luther-king-mandela-opposent-la.» 29/6/2017.

Crouch, Ian. “Why Don’t More Americans Win the Nobel Prize?” *The New Yorker*. “https://www.newyorker.com/books/page-turner/why-dont-more-americans-win-the-nobel-prize.” 8/10/2013.

Davis, Jess. “‘It’s not a cute little Bambi’: Farmers frustrated as deer wreak havoc across the landscape.” *ABC*. “https://www.abc.net.au/news/rural/2018-10-06/deer-wreaking-havoc-farms-environment/10343406.” 5/10/2018.

Dominiczak, Peter. “Britain to become a ‘nation of shopkeepers again,’ ministers pledge.” The *Telegraph*. “https://www.telegraph.co.uk/news/politics/conservative/11795160/Britain-to-become-a-nation-of-shopkeepers-again-ministers-pledge.html.” 10/8/2015.

Doyle, Patrick. “How Hunter S. Thompson Became a Legend.” *Rolling Stone*. “https://www.rollingstone.com/culture/culture-news/rolling-stone-at-50-how-hunter-s-thompson-became-a-legend-115371/.” 18/7/2019.

The Editorial Board. “‘Antifa’ groups only help the hateful forces they claim to oppose.” The *Washington Post*. “https://www.washingtonpost.com/opinions/antifa-groups-only-help-the-hateful-forces-they-claim-to-oppose/2017/08/29/d7c900b4-8cca-11e7-8df5-c2e5cf46c1e2_story.html?utm_term=.735592ace10f.” 29/8/2017.

Elgot, Jessica. “Beyoncé unleashes Black Panthers homage at Super Bowl 50.” *The Guardian*. “https://www.theguardian.com/music/2016/feb/08/beyonce-black-panthers-homage-black-lives-matter-super-bowl-50.” 8/2/2016.

Elks, Sonia. "Arctic cruises accused of leaving Indigenous communities in the cold." *Arctic Today*. "https://www.arctictoday.com/arctic-cruises-accused-of-leaving-indigenous-people-in-the-cold/." 14/8/2019.

Elsmoor, James. "Cruise Ship Pollution Is Causing Serious Health And Environmental Problems." *Forbes*. «https://www.forbes.com/sites/jamesellsmoor/2019/04/26/cruise-ship-pollution-is-causing-serious-health-and-environmental-problems/#49eecf8637db.» 26/4/2019.

Embury-Dennis, Tom. "US states consider laws allowing Creationism to be taught by science teachers." *Independent*. "https://www.independent.co.uk/news/world/americas/us-states-texas-creationism-science-teacher-state-law-evolution-religion-a7632931.html." 16/3/2017.

Farley, Robert. "Fact Check: The Truth About White Nationalism." *USA Today*. "https://eu.usatoday.com/story/news/politics/2019/03/20/fact-check-truth-white-nationalism/3226378002/." 20/3/2019.

Farrow, Ronan. "From Aggressive Overtures to Sexual Assault: Harvey Weinstein's Accusers Tell Their Stories." *The New Yorker*. "https://www.newyorker.com/news/news-desk/from-aggressive-overtures-to-sexual-assault-harvey-weinsteins-accusers-tell-their-stories." 10/10/2017.

Fawbert, Dave. "'Eco-anxiety': how to spot it and what to do about it." *BBC*. „https://www.bbc.co.uk/bbcthree/article/b2e7ee32-ad28-4ec4-89aa-a8b8c98f95a5." 27/3/2019.

Fehrenbacher, Katie. "A Lithium Gamble that Could Win Big for Tesla." *Fortune*. "http://fortune.com/2016/03/29/lithium-tesla-mine-nevada/" 29/3/16.

Ferguson, Benjamin. "Ideas don't exist, except within emotions: Interview with Joshua Cohen." *La clé des langues*. «http://cle.ens-lyon.fr/anglais/litterature/litterature-americaine/interview-with-joshua-cohen.» 20/12/2019.

Filkins, Dexter. "Did George W. Bush Create Isis?" *The New Yorker*. "https://www.newyorker.com/news/news-desk/did-george-w-bush-create-isis." 15/5/2015.

Frank, Priscilla. "In The #MeToo Era, Do These Paintings Still Belong In A Museum?" *Huffington Post*. "https://www.huffpost.com/entry/museums-me-too-sexual-harassment-art_n_5a2ae382e4b0a290f0507176." 14/12/2017.

Frankel, Todd C. and Peter Whoriskey. "Tossed Aside in the White Gold Rush: Indigenous people are left poor as tech world takes lithium from under their feet." *The Washington Post*. "https://www.washingtonpost.com/graphics/business/batteries/tossed-aside-in-the-lithium-rush/??noredirect=on" 19/12/16.

Frick-Wright, Peter, Robert Carver, and Ryan Haas (Producers). Bundyville Podcast. *Oregon Public Broadcasting*. "https://www.opb.org/news/article/bundyville-occupation-podcast/." Viewed on 9/7/2019.

Geier, Ben. "What Did We Learn From the Dotcom Stock Bubble of 2000?" *Time*. "http://time.com/3741681/2000-dotcom-stock-bust/." 12/3/2015.

Ghose, Aurobindo. "The Superman." *Site of Sri Aurobindo and the Mother*. «https://www.aurobindo.ru/workings/sa/16/0025_e.htm.» 4/1915.

Gibbens, Sarah. "Plastic proliferates at the bottom of world's deepest ocean trench." *National Geographic.* "https://news.nationalgeographic.com/2018/05/plastic-bag-mariana-trench-pollution-science-spd/." 13/5/2019.

Gladstone, Rick. "China and Russia Move to Cut Human Rights Jobs in U.N. Peacekeeping." *New York Times.* "https://www.nytimes.com/2018/06/27/world/africa/china-russia-un-human-rights-cuts.html." 27/6/2018.

Glenn, Alice Qannik. *Coffee and Quaq.* "www.coffeeandquaq.com." Viewed on 27/7/2019.

Gordon, Benjamin. "Social Media is Ruining Everything." *Huffington Post.* "https://www.huffpost.com/entry/social-media-is-ruining-everything_b_7253558" 11/5/2015.

Groves, Pamela. "Pleistocene Megafauna in Beringia." *National Park Service.* "https://www.nps.gov/articles/aps-17-1-4.htm." Viewed on 6/2/2019.

Gupta, Prachi. "A Brief Timeline of Palin Family Drama." *Cosmopolitan.* "https://www.cosmopolitan.com/politics/news/a52429/a-brief-timeline-of-palin-family-controversies/.» 20/1/2016.

Gupta, Rupa. "Indigenous Peoples and the International Environmental Community: Accomadating Claims Through a Cooperative Legal Process." *New York University Law Review: Vol. 74:1741.* "https://www.nyulawreview.org/wp-content/uploads/2018/08/NYULawReview-74-6-Gupta.pdf." 12/1999.

Harball, Elizabeth. "Alaska voters strike down 'Stand for Salmon' ballot initiative." *KTOO Public Media.* "https://www.ktoo.org/2018/11/06/ballot-measure-1-additional-protections-for-salmon-habitats/." 6/11/2018.

Harris, Kate. "Why the World Needs Barry Lopez." *Outside.* "https://www.outsideonline.com/2391391/barry-lopez-horizon." 19/3/2019.

Hillman, Anne. "Traditional gut sewing at the Anchorage Museum." *Alaska Public Media.* "https://www.alaskapublic.org/2014/12/03/traditional-gut-sewing-at-the-anchorage-museum/." 3/12/2014.

Holpuch, Amanda and Justin McCurry. "You would have to be a lunatic: Tourists to North Korea describe risks and rewards." *The Guardian.* "https://www.theguardian.com/world/2017/jul/22/you-would-have-to-be-a-lunatic-tourists-to-north-korea-describe-risks-and-rewards." 22/2/2017.

Jacobs, Julia. "With 'Molly of Denali,' PBS Raises its Bar for Inclusion." *The New York Times.* "https://www.nytimes.com/2019/07/15/arts/television/molly-of-denali-alaska-native-pbs.html." 15/7/2019.

Kalogerakis, George. "Change Is in the Air at the E. B. White Farm." *The New Yorker.* "https://www.nytimes.com/2017/09/07/arts/e-b-white-house-charlottes-web.html." 7/9/2017.

Kaskey, Jack and Simon Casey. "Tesla to Use North American Material Amid Pollution Worry." *Bloomberg.* „https://www.bloomberg.com/news/articles/2014-03-28/tesla-to-use-north-american-material-amid-pollution-worry" 31/3/14.

Kommenda, Niko. "How your flight emits as much CO2 as many people do in a year." *The Guardian*. "https://www.theguardian.com/environment/ng-interactive/2019/jul/19/carbon-calculator-how-taking-one-flight-emits-as-much-as-many-people-do-in-a-year." 19/7/2019.

Korte, Gregory. "Obama Administration renames Mount McKinley to Denali." *USA Today*. "https://eu.usatoday.com/story/news/politics/2015/08/30/obama-rename-nations-tallest-mountain/71426656/." 31/8/2015.

Koster, Kyle. "Did the Pace Picante commercials of the 1990s predict our current culture wars?" *The New York Times*. "https://www.dailydot.com/layer8/pace-picante-salsa-new-york-city-2016-election/." 4/9/2018.

Kurutz, Steven. "Face Tattoos Go Mainstream." *New York Times*. "https://www.nytimes.com/2018/08/04/style/face-tattoos.html." 4/8/2018.

Lartey, Jamiles. "Why the days of 'stand your ground' self-defense laws may be numbered." *The Guardian*. "https://www.theguardian.com/us-news/2018/aug/07/why-the-days-of-stand-your-ground-self-defense-laws-may-be-numbered." 8/8/2018.

Leahy, Stephen. "Polar Bears Really Are Starving Because of Global Warming, Study Shows." *National Geographic*. "https://news.nationalgeographic.com/2018/02/polar-bears-starve-melting-sea-ice-global-warming-study-beaufort-sea-environment/." 1/2/18.

Levin, Sam and Will Parrish. "'Treating protests as terrorism': US plans crackdown on Keystone XL activists." *The Guardian*. "https://www.theguardian.com/environment/2018/sep/20/keystone-pipeline-protest-activism-crackdown-standing-rock." 20/9/2018.

Lewis, Jeremy. "Colin Thubron: 'I wanted to vanish into the background'." *The Independent*. "https://www.independent.co.uk/arts-entertainment/books/features/colin-thubron-i-wanted-to-vanish-into-the-background-182816.html." 6/7/2002.

Leyshon, Cressida. "New Yorker Novella: Callan Wink on Fiction and Fly-Fishing." *The New Yorker*. "https://www.newyorker.com/books/page-turner/new-yorker-novella-callan-wink-on-fiction-and-fly-fishing." 20/11/2015.

Lopate, Leonard. "Phillip Lopate on the greatest essays the world has every known." *WBAI*. "leonardlopateatlarge.com." 3/6/19.

Lord, Nancy. "Well-known novelist's latest offers Alaska backdrop, but don't mistake it for the real deal: Heroes of the Frontier, by Dave Eggers." *Alaska Dispatch News*. https://www.adn.com/alaska-life/we-alaskans/2016/09/04/well-known-novelists-latest-offers-alaska-backdrop-but-dont-mistake-it-for-the-real-deal/. 4/9/2016.

Madsen, Peter. "Deep Ecology." *Britannica*. "https://www.britannica.com/topic/deep-ecology." Viewed on 2/4/2020.

Marantz, Andrew. "The Alt-Right Hails Its Victorious God-Emperor." *The New Yorker*. "https://www.newyorker.com/news/news-desk/the-alt-right-hails-its-victorious-god-emperor." 12/11/2016.

Marin, Rick. "Grunge: A Success Story." *The New York Times*. "https://www.nytimes.com/1992/11/15/style/grunge-a-success-story.html." 15/11/1992.

Martinson, Erica. "Murkowski voted no on her party's health care motion. Here's why. *Anchorage Daily News*. "https://www.adn.com/politics/2017/07/25/lisa-murkowski-voted-no-on-her-partys-health-care-motion-heres-why/." 25/7/2017.

McVeigh, Karen. "Rick Perry signs wide-ranging Texas bill to limit access to abortion." *The Guardian*. "https://www.theguardian.com/world/2013/jul/18/rick-perry-texas-abortion-bill." 18/7/2013.

Menand, Louis. "Mystery Man." *The New Yorker*. "https://www.newyorker.com/magazine/2005/02/14/mystery-man-2." 6/2/2005.

Merica, Dan. "Trump asked about reversing Obama to rename Denali as Mt. McKinley." *CNN*. „https://edition.cnn.com/2017/10/24/politics/trump-denali-mt-mckinley/index.html.» 25/10/2017.

Michaud, Jon. "Malcolm X and Elijah Muhammad." *The New Yorker.* "https://www.newyorker.com/books/double-take/malcolm-x-and-elijah-muhammad." 15/4/2011.

Muller, Carol B. "The Underrepresentation of Women in Engineering and Related Sciences: Pursuing Two Complementary Paths to Parity." *National Center for Biotechnology Information*. "https://www.ncbi.nlm.nih.gov/books/NBK36356/." Viewed on March 5, 2018.

Nijhuis, Michelle. "Fishers of the Yakama Nation." *The New Yorker*. "https://www.newyorker.com/tech/annals-of-technology/fishers-of-the-yakama-nation." 3/12/2015.

O'Harra, Doug and Peter Porco. "Halt salmon fishing, Peta asks Alaska." *The Seattle Times*. "https://www.seattletimes.com/nation-world/halt-salmon-fishing-peta-asks-alaska/" 20/2/2005.

Pace, Eric. "William Shawn, 85, Is Dead; New Yorker's Gentle Despot." *The New Yorker*. "https://www.nytimes.com/1992/12/09/obituaries/william-shawn-85-is-dead-new-yorker-s-gentle-despot.html." 9/12/1992.

Pemberton, Jennifer. "The lure of John McPhee's 'Coming into the Country,' 40 years later." *Alaska Public Media*. "https://www.alaskapublic.org/2017/01/23/the-lure-of-john-mcphees-coming-into-the-country-40-years-later/." 23/1/2017.

Peraldo, Emmanuelle. "From Mercantilist to Utilitarian Crusoe: the Transformative Impact of the Desert Island." *Études Anglaises. 2019/2* (*Vol. 72*). «https://www.cairn.info/revue-etudes-anglaises-2019-2-page-151.html?contenu=article.» 151–166.

Prendergast, Renee. "The Political Economy of Swift's Satires and other Prose Works." *Oeconomia: Economics and Literature*. 4–4/2014. "https://journals.openedition.org/oeconomia/1021." 593–619.

Preston, John. "Colin Thubron interview: 'Travel has always been an addiction for me.'" *The Telegraph*. "https://www.telegraph.co.uk/books/what-to-read/colin-thubron-interview-travel-has-always-been-an-addiction-for/." 12/8/2016.

Remnick, David and John McPhee. "Episode 38: The Wisdom of John McPhee, and the Agony of an iPod Lockout." *The New Yorker Radio Hour*, 2016. http://www.newyorker.com/podcast/the-new-yorker-radio-hour/episode-38-the-wisdom-of-john-mcphee-and-the-agony-of-an-ipod-lockout.

Remnick, David. "Episode 46: Gary Johnson, Angel Olsen, and A Bee Stylist" (podcast). *The New Yorker*. "https://www.newyorker.com/podcast/the-new-yorker-radio-hour/episode-46-gary-johnson-angel-olsen-and-a-bee-stylist." 2/9/2019.

Ritschel, Chelsea. "The Reasons Why Palm Oil is so Controversial." *Independent*. "https://www.independent.co.uk/life-style/palm-oil-health-impact-environment-animals-deforestation-heart-a8505521.html." 24/8/2018.

Robins, Peter. "The New Yorker's grammar rules (and how to break them)." *The Spectator*. "https://www.spectator.co.uk/2015/05/the-new-yorkers-grammar-rules-and-how-to-break-them/." 16/5/2015.

Rosenthal, Max J. "The United States isn't the only country downplaying climate change." *The Washington Post*. "https://www.washingtonpost.com/energy-environment/2018/12/11/united-states-isnt-only-country-downplaying-climate-change/?utm_term=.7b103d92f8eb." 11/12/2018.

Salter, James. "Postscript: Peter Matthiessen (1927–2014)." *The New Yorker*. "https://www.newyorker.com/books/page-turner/postscript-peter-matthiessen-1927-2014." 14/4/2014.

Sandberg, Eric. "Migration in Alaska." *Alaska Economic Trends*. "http://labor.alaska.gov/trends/mar18.pdf." 2/2018.

Salter, James. "Postscript: Peter Matthiessen (1927–2014)." *The New Yorker*. "https://www.newyorker.com/books/page-turner/postscript-peter-matthiessen-1927-2014." 14/4/2014.

Sarkar, Ash. "The colonial past is another country. Let's leave Boris Johnson there." *The Guardian*. "https://www.theguardian.com/commentisfree/2018/jul/10/past-boris-johnson-colonial-brexit-colony." 10/7/2018.

Sadasivam, Naveena. "Report: Going 100% renewable power means a lot of dirty mining." *Grist*. „https://grist.org/article/report-going-100-renewable-power-means-a-lot-of-dirty-mining/." 17/4/2019.

Seabrook, John. "Free." *The New Yorker*. "https://www.newyorker.com/magazine/2015/02/02/free-6." 26/1/2015.

Sidner, Sara. "The rise of Black Lives Matter: Trying to break the cycle of violence and silence." *CNN*. „https://edition.cnn.com/2015/12/28/us/black-lives-matter-evolution/index.html.» 28/12/2015.

Silva, Cristina. "People no longer want to move to Alaska for the first tie since World War II; Here's why." *Newsweek*. "https://www.newsweek.com/people-no-longer-want-move-alaska-first-time-world-war-ii-heres-why-843609." 13/3/2018.

Solomon, Deborah. "America, Illustrated." *The New York Times*. "https://www.nytimes.com/2010/07/04/arts/design/04rockwell.html." 1/7/2010.

Steinmetz, Katy. "'Free Speech Week' Is Coming to Berkeley. Things Could Get Heated Again." *Time*. "https://time.com/4924477/berkeley-yiannopoulos-antifa-free-speech-week/." 5/9/2017.

Sulleyman, Aatif. "Facebook is destroying society and making users feel vacant and empty," *Independent*."https://www.independent.co.uk/life-style/gadgets-and-tech/news/facebook-society-destroy-social-media-network-users-damage-communicate-connect-chamath-palihapitiya-a8105131.html.'» 12/12/2017.

Traff, Thea. "Gone Fishing." *The New Yorker*. "https://www.newyorker.com/culture/photo-booth/photographer-corey-arnold-gone-fishing." 21/8/2014.

Trufelman, Avery. Ed. "Punk Style: Articles of Interest." *99% Invisible*. "https://99percentinvisible.org/episode/punk-style-articles-of-interest-6/." 12/10/2018.

Unknown Author

"Alaskans descend on the Kenai River to dipnet for salmon." *Anchorage Daily News*. "https://www.youtube.com/watch?v=AL_fzdpmvo4." 19/7/2011.

"Alaska Purchase." *Britannica*. "https://www.britannica.com/event/Alaska-Purchase." Viewed on 27/11/2019.

"The Alaskan Rainforest." *Alaskan Department of Fish and Game*.

"http://www.adfg.alaska.gov/static/education/educators/curricula/pdfs/the_alaskan_rainforest.pdf." viewed on 21/8/2018.

"Alaska Statehood Act." *The Alaska State Legislature*. "https://www.akleg.gov/basis/get_documents.asp?session=29&docid=29890." Viewed on 27/11/2019.

"All Nobel Prizes in Literature." *The Nobel Prize*. "https://www.nobelprize.org/prizes/lists/all-nobel-prizes-in-literature/." Viewed on 23/6/2019.

"Animal Love / L'amour des animaux." *Université de Toulouse*. https://www.univ-tlse2.fr/accueil/navigation/recherche/animal-love-l-amour-des-animaux-604381.kjsp. Viewed on 5/6/2019.

"Bundyville." *Bundyville Podcast*. Oregon Public Broadcasting. Peter Frick-Wright, Robert Carver, and Ryan Haas (Producers). "https://www.opb.org/news/article/bundyville-occupation-podcast/." Viewed on 9/7/2019.

"Champions & Record Holders." *Iditarod*. "http://iditarod.com/about/champions-records/" Viewed on 22/7/18.

"Contributors: John McPhee." *New Yorker*. "https://www.newyorker.com/contributors/john-mcphee." Viewed on 9/9/2019.

"Contributors: Rebecca Solnit." *The New Yorker*. "https://www.newyorker.com/contributors/rebecca-solnit." Viewed on 19/11/2019.

"Cueillette de champignons: près de 500 cas d'intoxication en deux semaines." *Le Figaro*. «http://sante.lefigaro.fr/article/cueillette-de-champignons-pres-de-500-cas-d-intoxication-en-deux-semaines/.» 24/10/2019.

"Defending the faithful," *charleskochfoundation.org*. "https://www.charleskochfoundation.org/story/stanford-law-religious-liberty/." 22/1/2019.

"Habitat for Humanity." *Habitat for Humanity*. "https://www.habitatnashville.org/cwp." Viewed on 1/7/2019.

"Hamilton." Hamilton. "hamiltonmusical.com." *Lin-Manuel Miranda*. Viewed on 9/7/2019.

"History." *Walt Disney Studios*. "https://www.waltdisneystudios.com/history/." Viewed on 6/11/2019.

"How To Deal With Ethical Challenges In Documentary Filmmaking." *New York Film Academy*. "https://www.nyfa.edu/student-resources/documentary-filmmaking-how-to-deal-ethical-challenges/." 23/9/2014.

"Lake Hood Floatplane Base Walking Tour." *alaska.org*. http://www.alaska.org/guide/lake-hood-walking-tour-a-look-around-the-worlds-busiest-seaplane-base. Viewed on 5/6/2019.

"Nancy Lord." *NancyLord.com*. "http://www.writernancylord.com/bio.html." Viewed on 10/12/2019.

"The New Yorker." *The New Yorker*. "https://www.britannica.com/topic/The-New-Yorker." Viewed on 16/11/2019.

"Obituaries: Martin Louis Murie'50." *Reed College*. http://www.reed.edu/reed-magazine/in-memoriam/obituaries/december2013/martin-louis-murie-1950.html. December 2013.

"Oceangoing Kayak and Wooden Paddle." *Anchorage Museum*. "https://www.anchoragemuseum.org/exhibits/death-in-the-ice-the-mystery-of-the-franklin-expedition/gallery/oceangoing-kayak-and-wooden-paddle/." Viewed on 20/7/2019.

"Oomingmak." *Burns Bizarre*. "https://www.burnsbizarre.com/product/596/Oomingmak-Matthiessen-Peter." Viewed on 6/12/2019.

"Peta's empathy is misplaced: fishing is about survival, not cruelty." *Alaska Journal of Commerce*. "http://alaskajournal.com/community/2004-12-05/petas-empathy-misplaced-fishing-about-survival-not-cruelty#.WoL-dy97GRs" 4/12/2004.

"Peter Matthiessen." *The New Yorker*. "https://www.newyorker.com/contributors/peter-matthiessen/page/3" viewed on 22/8/2018.

"Peter Matthiessen 1927–2014." *The Paris Review*. "https://www.theparisreview.org/miscellaneous/6320/peter-matthiessen-1927-2014-terry-mcdonell." viewed on 22/8/2014.

"President Bush Expanded And Enforced Trade Agreements To Open New Markets For American Products." *The Bush Record*. "https://georgewbush-whitehouse.archives.gov/infocus/bushrecord/factsheets/tradeagreements.html." Viewed on 27/10/2019.

"Presidential Election Votes: Donald J. Trump Wins." *nytimes.com*. "https://www.nytimes.com/elections/2016/results/president." 9/8/2017.

"This Changes Everything." *This Changes Everything*. "thischangeseverything.org." Viewed on 30/6/2019.

"Tom Lowenstein." *Wordpress*. "https://tomlowenstein.wordpress.com/about/." Viewed on 29/7/2019.

"Top 10 Reasons Not to Eat Salmon." *Peta*. "https://www.peta.org/living/food/top-10-reasons-eat-salmon/» 25/6/2010.

"Trans-Alaska Pipeline." *Britannica*. "https://www.britannica.com/topic/Trans-Alaska-Pipeline." Viewed on 27/11/2019.

"Understanding Predator Management in Alaska." *Alaska Department of Fish and Game*. "https://www.adfg.alaska.gov/static/research/programs/intensivemanagement/pdfs/predator_booklet.pdf." Viewed on 7/11/2019.

"What is Ecotourism?" *The International Ecotourism Society*. "https://ecotourism.org/what-is-ecotourism/." Viewed on 29/10/2019.

"Whitetail Deer Becoming Major Pest." *The New York Times*. "https://www.nytimes.com/1983/11/27/nyregion/whitetail-deer-becoming-major-pest.html." 27/11/1983.

"Wyatt Earp dies in Los Angeles." *History*. "https://www.history.com/this-day-in-history/wyatt-earp-dies-in-los-angeles." 16/11/2009.

"2019–2020 Alaska Hunting Regulations." *Alaska Department of Fish and Game*. "https://www.adfg.alaska.gov/static/applications/web/nocache/regulations/wildliferegulations/pdfs/general.pdfCDCAD3F13A6A206A66962AAB2B1D6874/general.pdf." Viewed on 7/11/2019.</ENL>

Urban, Amanda. "Biography." *cormacmccarthy.com*. "https://www.cormacmccarthy.com/biography/," Updated 4/8/2013.

Vail, Sharon. "Sourdough: More than a Bread." *NPR*. „https://www.npr.org/templates/story/story.php?storyId=6061648.» 12/9/2006.

West, Karl. "Carmakers' electric dreams depend on supplies of rare minerals." *The Guardian*. "https://www.theguardian.com/environment/2017/jul/29/electric-cars-battery-manufacturing-cobalt-mining," 29/7/18

Wheelwright, Jeff. "A Writer's Controversial Past That Will Not Die." *The New York Times*. "https://www.nytimes.com/2018/02/02/books/review/peter-matthiessen-paris-review-cia.html." 2/2/2018.

Williams, Margaret. "Polar bear population decline a wake up call for climate action." *World Wildlife Fund*. "https://www.worldwildlife.org/stories/polar-bear-population-decline-a-wake-up-call-for-climate-change-action." Viewed on 6/10/18.

Index

www.ingramcontent.com/pod-product-compliance
Lightning Source LLC
Chambersburg PA
CBHW071248230926
51849CB00001B/270
9789004734753